Self-Sustaining Solutions for Streams, Wetlands, and Watersheds

Proceedings of the
12 - 15 September 2004 Conference
Radisson Riverfront Hotel
St. Paul, Minnesota USA

Jessica L. D'Ambrosio, Proceedings Editor

Published by

American Society of Agricultural Engineers (ASAE)
2950 Niles Road, St. Joseph Michigan 49085-9659 USA
www.asae.org

The Society for engineering
in agricultural, food, and
biological systems

FOREWORD

For thousands of years humanity has been modifying wetlands, streams and watershed. Most Roman cities were built next to rivers and roman bridges, roads, and aqueducts that date back to 312 BC are still in use today. There are also many examples of water engineering projects performed by earlier civilizations. However, most modifications to wetlands, streams and watersheds have occurred in the last 200 years. Perhaps this is not surprising as in the year 1800 the population of the world was about one billion people. The current population of the world is about seven billion people and it is possible that some time this century the global population will exceed 10 billion people. Throughout the world we have many examples of large dams, bridges, and river lock systems that are truly astonishing engineering feats. We also have countless examples of bridges and culverts that allow us to safely cross streams and rivers without fear of collapse or flooding. However, we also have countless situations where these same structures have adversely impacted on these channel system, where wetlands have been eliminated, and were streams and rivers have been straighten, dredged, and their banks artificially hardened. Often the focus has been on hydraulics, with little regard to ecology or natural stream processes, and limited understanding of the influence of rapid land use change on many of our watersheds. Wetland losses in the United States and worldwide exceed 50 percent and continue to grow.

The last few decades have seen increasing interest in enhancing, restoring, and protecting the ecology of wetlands, streams, and watersheds. To achieve these goals will require sound fundamental and applied knowledge, close interaction between scientists and engineers, a systems approach, and a good understanding of spatial and temporal scales. Often a better approach, than has been the norm, might be to reduce the amount of initial engineering of the system and to assist nature in developing a more self-sustaining system. To help in achieving these goals the American Society of Agricultural Engineers is pleased to host this Specialty Conference that will:

- **Explore innovative solutions, for stream and wetland restoration and management that seek to re-establish and maintain self-sustaining ecosystems.**
- **Expand our understanding of stream and wetland processes and how knowledge of these processes can be used in water resource protection, design, and management.**
- **Have experts provide cutting-edge knowledge on theory, methods, and tools that will aid in maintaining or developing self-sustaining high quality stream and wetland systems.**
- **Enhance the capabilities of professionals engaged in natural channel design.**
- **Through workshops, provide knowledge and training that will better position participants to protect and enhance water resources.**

We would like to express our gratitude to the following sponsor organizations that helped to make the 2004 Self-Sustaining Solutions for Streams Wetlands and Watersheds Conference a success: American Society of Agricultural Engineers, The Ohio State University, Ohio Department of Natural Resources, Environmental Protection Agency Great Lakes Grants Program, National Fish and Wildlife Foundation, and U.S. Environmental Protection Agency.

We are very grateful and would like to thank the many experts who have graciously offered to participate in this conference. Workshop instructors and keynote speakers include some of our leading international and national experts who have been in the forefront in developing methods to further the advance of fully functional and self-sustaining stream, wetland, and watershed systems. We would like to extend a heartfelt thank you for their special effort. We are also very excited by the huge response and high caliber of the oral and poster presentations made during this four day conference and greatly appreciate all the authors and co-authors, some of whom volunteered their time to moderate the technical sessions, who shared their research, projects, designs, and ideas in the more than 80 technical and poster presentations.

This Conference would not have been possible without the tireless work of staff members from The Ohio State University, the University of Minnesota, and the American Society of Agricultural Engineers. In particular, we are grateful to Sharon McKnight and Kelly Sernau of ASAE headquarters for their help with the conference organization, and Sandy Rutter, also of ASAE headquarters, for her help with organizing and printing the Conference Proceedings.

Special thanks also goes to the staff at the Radisson Riverfront Hotel in St. Paul, Minnesota for collaborating with the Coordinating Committee to ensure the smooth running of technical sessions, social events, and accommodation of conference participants.

We thank you for joining us for this exceptional educational opportunity.

Andy Ward & Bruce Wilson, Conference Co-Chairs

KEYNOTE SPEAKERS

Edwin E. Herricks, Professor of Environmental Biology in the Environmental Engineering and Science Program and member of the faculty of the Department of Civil and Environmental Engineering since 1975. His research analyzes and interprets the effects of environmental change on aquatic and terrestrial species, populations, and communities, with a particular emphasis on the development of methods to improve environmental decision-making and ecologically relevant engineering design. His recent research includes an evaluation of the effect of climate change scenarios on fisheries; the restoration of streams in urban areas, including the development of ecological engineering concepts for watershed management; and development of an integrated hydrologic, geomorphic, and ecological classification system for watershed management. Professor Herricks has published widely and frequently serves as an advisor to local, state, and federal government organizations. http://ux6.cso.uiuc.edu/~herricks/

Richard Hey, Professor of Applied Environmental Sciences in the School of Environmental Sciences at the University of East Anglia, Norwich, UK specializing in River Mechanics and Engineering. His research has focused on the processes controlling natural river morphology and the development of design equations for predicting their dimension, pattern and profile. He has conducted applied research on the design of flood alleviation schemes and the restoration of natural rivers, restoration of lowland river habitats, the effect of river engineering works on channel stability, instream structures to control erosion and deposition, bridge pier and abutment scour control and on channel forming discharge. Dr. Hey has published seven books and numerous refereed papers relating to his research and consulting work. He is a recipient of the Marshall Nixon Award from the Chartered Institution of Water and Environmental Management and the J. C. Stevens Award from the American Society of Civil Engineers. http://www.uea.ac.uk/env/faculty/heyrd.htm

Dr Bill Mitsch is Distinguished Professor of Natural Resources, Environmental Science and Ecological Engineering and Director of the Olentangy River Wetland Research Park at The Ohio State University. His research and teaching have focused on wetland ecology and biogeochemistry, wetland creation and restoration, and ecosystem ☐odeling. Bill Mitsch has published over 200 papers on ecological/wetland matters and has edited or written 12 books including *Wetlands* and *Ecological Engineering and Ecosystem Restoration*. Dr. Mitsch is Past President of the Society of Wetland Scientists (SWS) and Past President of the American Ecological Engineering Society (AEES). Bill Mitsch and Sven-Erik Jørgensen are co-recipients of the Stockholm Water Prize 2004 for

their pioneering development and global dissemination of ecological models of lakes and wetlands, widely applied as effective tools in sustainable water resource management. http://swamp.ag.ohio-state.edu/

Dave Rosgen is a registered Professional Hydrologist and principal hydrologist of Wildland Hydrology Consultants. He has 42 years of experience in stream morphology, restoration, sedimentology, stream classification development and applications, grazing and riparian systems management, cumulative water resource impact assessment and modeling, fish habitat enhancement, and conducts research in river studies. He designs, supervises, contracts and monitors a variety of large scale river restoration projects throughout the United States. Dave conducts short courses throughout North America for government agency personnel, universities, and consulting firms in river morphology, restoration, and Wildland hydrology. Dave has published widely and is the author of the book *Applied River Morphology*. http://www.wildlandhydrology.com/

Andrew Simon is a Research Geologist at the ARS-National Sedimentation Laboratory in Oxford Mississippi. He has 24 years of research and applied experience (16 with the USGS) in sediment transport and unstable landscapes, particularly incised channels and streambank processes. He is the author of more than 100 technical publications and has edited several books and journals. Dr. Simon is an adjunct Professor at the University of Mississippi, Special Professor in the School of Geography, University of Nottingham and is on the Editorial Board of the journal *Geomorphology*.
Msa.ars.usda.gov/ms/oxford/nsl/personal_pages/Simon.html

Stanley W. Trimble, Professor in the Department of Geography at UCLA and a member of the faculty since 1975. His interests include historical geography of the environment and especially human impacts on hydrology including soil erosion, stream and valley sedimentation, and stream flow and channel changes. He is the joint editor of CATENA, an Elsevier international journal of soils, hydrology, and geomorphology. He has taught courses in environmental geology/hydrology for the US Army Corps of Engineers and he is a hydrologic/geomorphologic consultant for several agencies. Professor Trimble is the co-author of *Environmental Hydrology (2nd edition)* and has published numerous scientific articles including several in *Science*. http://www.geog.ucla.edu/People/Faculty/trimble/trimble.htm

Sponsors and Endorsing Organizations

Sponsored By

The Society for engineering
in agricultural, food, and
biological systems

2950 Niles Road, St. Joseph, MI 49085-9659, USA
269.429.0300   fax 269.429.3852   hq@asae.org   www.asae.org

The Ohio State University
Ohio Department of Natural Resources
Environmental Protection Agency Great Lakes Grants Program
National Fish and Wildlife Foundation
U.S. Environmental Protection Agency

Endorsing Organizations

American Water Resources Association
American Society of Agronomy
Crop Science Society of America
Soil Science of America

Coordinating Committee

Andy Ward, The Ohio State University, ward.2@osu.edu
Bruce Wilson, University of Minnesota, wilson@umn.edu
Jessica D'Ambrosio, The Ohio State University, dambrosio.9@osu.edu
Greg Jennings, North Carolina State University, Greg_Jennings@ncsu.edu
Tess Wynn, Virginia Tech, thwynn@vt.edu

Table of Contents

Stream Restoration and Urban Impacts on Streams

Watershed Evaluations

Stream System Ecology and Water Quality

Design Considerations and Applications

Monitoring and Modeling

Wetland Systems

Sediment Transport

Stream Restoration and Urban Impacts on Streams

USDA-NRCS NEH 654 Stream Restoration Handbook and Selected Topics

Water Quality Considerations

Stream Geomorphology

Poster Sessions

Restoration WARSSS

David Rosgen

The four C's of river restoration…Cause, Consequence, Correction and Communication are better understood through the Watershed Assessment and River Stability and Sediment Supply (WARSSS) methodology. The WARSSS procedure involves a cumulative effects analysis of the watershed including the change in hillslope, hydrologic and channel processes influenced by various land use activities. Causes and consequences of instability related to specific processes are important to understand in order to "correct" the problem(s) rather than "patch" symptoms. Applying a geomorphic approach for river restoration and natural channel design, is ***not*** a "simple cookbook" procedure contrary to statements made by those unfamiliar with the method. Many river restoration projects that ignore the complexities inherent in assessment and geomorphic-based design, are often the central cause of failure.

Conceptual outlines and examples are presented showing the watershed linkage to sedimentological, hydrological, morphological and biological process integration into river restoration.

EVALUATION OF NATURAL STABLE RIVER DESIGN PROCEDURES

Dr. Richard Hey and Dave Rosgen

Several design procedures are available for predicting the three dimensional morphology of natural stable rivers: analytical, empirical and analogue. First analytical, or rational procedures are the most theoretically sound as they are based on the simultaneous solution of a set of process equations which link the three dimensional morphology of the river (bankfull width, mean and maximum depths, slope, sinuosity, meander arc length and velocity) with the controlling boundary conditions (bankfull discharge and bed material load, bed material calibre, bank sediment and vegetation and valley slope). Second, empirical regime equations have been developed from field data using statistical methods. They simply link the variables defining channel form with one or more of the controlling variables. Third analogue, or geomorphological procedures enable designs to be scaled from the morphology of a natural stable reference reach of the required stream type.

The basis of these three distinctly different procedures are critically reviewed and compared. Field data are then used to evaluate their predictive capability. This indicates that significant errors can occur with both rational and regime equations even when used appropriately. Provided that reference reaches are correctly identified, the geomorphological procedure enables the three dimensional morphology to be specified within +/- 10% of the true value. Unlike the other procedures, which only predict reach average conditions, it also enables the morphology of riffles, runs, pools and glides to be prescribed. The rules for correctly identifying an appropriate reference reach are identified.

\\Server\data (z)\My Documents - Dave\Abstracts 2004\Evaluation of Stable River Design Procedures.doc

An Integrated Process-Based Approach to Regional Sediment Issues Using Geomorphic and Numerical-Modeling Analyses

Andrew Simon, Robert Wells, Eddy Langendoen, and Ron Bingner

Sediment is listed as one of the principle pollutants of surface waters in the United States, both in terms of sediment quantity ("clean sediment") and sediment quality due to adsorbed constituents. Sediment-transport rates and bed-material conditions can be interpreted as (1) "natural" or background, resulting from generally stable channel systems, (2) "impacted", with greater transport rates and amounts, reflecting a disturbance of some magnitude and more pervasive erosion, and (3) "impaired", where erosion and sediment transport rates and amounts are so great (or in a some cases so low) that biologic communities and other designated stream uses are adversely effected. Impairment of designated stream uses by clean sediment (neglecting adsorbed constituents) may occur through processes that occur on the channel bed or by processes that take place in the water column. To identify those sediment-transport conditions that represent impacted conditions it is critical to first be able to define the non-disturbed, stable, or "reference" condition for the particular waterbody.

Stream channels integrate the transfer of energy and materials from a watershed and continually adjust their morphology according to imbalances between flow and sediment load delivered from upstream. Because of this, channel morphologies can be used to determine dominant processes and, therefore, relative channel stability. A conceptual models of channel evolution permits identification of stable and unstable, or "reference" conditions. In combination with historical flow and sediment-transport data, background transport rates are, therefore identified for streams on a regional basis. The ecoregion concept has been used to separate stream systems on a regional basis with data from more than 2,900 sites nationwide used to determine ranges of suspended-sediment transport. Bed-material characteristics for these stable streams are also used to identify threshold levels of embeddedness (for coarse-grained streambeds) for use in determining impacted habitat conditions.

With greater resources, numerical-modeling tools can be used to identify source areas of sediment and potential mitigation measures. These models cover a range of complexity and processes ranging from an Excel-based Bank-Stability Model, to an empirical upland model (AnnAGNPS), to a deterministic channel-evolution model (CONCEPTS). The latter two models have been used successfully in tandem in a variety of environmental settings to determine sources of suspended sediment and the relative contributions of upland and channel processes.

Changes in the Sediment Budget and in Stream Morphology, Coon Creek Wisconsin, 1975-1993

Stanley W. Trimble

Department of Geography, University of California, Los Angeles 90095-1524 USA

Coon Creek Wisconsin is a stream that has undergone significant morphological changes over historical time. These changes are due to changes of hydrology and sediment loads which, in turn, are due to changes of land use and land treatment. Considered here are changes for the period 1975-1993 when, due to implemented soil conservation measures, storm hydrologic response was mild and sediment loads were very low (1). Tributaries, now sediment sinks, featured continued vertical accretion on new (post-1930s), lower, inset floodplains created since the 1930s when soil conservation measures were first installed. The upper main valley had been a strong sediment source, primarily from eroding cut banks, since the 1930s, but vertical accretion on new (post 1930s), lower, inset floodplains offset the loss by cut banks in the recent period. This was the greatest surprise of the study. The floodplain of the lower main valley continued to aggrade but at only about 6% of the maximum rate of the 1930s. However, areas adjacent to the streams aggraded significantly more than backswamp areas more distant from the stream. This appears to be due to a coarser sediment load created by the stream reworking old upstream alluvium.

Considering Urban Stream Naturalization – Ecosystem Criteria for Habitat Quality and Quantity

Edwin E. Herricks

There is a recent emphasis in urban watershed management on the development of self-sustaining ecosystems particularly in stream channels. Although there are a wide range of practices (BMPs) available for channel management, there is little guidance on selection criteria, and almost no information on the ecological performance of in-stream management practices operating within a single stream reach or at several locations throughout a watershed. To develop selection guidance, it will be necessary to first examine the ecological performance of BMPs and then develop a protocol for watershed-wide evaluation of the contribution of both in-channel and landscape BMPs to ecosystem sustainability. This paper will review methods for integrating hydraulic, geomorphological, and ecological information to derive measures of ecological performance that are appropriate for different scales. A particular focus will be on the ecological performance of BMP types and the appropriateness of particular practices for system-wide improvements in ecological resources.

Celebrating Stormwater at Coffee Creek Center, Chesterton, Indiana

Andrew Bender, PE
JFNew

This development highlights environmental restoration as an integral part of a 640-acre "sustainable" development. When fully developed, the site will feature 1,200 town homes, 1,000,000 square feet of office/retail space, including a 180-acre corridor of woodlands and prairies along Coffee Creek.

JFNew was involved with all environmental issues, including restoration of a trout stream, native prairies, and restoration of a bottomland forest. As a key member of the site planning team, JFNew evaluated the topography, soils and hydrology of the prairies, wetlands, woodlands, determined species appropriate and species best adapted to wastewater and stormwater absorption. Plant selections were based on native species composition of remnant prairies and riparian areas.

JFNew prepared a wastewater treatment master plan of subsurface flow wetlands and absorption fields based on site characteristics and estimated wastewater volume from each phase of the development. Treated wastewater will be piped underground to biofields planted with deep-rooted native grasses and flowers. Stormwater is treated by prairies and wetlands through more than a mile of level spreaders, a series of underground pipes and gravel infiltration areas under the restored prairies.

Integrated Drainage-Wetland Systems for Reducing Nitrate Loads from Tile Drained Landscapes

William G. Crumpton and Matthew Helmers

In addition to raising local water quality concerns, nitrate loads from Midwest agriculture are suspected as a primary contributor to hypoxia in the Gulf of Mexico. Over-application of fertilizer can exacerbate the problem, but the major causes are hydrological and land-use changes that came with tile drainage. Subsurface drainage creates very productive croplands and reduces water quality problems associated with surface runoff, but subsurface flow and nitrate transport are substantially increased. A permanent solution to the environmental problem of hypoxia in the Gulf of Mexico will likely require more than improved nitrogen management and tillage practices. We present results of simulations integrating nitrate-removal wetlands, as a proven technology, with the emerging technologies of drainage modification. Relatively small areas of wetlands intercepting tile drainage can remove over 50% of the nitrate in tile drainage water. Controlled drainage and shallow drainage can reduce subsurface flow and nitrate export by as much as 50%. The integration of shallow and controlled drainage systems with nitrate-removal wetlands has the potential to simultaneously decrease the volume of subsurface drainage, increase the number of wetland sites, push those sites closer to the nitrate source, and enhance wetland performance by increasing the average residence time in the wetlands.

EFFECT OF MACROPHYTE SPECIES ON SUBSURFACE FLOW WETLAND PERFORMANCE IN COLD CLIMATE

C. M. Ouellet-Plamondon [1], J. Brisson[1] and Y. Comeau[2]

ABSTRACT

Horizontal subsurface flow constructed wetlands (HSSCW) allows organic matter and nitrogen removal of fish farm effluent prior to streams discharge. The effect of macrophyte species on HSSCW efficiency was tested in ten units in a greenhouse experiment, in summer and winter. Eight units were individually planted with *Phragmites australis*, *Typha angustifolia*, *Phalaris arundinacea* and *Calamagrostis canadensis* (two units per species) and the remaining two units were left unplanted. The units were fed with a reconstituted effluent made from trout farm sludge. The sludge was diluted to obtain an average of 15 g $COD/m^2/d$, 3 g $BOD_5/m^2/d$, 6 g $TSS/m^2/d$, 0.50 g $N/m^2/d$, 0.15 $gP/m^2/d$ and a resulting hydraulic loading of 3 $cm/m^2/d$. Water quality was analysed in summer 2002 and winter 2003 for COD, BOD, TSS, TNK, NH_4^+, NO_2^- + NO_3^-, TP and o-PO_4. Planted units were at least 5% more efficient in pollutant removal than unplanted units in summer and at least 10% in winter. The increase in removal efficiency for planted units was small, mainly because of the low loading conditions of the fish farm effluent. TKN (96% in summer and 88% in winter) and COD (96% in summer) removal were more efficient for *Phragmites* and *Typha*, the two species with large rhizomes. *Phalaris* was more efficient than the others with COD and BOD removal at 95% in winter. *Calamagrostis* was the least efficient species, with the largest difference being for winter nutrient removal. It was also the species with the lowest belowground: aboveground biomass ratio, around 0.25 compared to above 2 for *Typha*. Units planted with macrophytes with large belowground biomass showed less seasonal variability.

KEYWORDS. Subsurface flow wetland, Macrophytes, Belowground biomass, Cold climate, Aquaculture effluent treatment

INTRODUCTION

Macrophytes are an active component of horizontal subsurface flow constructed wetlands (HSSFCW). They distribute and decrease current velocities. They increase contact time between water and plant surface area and surface area for attached microbial growth. They release oxygen and organic compounds to the rhizosphere. Macrophytes also assimilate nutrients a small fraction of nutrient. Their presence make constructed wetland a suitable solution for decentralised wastewater treatment applications (IWA, 2000). Moreover, HSSFCW can be integrated in the landscape and create wildlife habitat (Kadlec and Knight, 1996). Most studies comparing planted versus non-planted system showed a significant positive effect of plants on nutrients removal (IWA, 2000; Allen *et al.*, 2002, Jing *et al.*, 2002).

There is a need to diversify macrophytes species used in HSSFCW. Fast growth rate and establishment, tolerance to anoxic conditions, ability to form large monoculture and adaptation to

[1] Institut de recherche en biologie végétale, University of Montréal, 4101 E Sherbrooke, Montreal (QC) HIX 2B2 Canada

[2] Department of Civil, Geological and Mining Engineering, Ecole Polytechnique, C.P. 6079, Succ. Centre-ville, Montreal (QC) H3C 3A7 Canada

local conditions (ex. cold hardiness) are biological attributes for HSSFCW application (IWA, 2000). There are some comparison studies, for different set of species, but they are not always conclusive (Bachand and Horne, 2000; Jing et al., 2002). Also, species applicability can change with latitude and local climate. Therefore, rigorous comparison studies with statistical replicates under controlled conditions are necessary to choose appropriate species for HSSFCW construction.

This study is part of a research project aiming to develop an appropriate treatment for fish farm effluent prior to stream discharged. The large quantity of phosphorus released in fresh bodies of water forces fish farm to adapt to new water quality standard. The system we are proposing to treat the concentrated effluent from the raceway decantation basin is a two-step procedure. First, HSSFCW removes organic matter and nitrogen. Secondly, a smaller non-planted unit filled with steel slag, a highly adsorbing substrate, removes phosphorus (Naylor et al., 2003).

The present experiment investigates the effect of species on HSSFCW efficiency. We compared the performance of Phragmites, Typha, Phalaris and Calamagrostis, in the treatment of a fish farm effluent in a greenhouse experiment, both during active and dormant growth season. We selected Phragmites, Phalaris and Typha, because they are the more common macrophytes used in constructed wetland (IWA, 2000). While Calamagrostis is not currently used, we selected it because it has the ability to form large monoculture looked for in HSSFCW applications.

MATERIAL AND METHODS

Research took place in a greenhouse at the Botanical garden of Montreal. Controlled environment limited the number of variables interacting with the objectives of the study. The experimental system consisted of two 1500 L refrigerated bulk tanks to store the effluent, a central peristaltic pump, two redistribution basins with mixers, four peristaltic pumps, ten 1 m^2 wetland units (1.23 m x 0.78 m x 0.32 m) and two buckets after each wetland unit to collect the treated effluent. The wetland units were filled with a neutral 10-15 mm neutral substrate up to 3 cm from the edge. A 30-40 mm neutral river stone gabion at the inlet and the outlet facilitated water distribution and evacuation. Water table was kept 4 cm under the substrate surface. Eight units were individually planted with Phragmites australis, Typha angustifolia, Phalaris arundinacea and Calamagrostis canadensis (two units per species) and the remaining two units were left unplanted. Beds were planted on May 3, 2002 from collected rhizomes in fields on the south shore of the St-Lawrence River near Montreal. They were fertilized for the first month to ensure a proper start up.

A reconstituted effluent from sludge collected in a silo acting as an anaerobic digester at a through-flow trout farm was used in the experiment. The sludge was diluted 50 times to obtain an average of 15.0 g $COD/m^2/d$, 3.00 g $BOD_5/m^2/d$, 6.00 g $TSS/m^2/d$, 0.50 g $N/m^2/d$ and 0.15 g $P/m^2/d$, comparable to the silo supernatant. Fifteen litres were fed twice a day in each bed starting in June. The resulting hydraulic loading rate was 3.0 $cm/m^2/d$ and the resulting voids hydraulic retention time was 3.6 days.

Results were collected during two periods, summer and winter. In summer, measurements were taken from July to October 2002 (macrophytes fully active, $T_{avg} = 22$ °C). In winter, measurements were taken from January to end of March 2003 (dormant season, above-ground portion previously harvested, $T_{avg} = 7$ °C). During these periods, the volume of the final effluent was measured three to four times a week to measure evapotranspiration. Efficiency calculation was based on mass balance between a sampling point in the storage bulk tank and a second one at the end of the wetland unit. The eight following water quality parameters were measured in compliance with the Standard methods 1998 (APHA, 1998): COD, BOD_5, TSS, TNK, NH_4^+, $NO_2^- + NO_3^-$, TP and o-PO_4. Analyses of variance (two-way-ANOVA) followed by multiple comparison of means according to Tukey's method were performed to test differences between treatments within each growth season.

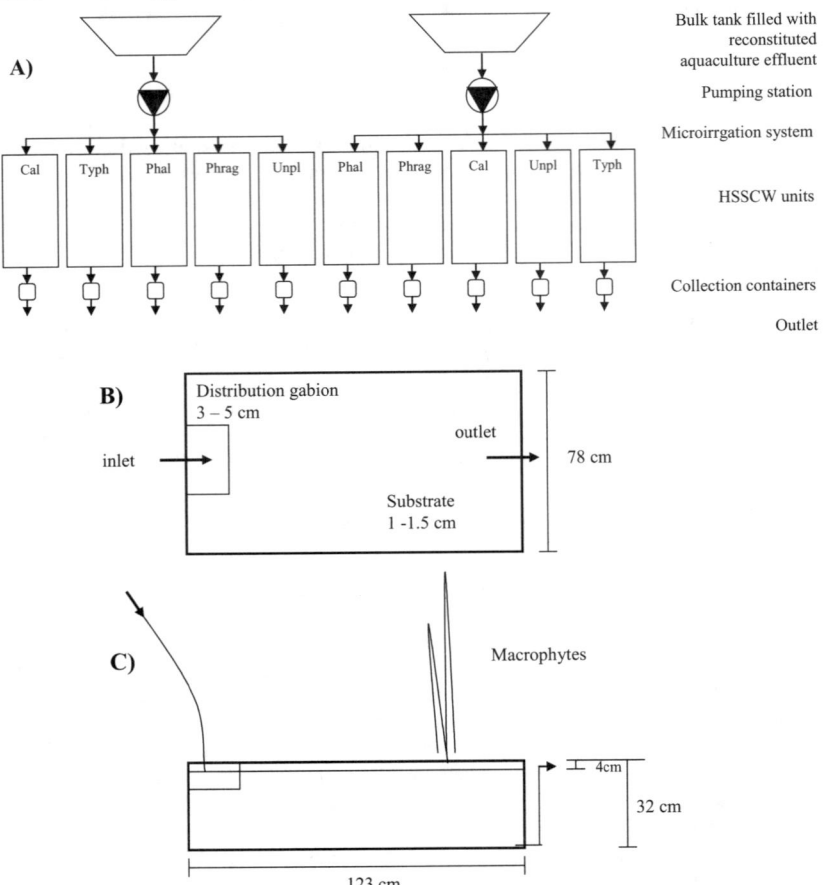

Figure 1. Experimental set-up A) Experimental layout (plan view) (Cal=*Calamagrostis*, Typh=*Typha*, Phal=*Phalaris*, Phrag=*Phragmites*, Unpl=Unplanted), B) Unit plan (view design), C) Unit design (cross section).

At the end of the summer period in 2002, the entire aboveground portion of the macrophytes was harvested and their foliage was analysed for nutrient content. Also, two 30 cm diameter cores of substrate were taken along the planted basins central axis, at 30 cm from the inlet and 30 cm of the outlet, to estimate belowground biomass. Both aboveground and belowground biomass were dried at 60°C for 72 h and weighted.

RESULTS

Biomass

The aboveground biomass was approximately 1 kg/m² for the first summer growing season for all species. Belowground biomass and, consequently, belowground: aboveground biomass ratio varied among species (Table 1). Although the variability of the results, *Typha* developed significantly more belowground biomass, followed by *Phragmites* and *Phalaris*. *Calamagrostis* developed considerably belowground biomass and it was concentrated near the surface of the bed. *Typha* produced significantly more rhizome compared to the three other species ($p=0.0013$).

Table 1. Biomass data

Species	Aboveground biomass	Belowground biomass	Belowground: Aboveground biomass ratio
	g/m²	g/m²	BG:AG
Phragmites	1115 +/- 36	946 +/- 356	0.85 +/- 0.35
Typha	988 +/- 51	2461 +/- 817	2.47 +/- 0.70
Phalaris	1187 +/- 164	1657 +/- 954	1.35 +/- 0.62
Calamagrotis	1057 +/- 12	256 +/- 147	0.24 +/- 0.14

<u>Water Loss and Evapotranspiration</u>

During summer, there was a significant difference in evapotranspiration (Et) between unplanted and macrophytes units (Figure 1). *Phalaris, Phragmites, Typha* had significantly more evapotranspiration, followed by *Calamagrostis* and unplanted unit. There was no statistical difference in evapotranspiration in winter. It must be noted that evapotranspiration was probably higher than expected during winter due to a change in relative humidity between summer and winter. In the controlled greenhouse environment, vaporisation was used to decrease the ambient temperature in the summer. The relative humidity was higher in summer (60 % average) than in winter (RH 46 %).

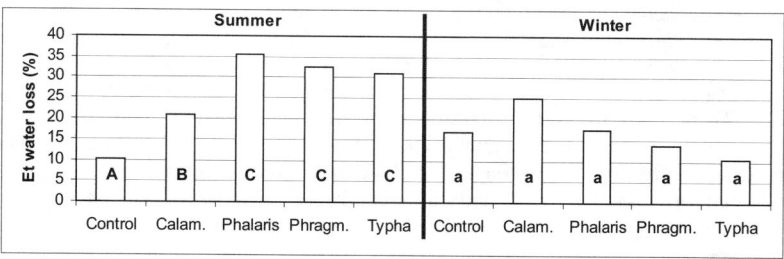

Figure 1. Effect of macrophyte species on summer and winter evapotranspiration [3]

<u>Organic matter</u>

During summer, macrophytes had a significant effect on COD removal ($p_s = 0.0116$) (Figure 2a). *Phragmites, Typha* and *Phalaris* were the most efficient and were statistically different from unplanted unit. *Typha* and *Phragmites* were also different from *Calamagrostis*. There was no significant different in summer BOD_5 removal (Figure 2b). During winter, COD and BOD_5 performance decreases were less important for planted units ($p_wCOD = 0.0263$; $p_wBOD_5 = 0.028$). *Phalaris* was the best treatment and unplanted was the worst.

---

[3] Treatments with different letter code are significantly different according to multiple comparisons of means using Tukey's method. Winter statistical tests were conducted independently from the summer.

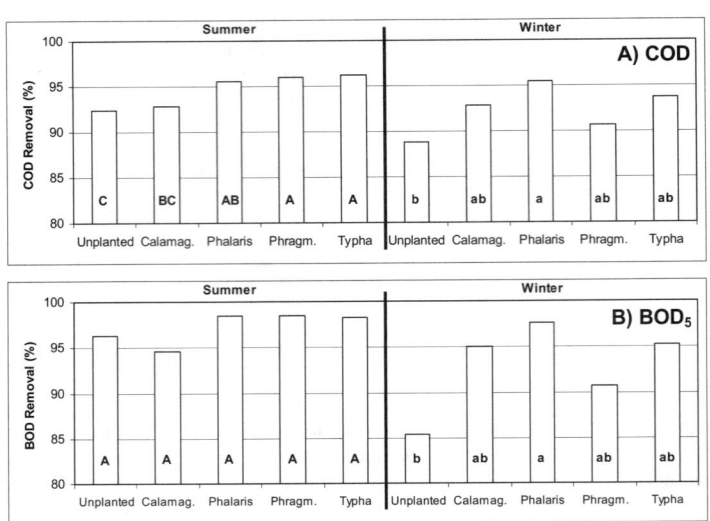

Figure 2. Effect of macrophyte species on summer and winter organic matter removal A) COD, B) BOD₅.

Nutrients

Macrophytes showed their most significant effect for nitrogen removal (p_s = 0.0024; p_w = 0.0255). *Typha, Phragmites* and *Phalaris* were more efficient than unplanted in both seasons (Figure 3a). *Phalaris* was not significantly different from *Calamagrostis* and from unplanted in winter also. *Calamagrostis* was more efficient than unplanted units, but the difference was not significant. There was a 10% efficiency drop in the winter. There was a statistical difference in macrophytes effect on phosphorus removal (p_w=0.0068) (Figure 3b). *Phalaris* and *Typha* were different from *Calamagrostis* and unplanted. *Phragmites* was different from unplanted only, while *Calamagrostis* was not significantly different from all treatment. The winter performance decrease was more important for unplanted and *Calamagrostis* units.

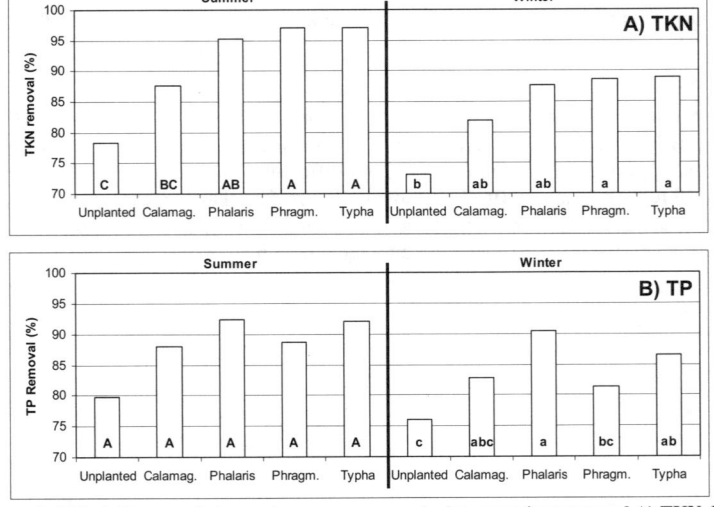

Figure 3. Effect of macrophyte species on summer and winter nutrient removal A) TKN, B) TP.

DISCUSSION

Effect of macrophytes

This experiment confirmed the positive role of macrophytes on organic matter and nutrient removal (Kadlec and Knight, 1996). During summer, macrophytes increased evapotranspiration water lost, COD removal and nitrogen removal. During winter, macrophytes had an effect on COD, BOD_5, phosphorus, TKN and NH_4. TSS removal remained a physical process. Our results are consistent with the role of macrophytes for nitrogen removal and initial phosphorus removal (Tanner, 2001). HFSSCW fed with fish farm effluent are usually more efficient than average and the macrophytes increase in removal efficiency remained small. Lin *et al.* (2001) measured similar efficiencies in a HSSFCW experiment with similar loading and hydraulic retention time (HRT). Schulz *et al.* (2003) has similar results for TSS removal, but lower organic matter and nutrient removal for a HRT of less than a day. The observed high efficiency can also be explained by the controlled environment and by possible previous processing in the distribution tank. During summer, higher HRT caused by evapotranspiration improved the performance too (Bachand and Horne, 2000). The theoretical HRT was 3.6 day, but 5.4 day with the measured 33% evapotranspiration (relative humidity 60%) in planted beds.

Species differentiation

Overall, we can say *Typha, Phragmites* and *Phalaris* had a similar removal and *Calamagrostis* was less efficient. The ranking of species is not definitive in the literature. Plant species does not make a difference in many studies (Bachand and Horne, 2000; Jing *et al.*, 2002). *Typha latifolia* out-performed in some species comparison (Coleman *et al.*, 2001), while in others its efficiency may decreased during winter (Allen *et al.*, 2002). *Typha angustifolia* was very performant in our study. *Typha angustifolia* has also more reserve in rhizomes than *Typha latifolia* (Grace and Wetzel, 1982). It was also observed that *Typha angustifolia* was more tolerant to high ammonia level (Clarke and Balwin, 2002). There are other positive results with *Phalaris* trials (Vymazal *et al.*, 2001b) Species differentiated more during winter.

The enhancement of species differentiation in winter is consistent with other studies (IWA, 2000; Allen *et al.*, 2002). Because the aboveground portion of the biomass was harvested, the difference is at the root level, largely attributed to the root structure and biomass allocation. The reliance on root biomass for nitrogen removal efficiency has been observed (Farahbakhshazad and Morrison, 1997). Our results showed a correlation between species efficiency and belowground biomass, even though belowground biomass results showed a greater variability due to sampling. *Calamagrostis* was the least efficient species, with the largest difference being for winter TKN and TP removal. It was also the species with the shallowest and the lowest belowground biomass. The low belowground biomass may also be explained by the large granulometry of the media (10-15mm). Larger biomass was found in a sandy loam potting soil (Powelson and Lieffers, 1992). The belowground: aboveground biomass ratio was the lowest, around 0.25 compared to above 2 for *Typha*. This ratio has a potential to become a criteria for species selection. There were little differences between the three other species efficiency, all with larger belowground biomass. TKN (summer and winter) and COD (summer only) removal was more efficient for *Phragmites* and *Typha*. The belowground biomass of these two species consisted of large rhizomes. *Typha* had significantly more rhizome than *Phragmites* in our study. *Phragmites* was also reported to have more microorganisms per root surface area than *Phalaris* in bacterial count (Vymazal *et al.*, 2001a). *Phalaris* was more efficient than the others at COD and BOD_5 removal in winter. It developed a different root structure, more fibrous, less deep, less large rhizomes and its belowground biomass was between that of *Typha* and *Phragmites*. *Phalaris* is also known for a longer active period in number of days (USDA, 2004). Thus, our study suggests a relationship between the macrophyte function for the removal of organic matter and nutrient and its physiological root structure.

Dormant season

Temperature and macrophytes activity changed the seasonal performance. TKN removal had the more important seasonal decrease, followed by TP and BOD_5 and COD. TKN removal efficiency was lower during winter for all treatments as reported by Werker *et al.* (2002). There was no uptake of nutrient in the dormant season. Temperature affects several biogeochemical processes which regulate nutriment removal in wetlands (Reddy and Burgoon, 1996; IWA, 2000). Bigger performance decreases related to temperature have been observed in outdoor facilities (IWA, 2000). The cause for the greater effect of macrophytes in winter is still a controversy. Allen *et al.* (2002) suggests it is because oxygen consumption for respiration decreases with temperature. The increased of oxygen solubility in winter provided more oxygen which improved transformation, while kinetics decreases (Kadlec and Knight, 1996).

Ecological considerations

The four species studied here have a potential to be invasive. Some region in the world, such as Australia, banned *Phragmites* introduction (Chambers and McCombs, 1994). *Phragmites* and *Phalaris* are known to be very active in the vegetation dynamics in the South west of Quebec (Lavoie *et al.*, 2003). *Phalaris* and *Typha angustifolia* were also reported to replace native, perennial herbaceous, such as *Calamagrostis canadensis*, (Galatowitsch *et al.*, 2000). *Calamagrostis canadensis* also has a potential to be invasive in disturbed open forest sites (Lieffers *et al.*, 1993). Before selecting a species, local invasive status of the species should be determined.

CONCLUSION

Macrophytes affect organic matter, nutrient removal and water balance, especially in winter. Our study showed an important link between physiological structure (belowground biomass, belowground: aboveground biomass ratio, rhizomes) and efficiency for subsurface flow wetland. We recommend these structures as criteria for species selection among native one. More studies are needed on the role of belowground plant structures on pollutant removal, especially in winter. We found no important differences between *Typha angustifolia*, *Phragmites australis* and *Phalaris arundinacea* and depending on local availability and invasiveness, all three species could be used. We do not recommend *Calamagrostis canadensis* under conditions comparable to ours because of its lower efficiency. This research confirms the important role of macrophytes in organic matter removal and nitrogen transformation in SSW, even in winter conditions when the plants are dormant.

Acknowledgements

We thank Denis Bouchard, Mourad Kharoune, Chris Donka, Laurent Côté, Martine Provost, Christine Galipeau, Sabina Tigges and Vincent Gagnon for technical support, Stéphane Daigle for statistical analysis, Florent Chazarenc for manuscript review and NSERC for financial support.

REFERENCES

Allen, W. C., P.B. Hook, J.A. Biederman and O.R. Stein. 2002. Temperature and wetland plant species effects on wastewater treatment and root zone oxidation. Journal of Environmental Quality 31(3): 1010-1016.

APHA. 1998. Standard Methods For the Examination of Water and Wastewater: including bottom sediments and sludges18[th] Edition . New York, N.Y.: American Public Health Association.

Bachand, P.A.M. and A.J. Horne. 2000. Denitrification in constructed free-water surface wetlands II: Effects of vegetation and temperature. Ecological Engineering 14: 17-32.

Chambers, J.M. and A.J. McCombs. 1994. Establishing wetlands plants in artificial systems. Water Science and Technology 29 (4): 79-84.

Clarke, E. and A. H. Baldwin. 2002. Responses of wetland plants to ammonia and water level. Ecological Engineering 18 (3): 257-264.

Coleman, J., K. Hench and K. Garbutt. 2001. Treatment of domestic wastewater by three plant species in constructed wetlands. Water, Air, and Soil Pollution 128 (3-4): 283-295.

Farahbakhshazad, N. and G. M. Morrison. 1997. Ammonia Removal Processes for Urine in a Upflow Macrophyte System. Environmental Science and Technology 31: 3314-3317.

Galatowitsch, S.M., D.C. Whited, R. Lehtinen, J. Husveth and K. Schik. 2000. Environmental Monitoring and Assessment 60: 121-144.

IWA. 2000. Constructed Wetland for Pollution Unplanted Processes, Performance, Design and Operation. Scientific and Technical Report No. 8.g London, UK: IWA Publishing.

Jing, S., Y. Lin, T. Wang and D. Lee. 2002. Microcosm Wetlands for Wastewater Treatment with Different Hydraulic Loading Rates and Macrophytes. Journal of Environmental Quality 31(2): 690-696.

Kadlec, R.H. and R.L. Knight. 1996. Treatment Wetlands. Boca Raton, FL: CRC Press, Inc.

Lavoie C., M. Jean, F. Delisle and G. Létourneau . 2003. St. Lawrence River wetlands: A spatial and historical analysis. Journal of Biogeography 30: 537-549.

Lin, Y., S. Jing and D. Lee. 2002. Nutrient removal from aquaculture wastewater using a constructed wetlands system. Aquaculture, 209: 169-184.

Naylor, S., J. Brisson, M.A. Labelle, A. Drizo and Y. Comeau. 2003. Treatment of fresh fish farm effluent using CWSs – The role of plants and substrate. Water Science and Technology 48(5): 215-222.

Powelson, R.A. and V. J. Lieffers. 1992. Effect of light and nutrients on biomass allocation in *Calamagrostis canadensis*. Ecography 15: 1 (1992).

Reddy, K.R. and P.S. Burgoon. 1996. Influence of temperature on biogeochemical processes in constructed wetlands – Implications to wastewater treatment. Paper presented at the Symposium on Constructed wetlands in Cold Climates, June 4-5, 1996, Niagara-on-the-Lake, Ontario.

Schulz, C., J. Gelbrecht and B. Rennert. 2003. Treatment of rainbow trout farm effluents in constructed wetland with emergent plants and subsurface horizontal water flow. Aquaculture 217: 207-221.

Tanner, C.C. 2001. Plants as ecosystem engineers in subsurface-.treatment wetlands. Water Science and Technology 44(11-12): 9-17.

USDA, NRCS, 2004. The Plant Database, Version 3.5. National Plant Data Center, Baton Rouge. Available on http://plants.usda.gov (search on 2003-04-10).

Vymazal, J., J. Balcarová and H. Doušová. 2001a. Bacterial dynamics in the sub-surface constructed wetland. Water Science and Technology 44 (11-12): 207–209.

Vymazal, J. 2001b. Constructed wetlands for wastewater treatment in the Czech Republic. Water Science and Technology 44 (11-12): 369-374.

Werker, A.G., J.M. Dougherty, J.L. McHenry and W.A. Van Loon. 2002. Treatment variability for wastewater treatment design in cold climates. Ecological Engineering 19: 1-11.

Yang, L., H.T. Chang and M.L. Huang. 2001. Nutrient removal in gravel- and soil- based wetland microcosms with and without vegetation. Ecological Engineering 18: 91-105.

Agricultural Wetland and Pond Hydrologic Analyses Using the SPAW Model

Dr. K. E. Saxton[1] and Mr. P. H. Willey[2]

Abstract

A computer model, SPAW, has been developed to describe and evaluate the hydrology of wetlands and ponds which have agricultural watersheds as their water source. Wetland analyses include daily inundation depths and durations during the wetland growing season and multiple year statistics. The model can also be applied to evaluate the capacity and management of ponds for irrigation livestock water, and animal waste disposal. A comprehensive ponding routine added to the previously developed SPAW hydrologic field model provides a single hydrologic computational package with operational and data screens, output tables and graphs, manuals and examples. Documentation includes the SPAW model theory, typical applications, data requirements, operational details and example files. Program corroboration has been accomplished through research data, workshops and application assessments.

Keywords: Agricultural, Hydrology, Water budgeting, Wetland, Ponds, Reservoirs, Lagoons

Introduction

The SPAW model computes daily hydrologic budgets for agricultural fields with a moderate level of complexity to account for the most important hydrologic processes. Inputs describe the climate, soil and crops of a farm field in the one dimensional vertical plane. The principle hydrologic inputs are daily rainfall and potential evaporation, with optional daily maximum and minimum air temperatures to estimate snow and frozen soil processes. Soil and crop descriptions determine the daily disposition of this water into and out of the soil-plant-air-water (SPAW) system. Recent enhancements to SPAW were the addition of an irrigation field budget (scheduling) and an inundated pond (wetland/lagoon/pond/reservoir) budget.

The objective of the SPAW model development was to predict agricultural field hydrology and its interactions on crop production using mid-range technical complexity while minimizing input data and computation time. The SPAW model methodology achieved a "reasonable" and "balanced" approximation of real world hydrology using numerical solutions to address applicable physical, chemical and biological processes. SPAW model users have assisted with the model development which has resulted in improved accuracy, broader areas of application, and greater ease of use.

The SPAW model has been applied extensively to hydrologic simulations for both research and design. The original versions were focused on plant available soil water under rain-fed conditions (Saxton, 1980, 1981, 1985; Saxton et al., 1974; Saxton and McGuinness, 1982; Sudar et al., 1981). Subsequent versions have provided significant model additions, alterations, added capability and numerous application evaluations related to agricultural field hydrology (Saxton et al., 1992; Saxton and Bluhm, 1982; Rao and Saxton, 1995; DeJong and Zentner, 1985). The enhancement of soil water characteristic estimates was particularly useful to make the model applicable to a wide variety of field situations (Rawls, et al., 1982, Saxton et al., 1986). The most recent additions were an irrigation scheduling methodology and a POND module that utilizes daily SPAW field hydrology analysis to simulate daily pond inflow, outflow and storage. Auxiliary inflows and withdrawals from the pond allows for the simulation of a variety of pond

---

[1] Research Agricultural Engineer, USDA Agricultural Research Service (Retired), Pullman, WA 99164-6120
[2] Drainage and Wetland Engineer, USDA Natural Resources Conservation Service, National Water and Climate Center, Portland, OR

functions such as seasonal and permanent wetland ponds, irrigation storage reservoirs, animal waste storage ponds, and ponds used for livestock water supplies (Moffitt et al., 2003).

Hydrologic Processes Represented

Field model:
The principle hydrologic processes considered in the SPAW field model are the following as depicted in Figure 1 by a schematic of the vertical water movement of a field:

♦ Precipitation: daily totals including snow accumulation and melt when air temperature data are included.
♦ Runoff: Estimated by the USDA/SCS curve number method modified for daily soil moisture and vegetation conditions. Frozen soil effects are included if air temperature data are included. No stream routing is provided.
♦ Infiltration: A daily amount based on precipitation minus estimated runoff which infiltrates into the upper-most soil layers as storage is available.
♦ Redistribution within the soil profile: Infiltrated and existing soil water is moved between defined soil layers by a Darcy tension-conductivity procedure providing either downward or upward flow.
♦ Evapotranspiration: Daily estimates of plant transpiration, direct soil surface evaporation and surface interception evaporation estimated by a daily potential evaporation reduced by the current plant and soil water status. Potential evaporation data are derived externally by any one of several methods such as the Penman and/or Monteith method, daily pan evaporation, daily temperature or radiation methods or mean annual evaporation distributed as monthly mean daily values.
♦ Percolation: Daily water leaving the bottom of the described soil profile which will contribute to local ground water or horizontal inter-flow.

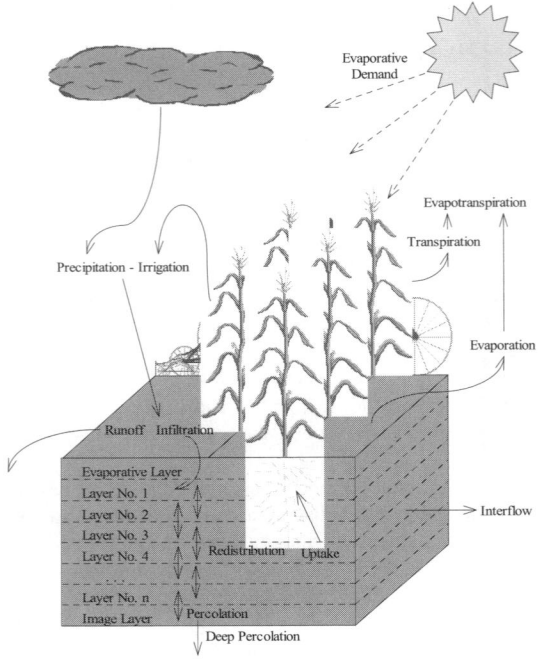

Figure 1: The principle hydrologic processes considered in the SPAW field model

POND Model:

The principle hydrologic processes of the POND model are the following as depicted in Figure 2 by inflows, withdrawals and losses representative of various inundated wetlands, lagoons, ponds or reservoirs:

- ♦ <u>Surface runoff inflow</u>: Water supplied to the pond by a watershed field(s) estimated by a SPAW field simulation.
- ♦ <u>Subsurface inter-flow</u>: Water supplied to the pond by a watershed field(s) deep seepage estimated by a SPAW field simulation.
- ♦ <u>Bank runoff</u>: Runoff from exposed pond banks above the current inundation level.
- ♦ <u>Input Pump</u>: A system delivering water from elsewhere such as an off-stream pump or an animal housing flush system.
- ♦ <u>Precipitation:</u> That falling directly on the currently inundated pond surface.
- ♦ <u>Evaporation:</u> Loss from the water surface estimated as the potential daily evaporation.
- ♦ <u>Infiltration:</u> An amount infiltrating into the dry pond bottom as it is initially inundated.
- ♦ <u>Seepage</u>: A constant seepage beneath the inundated area.
- ♦ <u>Pipe outlet</u>: Flow of a pipe outlet system having a defined stage-discharge relationship.
- ♦ <u>Spillway overflow</u>: An uncontrolled daily flow from the uppermost spillway or outlet.
- ♦ <u>Supply pump</u>: An amount pumped from the pond for designated rates and periods with a specified inlet pond depth, e.g. to supply an animal watering water tank.
- ♦ <u>Discharge pump</u>: An amount pumped from the pond for designated rates and periods with a specified inlet pond depth, e.g. to remove lagoon water to a disposal field.
- ♦ <u>Irrigation</u>: An amount supplied to one or more fields for an irrigation depth previously defined by a SPAW field simulation for each irrigated field.

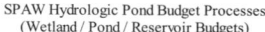

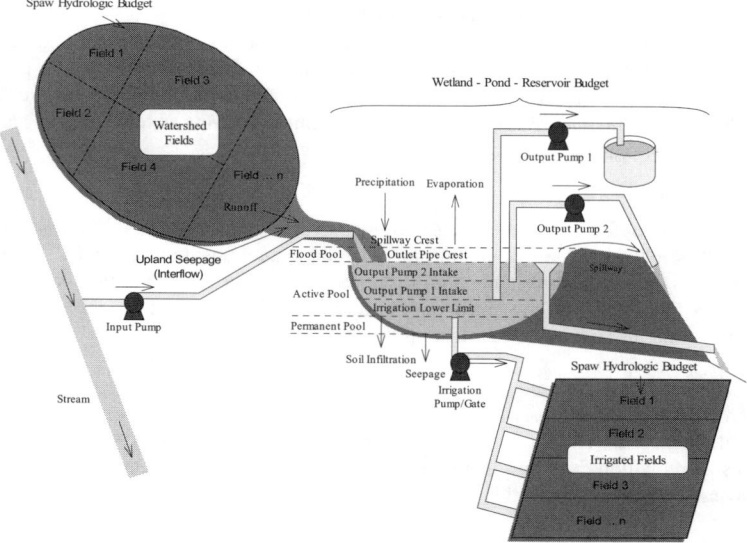

Figure 2: The principle hydrologic processes considered in the POND model

The pond volume is described by a depth-area table beginning with the pond bottom as a zero reference depth. Depths above the pond bottom are specified for permanent storage, pump inlets, pipe-outlet and the emergency spillway outlet. Each of these depths provides limits, or impacts the operation of the various budgeting processes such as the pumps, pipe outlet, or availability of irrigation water.

Limitations and Ranges

The SPAW model is a field scale, vertical water budget, given that the field can be considered spatially uniform in soil, crop and climate. These considerations limit the definition of a "field" depending on site conditions and the intended simulation accuracy. For many cases, the simulation will represent a typical farm field of tens to a few hundred acres growing a single crop with only moderate soil characteristic variations. In other cases, a single farm field may need to be divided into separate simulation regions because of distinct and significant differences of soil or crop characteristics.

The SPAW model is not intended for watershed hydrologic analyses, since each simulation is tied to a specific set of soil, crop and climate data. However, it could be utilized for relatively small watersheds composed of multiple farm fields, each simulated separately and the results combined. There is no stream flow routing or channel descriptors included and daily runoff is estimated as an equivalent depth over the simulated field. The POND model allows for a combined field concept to represent watershed input for inundation simulations of wetlands and ponds. The total combined area of fields used in POND simulations should adhere to maximum limits established for the SCS Runoff Curve Number process, approximately 800 ha (2000 acres).

The POND model has minimal limitations of size, shape or volumes. However, the climatic and watershed representations limit the practical application to maximum pond sizes in the range of 40 to 80 ha (100 to 200 ac). Interactions with ground water are included such that upward percolation is included if the surrounding water table is at a higher elevation than the pond water level.

Data Inputs and Sensitivities

Simulations are always an approximation of the real world, so it is quite logical to be concerned with the data description accuracy and their impact on the simulation results. This is particularly true when dealing with natural subjects such as soil, crops and climate which comprise the major inputs of hydrologic models. While there are few definitive answers as to what is an "adequate" accuracy, guidelines can be developed from simulation experiences.

Climatic Data:

The principal driving forces and most sensitive factors in the development of hydrologic budgets are precipitation data and the evaporation or potential evapotranspiration (PET). Daily values were selected as the most appropriate input for the SPAW model for a daily time-step hydrologic simulation. Values for each day are important and these data should be carefully checked for obvious errors and missing values, particularly for precipitation.

Climatic data must often be used from a nearby local source because they are not available at the simulation site. Thus the impact of spatial variation must be assessed and recognized to evaluate the simulation results. Evaporation and air temperature data are not as spatially or temporal variable as precipitation data, and therefore can often be transferred some distance or estimated by time-averaged weekly or monthly values and yet provide reasonable results.

Soil Profile Characteristics:

The soil profile comprises the primary hydrologic storage mechanism of a field and as such the specified water holding capacities play an important role in the water budget. Soil texture of each major profile layer is used to estimate water holding characteristics represented by average relationships previously determined from a very large data set of laboratory analyses (Saxton et al., 1986, Rawls et al., 1998). The field soil profile is described by a series of layers, each described as to its thickness and water holding and transmission characteristics. Limits for layer thickness are prescribed, but the uppermost soil layer is set to a thickness of 2.5 cm (1-inch) to facilitate the modeling of surface infiltration and evaporation processes. Typically, the second soil layer is specified as a 12.7 cm (5-inch) layer, with subsequent layers ranging in thickness from 15 to 30 cm (6 to 12 inches). The soil profile must be described to a depth below the maximum crop rooting depth.

Crop Characteristics:

Crop characteristics are described for a calendar year (or less if shorter simulations are being made) and include both growing and non-growing periods. Multiple crops during any one year would be described as a single cropping period. Two calendar year descriptions would be required for one crop for those that over-winter such as winter wheat, or southern hemisphere crops which grow over the calendar year end.

It is often difficult to describe a set of time-variable crop characteristics to sufficiently interact with the soil-water processes given the many variables and various configurations of stems, leaves and roots. However, reasonably accurate SPAW hydrologic simulations can be performed using a relatively simple set of crop descriptors. The crop definitions are four annual curves representing: 1) canopy cover, 2) greenness of the canopy, 3) root depth and 4) grain yield sensitivity to water stress. Given familiarity with the common crops in the region, the crop characteristic curves can usually be defined with sufficient accuracy to provide the general crop impact on the field water hydrology.

Field Management:

Fields may be managed with various crop rotations and inputs over multiple years. To define crop rotations, the previously defined annual crop data sets are selected and assigned a rotation year. If a single crop is used every year, such as pasture, only one year of crop rotation is needed. For multiple year rotations, several crops in a set sequence are defined. If irrigation is involved, an irrigation scheduling routine is specified with choices for the time, depth and method of irrigation for each year.

Pond Description:
The inundated pond area is described in terms of a depth-area relationship and maximum depth. If a constructed pond is being analyzed, inputs also include the discharge pipe inlet height and its depth-discharge relationship, and heights above the pond bottom to various pump inlets. An infiltration amount into the pond bottom soil at the beginning of each inundation cycle is specified such that the soil will absorb that amount before inundation begins. The user also provides an estimated constant daily seepage rate through the saturated pond bottom area, Net surface evaporation from the wetland pond area is determined from climate data.

Simulations and Results
The SPAW model may be used for hydrologic analyses of fields and ponds having unique and spatially uniform data of climate, soil and plant characteristics. In many applications this will generally be a designated farm field(s) providing water to an inundated pond. The SPAW data are input by user screens which provide flexibility and minimum duplication. Each input data file is stored for current and future simulations because these or similar data are often applicable to other fields and ponds.

Field Simulation:
A field simulation designates the appropriate previously defined data files and run criteria of: 1) climate, soil, and crop management files, 2) generation options, 3) optional observed data and 4) a simulation period. Runoff "curve numbers" by the USDA/SCS procedure (USDA/SCS-1985, USDA/NRCS-1997) are estimated from data specified in the crop and soil data screens. Given that this is an empirical approximation, the user may find it desirable to modify these first estimates to improve or calibrate the estimated runoff. Manual curve number values are optional, one for a fallow soil, and a second for the "seasonal average" of a cropped soil. Simulation results are viewed in either tabular (screen or printed) or graphic form.

Wetland/Pond/Reservoir Simulation:
A daily inundated wetland or pond budget is simulated subsequent to one or more SPAW field simulations. The simulations for one or more fields provide an estimate of runoff and subsurface flow into the pond. Pond definitions describe the physical size and characteristics. Additional descriptions are optional if pipe outlets or other water sources or applications are involved.

Results of current or previous POND simulations are viewed as graphics or text tables. Two statistical tables define the depth-duration and wetland inundation summaries of long-term simulations. The wetland statistics follow common criteria of wetland "growing periods" and "inundation period duration".

Example Wetland Analysis
An example analysis for a wetland ponding within a corn field in the vicinity of Jonesboro, AR demonstrates the simulation procedures and results. The hydrology of the contributing corn field was simulated for 30 continuous years using rainfall data from the USDA/NRCS climate center (AR3734), annual evaporation of 135 cm (53 inches) from a US evaporation map, a well drained silt loam soil, and continuous corn crop. The simulated wetland inundation pond was described to have a maximum depth of 1.8 m (6.0 ft.) before uncontrolled spillage, an area of 0.40 ha (1.0 ac) at inundation beginning (depth = 0.0 m.) and maximum area of 3.0 ha (7.5 ac) at the 1.8 m depth. A moderately high seepage rate of 25 mm/day (1.0 inch/day) was assumed, and the corn field drainage area set as 20.2 ha (50 ac).

The subsequent daily inundation simulations provided annual distributions of ponded depths as shown for one year in Figure 3. A summary and analyses routine provided wetland statistics according to the inundation criteria of being inundated a specified number of days within the "wetland" growing period of the year for more than 50% of the years. For this example, a ponded depth of approximately 60 cm (2.4 ft.) with an associated area of about 0.8 ha (2.0 acres) was defined to hydrologically qualify as a wetland.

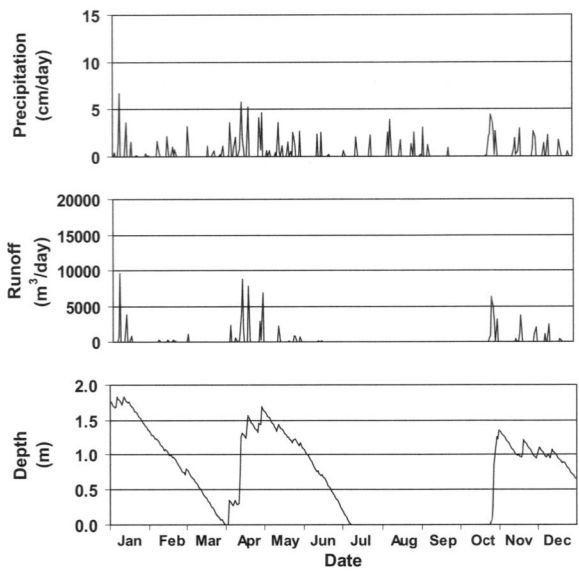

Figure 3: Daily rainfall, watershed runoff into the wetland, and inundation depth for 1991 near Jonesboro, AR.

CONCLUSIONS

An agricultural hydrologic field and pond simulation model, SPAW, has been described with emphasis and examples for wetland hydrologic analyses. The model consists of two linked routines, first a routine for the daily vertical hydrology of one or more agricultural fields, then a ponding routine which utilizes output from the field(s) simulation. The field model input includes daily climate data, annual crop definitions, and a layered soil profile with individual tension-conductivity soil water characteristics. The ponding routine utilizes the climatic data and runoff estimates from one or more previously simulated farm fields. The pond simulations may be applied to shallow wetlands, small ponds, or constructed lagoons and reservoirs. Outputs include annual, monthly or daily hydrologic budgets, graphics and wetland statistics. The SPAW model and documentation is publicly available by contacting the authors.

References

1. DeJong, R. and Zentner R. P. 1985. Assessment of the SPAW model for semi-arid growing conditions with minimal local calibration. Agric. Water Mgt., 10(1985):31-46.
2. Moffitt, D. C., Wilson, B. and Willey, P. 2003. Evaluating the design and management of waste storage ponds receiving lot runoff. Paper No. 03xxxx, ASAE Annual International Meeting, Las Vegas, NV, July 27-30, 2003.
3. Rao, A. S. and Saxton, K. E. 1995. Analysis of soil water and water stress for pearl millet in an Indian arid region using the SPAW model. Indian J. of Arid Environments 29:155-167.

4. Rawls, W. I., Brakensiek, D. L, and Saxton, K. E. 1982. Soil water characteristics. Trans. ASAE 25(5):1316-1328.
5. Saxton, K. E. 1980. Agricultural drought assessment by daily soil moisture predictions. Proc. of Climate and Risk Symposium, Washington, D. C., May, 1980. 22 pp.
6. Saxton, K. E. 1981. Mathematical modeling of evapotranspiration on agricultural watersheds. In Modeling Components of Hydrologic Cycle, a part of the Proc. Inter. Symp. On Rainfall-Runoff Modeling, Miss. State Univ., May 18-21, 1981. pp. 183-203.
7. Saxton, K. E. 1985. Soil water hydrology: Simulation for water balance computations. Proc. Int. Asoc. Hydro. Sciences General Assembly. IUGG, Hamburg, West Germany, Aug., 1983. IAHS Pub. No. 148, pp. 47-59.
8. Saxton, K. E. and Bluhm G. C. 1982. Regional prediction of crop water stress by soil water budgets and climatic demand. Trans. ASAE. 25(1):105-115.
9. Saxton, K. E. and McGuinness, J. L. 1982. Chapter 6--Evapotranspiration. In C. T. Haan, H. P. Johnson, and D. L. Brakensiek (eds) Hydrologic Modeling of Small Watersheds, Am. Soc. Agric. Eng. Monograph No. 5, pp. 229-273,
10. Saxton, K. E., Johnson, H. P. and Shaw, R. H. 1974. Modeling evapotranspiration and soil moisture. Trans. ASAE 17(4):673-677.
11. Saxton, K. E., Rawls, W. J., Romberger, J. S. and Papendick, R. I. 1986. Estimating generalized soil water characteristics from texture. Soil Sci. Soc. Amer. J. 50(4):1031-1036.
12. Saxton, K. E., Porter, M. A. and McMahon, T. A. 1992. Climatic impacts on dryland winter wheat yields by daily soil water and crop stress simulations. Agric. and Forest Meteor. 58(1992):177-192.
13. Sudar, R. A., Saxton, K. E. and Spomer, R. G. 1981. A predictive model of water stress in corn and soybeans. Trans. ASAE. 24(1):97-102.
14. USDA-SCS. 1985. National engineering handbook (NEH) Section 4, Hydrology. Soil Conservation Service, USDA, Washington D.C.
15. USDA-NRCS. 1997. Part 630 Hydrology, National Engineering Handbook. Natural Resources Conservation Service, Washington, D.C.

Two Stage Channel Design, A Comparison of Two Methodologies

Warren C. High
Senior Principal
MACTEC Engineering and Consulting
7209 E. Kemper Road
Cincinnati, Ohio 45249-1030
phone: 513.489.6611
fax: 513.489.6619
wchigh@mactec.com

Much attention has been given to two stage channels in recent years due to the need to convey small frequency storm events (i.e. 20 to 100 years floods) in a large channel while maintaining a small natural meandering stream that conveys bedload, maintains water quality, and provides habitat. In nature a floodplain and a natural stream would perform this function. In an urban environment, we have historically resorted to a trapezoidal ditch. A reasonable compromise is a two stage channel. A properly constructed two stage channel is self sustaining because it conveys bedload, it does not aggrade or degrade, and the only maintenance is the occasional removal of large woody vegetation to maintain the desired flow capacity.

This study involves the relocation of over 2000 feet of perennial stream in Southwestern Ohio. A two stage channel was the desired goal that would convey both a 100 year storm and a 1.5 year storm in a stable manner. A bridge in the center of the stream relocation provided a good place to divide the project. Both segments of stream had essentially the same flow, channel slope, and large trapezoidal channel. Calculations of channel cross section, meander width, meander length, and radius of curvature were calculated for the bankfull flow in order to create a C4 (Rosgen, 1994) type channel.

In the upper stream segment, a bankfull channel was cut and rock groins were installed to lock in the meander pattern. In the lower segment, an undersized channel was cut and allowed to erode to achieve equilibrium. It has been 5 years since construction was completed and both channel segments have achieved stable configurations.

The presentation will focus on the differences in the two channel segments including meander patterns, cross sectional area, bank stability, substrate size, and other features that differentiate the two segments. Additional discussion will include overall characterization of riparian habitat, vegetation, aquatic species, and water quality.

The unique features of this study are; 1) It is rare to have a side by side comparison with a large number of fixed parameters, 2) It is uncommon to use two methodologies on the same stream, and 3) The project has been in the ground long enough to allow useful comparison. The information that will be presented will have application to practitioners, especially in the developed environment where water quantity and water quality are a consideration.

OVERVIEW OF THRESHOLD AND ALLUVIAL CHANNEL DESIGN

Jon Fripp, Stream Mechanics Civil Engineer, USDA-NRCS National Design, Construction, and Soil Mechanics Center, Ft. Worth, TX.

and

Larry Goertz, Hydraulic Engineer, USDA-NRCS National Design, Construction, and Soil Mechanics Center, Ft. Worth, TX.

Abstract: The purpose of this paper is to outline a systematic hydraulic design methodology and design techniques for hydraulic engineers involved in design of stable streams and channels. The techniques that will be presented are divided into those that are for threshold channels and those for alluvial channels. The objective of the methodologies that will be described is to fit the channel design into the natural system within the physical constraints imposed by other project objectives and constraints.

A threshold channel has essentially rigid boundaries. Threshold channels include cases where the streambed is composed of very coarse material, erosion resistant bedrock, clay soil, or grass lining. In an alluvial channel, there is an exchange of material between the incoming sediment load and the bed and banks of an alluvial channel under design or normal flow conditions. Essentially, where a threshold channel is fixed under design conditions, an alluvial channel is one that is free to change its shape, pattern and planform in response to short or long term variations in flow and sediment.

The design approach for a threshold channel is to select a channel where the stress applied during design conditions is below the allowable stress of the channel boundary. This paper describes threshold channel design procedures based on the permissible velocity and tractive stress approaches. Channel design becomes more complicated in alluvial channels where the bed is mobile and where bed-material sediment inflow is significant. In addition to water-surface elevation, efficient transport of sediment becomes a focus in the hydraulic design of alluvial channels. The recommended design methodology for alluvial channels that will be briefly outlined includes analytical solutions of resistance and sediment transport equations in combination with application of geomorphic principles. When possible, alluvial channels are sized for the channel-forming discharge. However, the response of the design channel to the entire range of natural flow needs to be evaluated.

Natural Channel Design Applications for Restoring Streams in North Carolina

Greg Jennings, PhD, PE
Dan Clinton, PE
David Bidelspach, EI
Barbara Doll, PE
North Carolina State University
Greg_Jennings@ncsu.edu

Stream restoration can be defined as the application of engineering, geologic, and biological principles to improve hydrologic, habitat, and aesthetic functions of the stream corridor, considering current and future watershed conditions. Components of a successful stream restoration project may include: (1) adjusting the stream channel size and shape; (2) establishing a hydraulic connection between the channel and floodplain; (3) adding in-stream structures; (4) stabilizing streambanks; and (5) enhancing vegetation in the riparian corridor. The natural channel design approach makes use of reference stream morphology and biology information to devise a comprehensive project aimed at restoring and maintaining natural stream functions over the long term. The purpose of this paper is to describe lessons learned implementing natural channel design projects under various conditions in North Carolina.

REGIONAL CURVE DEVELOPMENT FOR KANSAS

Brock A. Emmert[1]

ABSTRACT

In 1999, a study to establish a baseline database of fluvial geomorphology characteristics for Kansas streams began. Data were employed to apply a uniform stream classification system and develop regional runoff curves. Regional runoff curve data are essential for estimating the discharge and dimensions of bankfull elevations in ungaged watersheds. The study has used the Rosgen (1994) stream classification. To date, 166 geomorphology surveys have been completed for Kansas. Of these surveys, 125 were at active and discontinued USGS gaging stations. The remainder were at reference reaches on ungaged watersheds. Data were collected by measuring the stream's dimension, pattern, and profile at the bankfull elevations. USGS data analysis of hydraulic geometry relationships and flood frequency data helped calibrate the bankfull stage. Study of each stream included surveys of channel cross sections, longitudinal profiles, channel patterns, and channel materials. Data analysis allowed classification of each stream survey as a Rosgen stream type, determination of the bankfull flow channel dimensions, and corresponding discharge. Further analysis developed relationships among hydraulic parameters at the bankfull stage to drainage area—helpful to produce regional curves. Regression analysis strongly correlated drainage area with bankfull discharge when surveys were grouped into hydrophysiographic provinces. This document presents results of the regional runoff curve analysis for identified hydrophysiographic provinces in Kansas.

KEYWORDS. fluvial geomorphology, Kansas streams, stream classification, regional runoff curves, bankfull discharge

INTRODUCTION

Kansas's streams and their associated riparian corridors are collectively a vital natural resource. For a large percentage of the State's population, streams provide water for drinking, irrigation, and recreation. Stream corridors serve as aquatic and terrestrial wildlife habitat for numerous species. Kansas stream corridors have undergone many disturbances. Fluvial systems are especially susceptible to disruption due to their numerous occurrences in the landscape and sensitivity to land use changes (Graf, 1977). Land use changes resulting from agriculture and urban development, channelization, irrigation, impoundment, removal of riparian corridors, over grazing of livestock in riparian corridors, and draining of riparian wetlands all have contributed to stream degradation in Kansas. All stages of the evolutionary sequence for channels developed by Schumm, Harvey, and Watson (1984) and Simon (1994) are found in Kansas. Some streams have reached a new steady-state or equilibrium, while many streams are still in the stages of adjustment.

Kansas has management strategies to help protect, enhance, or restore stream corridors. They involve primarily a site-specific or patch-in-place approach that improves the condition of a particular stream segment but may not change the overall health of the stream (Emmert and Hase, 2001). To improve the health of a stream system, the Kansas Riparian Technical Team (KRTT), a group of professionals in natural resource fields, recognized need for a watershed

---

[1]Geologist, Tetra Tech EM Inc. 1200 SW Executive Dr., Topeka, KS 66615-3850. brock.emmert@ttemi.com

approach to stream corridor protection and restoration. In 1998, the KRTT drafted and was awarded a Wetland Protection Grant entitled, "Geomorphic Assessment and Classification of Kansas Riparian System" from the Environmental Protection Agency (EPA), Region VII. The Kansas Water Office (KWO) administered this project that devoted three and one-half years to develop a fluvial geomorphology database for Kansas stream corridors. After completion of this project, KRTT won another EPA grant entitled, "Assessment, Geomorphic Definition, and Documentation of Kansas Stream Corridor Reference Reaches." This grant is administered by the State Conservation Commission (SCC) and performed by SCC and Tetra Tech EM Inc. personnel. Participants in these two projects have dedicated over six years to collecting fluvial geomorphology data. This paper represents part of the results from these two grants.

Fluvial geomorphology examines how water shapes the Earth's surface. Leopold, Wolman, and Miller (1964) identified eight major variables that influence stream morphology—width, depth, velocity, discharge, channel slope, roughness of channel materials, sediment load, and sediment size. To collect and organize stream morphology data clearly and meaningfully, KRTT used the Rosgen stream classification system and data collection protocol (Rosgen, 1994). Rosgen developed one of the most comprehensive and objective stream classification systems to date (Fig. 1) from measured morphologic characteristics and river-formed variables obtained from hundreds of actual river sites (Rosgen, 1996). Thorne (1997) stated that the Rosgen classification system incorporates all three dimensions of channel form while also accounting for differences in channel forming materials. This classification system provides natural resource professionals a common and universally applicable means of communicating about streams.

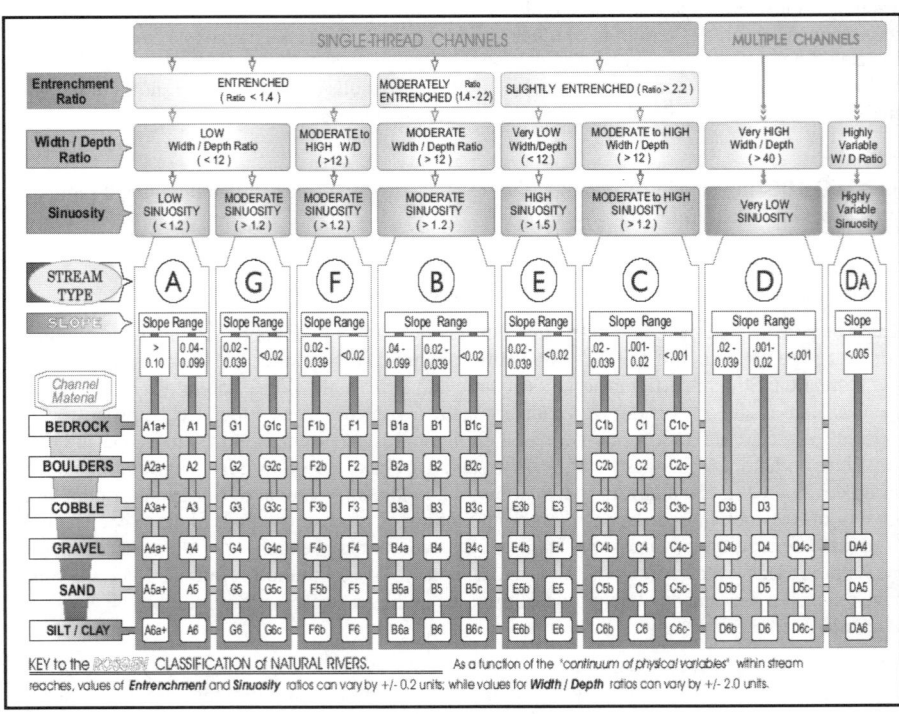

Figure 1. Rosgen Stream Classification System for Natural Rivers (Rosgen, 1994).

Basic to the Rosgen classification system are the dimension, pattern, and profile of the bankfull discharge. Dunne and Leopold (1978) define bankfull as the stage, "correspond[ing] to the discharge at which channel maintenance is the most effective, that is, the discharge at which moving sediment, forming or removing bars, forming or changing bends and meanders, and generally doing work that results in the average morphologic characteristics of channels." Wolman and Miller (1960) stated large flow events can cause major channel changes, but the frequency is too low to govern channel characteristics. They also concluded that low lows are too common and frequent, thus ineffective at changing channel characteristics. The bankfull discharge is the frequent, intermediate-sized event with an average return interval of 1.5 years (Dunne and Leopold, 1978). Rosgen (1996) concluded that bankfull discharge should recur every 1-2 years. Studies by Harman et al (1999), Harman et al (2000), Emmert and Hase (2001), and Odem and Gilman (2002) support the bankfull interval of 1-2 years. Odem and Gilman did have return intervals as high as 3.2 years in the arid southwest. Nonetheless, these studies confirm that bankfull discharges occur at similar frequencies in various types of streams across the United States.

Bankfull discharges from each survey were used to develop regional curves. Regional curves support estimates of bankfull discharge and related channel dimensions at ungaged sites (Rosgen, 1998). Regional curves are grouped by hydrophysiographic provinces that are areas of equivalent geology, climate, soils, and landuse which produce similar geomorphic and hydrologic conditions.

With few exceptions, all surveys included in this study have been in rural areas, and the flows reflect land uses. A few surveys occurred in urban areas where runoff processes completely differ from sites in rural areas. Therefore, regional runoff curves presented in this paper are intended to help determine bankfull in rural landscapes. Regional runoff curves identified are useful for professionals in natural resource fields. Uses include designing natural channels, determining bankfull dimensions in ungaged watershed, undertaking departure analysis in impaired watersheds, and helping to determine stream stability.

STUDY AREA

Kansas has a variety of conditions that influence the physical characteristics of its streams. Watersheds across the State vary greatly in precipitation, geology, soils, and landuse. Precipitation increases from west to east—from southwest Kansas receiving average rainfall of 14-16 inches to southeast Kansas with average rainfall over 40 inches. Kansas has a variety of ecoregions (Fig. 2). EPA classified ecoregions using many landscape features, including geology, and predominant landuse (Chapman et al., 2000).

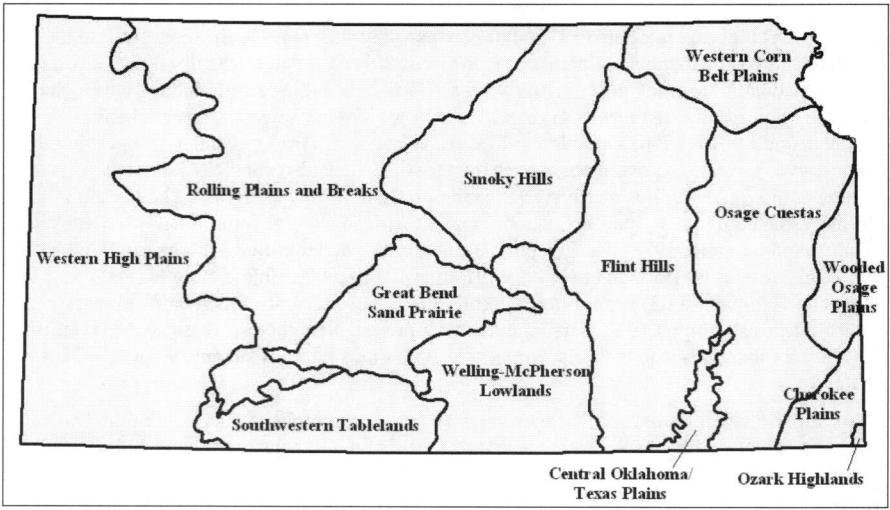

Figure 2. Major Ecoregions in Kansas (adapted from Chapman et al., 2000).

Field crews began surveying at USGS gaging stations. These included current and discontinued stations. USGS stations not included in this study were sites directly below federal impoundments and stations in highly disturbed watersheds or sites with modified channels. Between 1999 and 2001, 125 surveys at USGS gaging stations were completed (Fig. 3). Survey sites for the SCC grant are in ungaged drainages. We selected sites that represented healthy stream segments in watersheds not included in the USGS stream gaging network. Between 2002 and 2003, 41 surveys were completed (Fig. 3). The goals were to survey stream reaches typical for each hydrophysiographic province and to achieve an even geographic distribution across the State with a variety of drainage sizes.

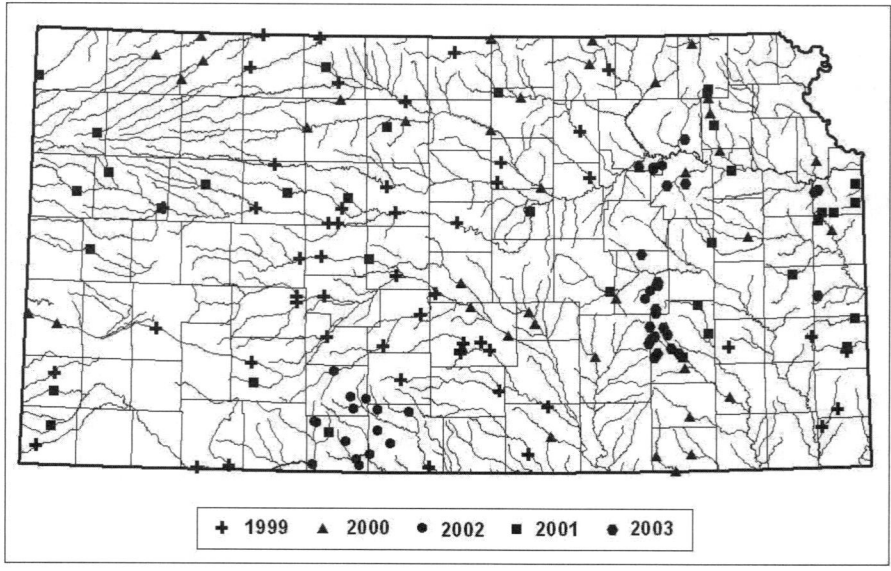

Figure 3. Geomorphology survey locations from 1999 – 2003.

METHODS

Surveys at USGS gaging stations were conducted to determine the bankfull dimensions and discharge, and assign a Rosgen stream type. Before each survey, USGS gage data were obtained for information on channel characteristics, flood events, and survey benchmarks. This information included a written gage description, 9-207 records, current expanded rating tables, and annual peak discharge data.

Written gage descriptions provided benchmark locations and elevations for referencing surveys. Hydraulic geometry relationships developed by Leopold and Maddock (1953) were determined from the 9-207 records. The relationship correlates parameters of width, depth, channel area, and mean velocity commensurate with an associated discharge (Rosgen, 1998). The expanded rating tables are used to relate a discharge to an elevation or gage height. Log-Person analysis was performed on annual peak flow data. A minimum of 10 years was used for this analysis.

Each survey used a Laser Alignment LB-10 as the surveying instrument. Survey techniques applied at each site were referenced from "Stream Channel Reference Sites: An Illustrated Guide to Field Technique" by Harrelson, Rawlins, and Potyondy (1994). Field crews referenced all elevations to USGS benchmarks and expressed the elevations as gage heights. At some discontinued and all ungaged sites, benchmarks were installed and a relative elevation used.

Cross sections were surveyed at riffles or cross-over reaches where a uniform cross section was along the channel. Each cross section was surveyed perpendicular to flow and, where possible, captured the channel from left top of bank to right top of bank. Survey crews noted various features in the cross section surveys—top of banks, terraces, edges of water, thalweg, and bankfull indicators. Indicators of the bankfull flow include top of point bars, change in vegetation, change in bank slope, change in bank materials, bank undercuts, and stain lines (Rosgen, 1996). When an indicator was identified, the expanded rating table was used to assign a discharge for the gage height elevation. Next, the return period of that discharge was checked from the peak flow analysis. If the return interval was in the proper range, the stage was used for the bankfull elevation. Once identified, the bankfull channel dimensions were extracted from the cross section survey.

The survey crew also conducted longitudinal profiles. Each profile was surveyed in the thalweg, and the length was normally 20 times the bankfull width. Some surveys fell short of the length due to landowner permission refusal and severe debris jams. Survey shots were taken in the thalweg, water surface, and bankfull indicators where possible throughout the entire profile.

Channel pattern measurements were collected from aerial photograph interpretation. Parameters measured included meander lengths, radius of curvatures, belt widths, and the channel sinuosity. An average was used for the first three measurements.

Channel materials collection or pebble counts first developed by Wolman (1954) and later modified by Rosgen (1993) were conducted for the entire survey reach. This pebble count procedure was used because the sampling is proportional to the survey reach. For instance, if 60% of a reach is riffles, 60% of the pebble count is sampled in riffles. Ten transects were sampled across the bankfull channel, with 10 measurements recorded in each transect. A sample was taken by looking away from the stream bed and reaching down until touching a sediment particle. The intermediate axis was measured and recorded by size classes based on the recommendations of the American Geophysical Union Subcommittee on Sediment Terminology.

For surveys conducted at stable ungaged sites, a different procedure was used to determine the bankfull discharge. First, a pebble count was conducted at the riffle cross sections. This pebble count contained at least 100 measurements of the cross section active channel bottom. Bankfull indicators were noted in the same fashion as the gage sites. The mean velocity was determined using the Darcy-Weisbach resistance coefficient (f), as expressed in the following equations:

31

$$u = \left(\frac{8gRS}{f}\right)^{\frac{1}{2}} \qquad\qquad (1)$$

where:

        u = mean velocity (ft/s)
        g = gravitational acceleration (ft/s^2)
        R = hydraulic radius (ft)
        S = bankfull average water surface slope (ft/ft)
        f = Darcy-Weisbach coefficient

The Darcy-Weisbach coefficient is related to the ratio of mean velocity to mean shear velocity by equation 2 (Bathurst, 1997):

$$\frac{u}{u^*} = \left(\frac{8}{f}\right)^{\frac{1}{2}} \qquad\qquad (2)$$

where:

        $u^* = (gRS)^{1/2}$ = mean shear velocity (ft/s)

The mean velocity then is calculated by the friction factor/roughness relationship (Rosgen, 1998) in equation 3:

$$u = u^*(2.83 + 5.7 \log R/D_{84}) \qquad\qquad (3)$$

where:

        R = bankfull hydraulic radius (ft)
        D_{84} = D_{84} from pebble count conducted at cross section (ft)

Once calculated, the mean velocity is used to determine the bankfull discharge. The discharge is checked with gage sites in the vicinity. If the bankfull discharge and dimensions correlate well with the gage sites, the bankfull elevation is used for the survey.

RESULTS

Every stream type (A…G) except A was classified in the geomorphic surveys. Stream types that imply instability in Kansas are D, F, and G. The bankfull information was not included in the regional runoff curves. Return intervals for bankfull discharge ranged from 1.06 to 1.77 years. The relationship between bankfull discharge/drainage area versus drainage area helped delineate hydrophysiographic provinces (Fig. 4). Hydrophysiographic provinces share similar stream properties based on the drainage size. The plots in Figure 4 assist in identifying hydrophysiographic provinces. Six different hydrophysiographic regions are identified to date in Kansas (Fig. 5). They correlate well with the ecoregions in Figure 2.

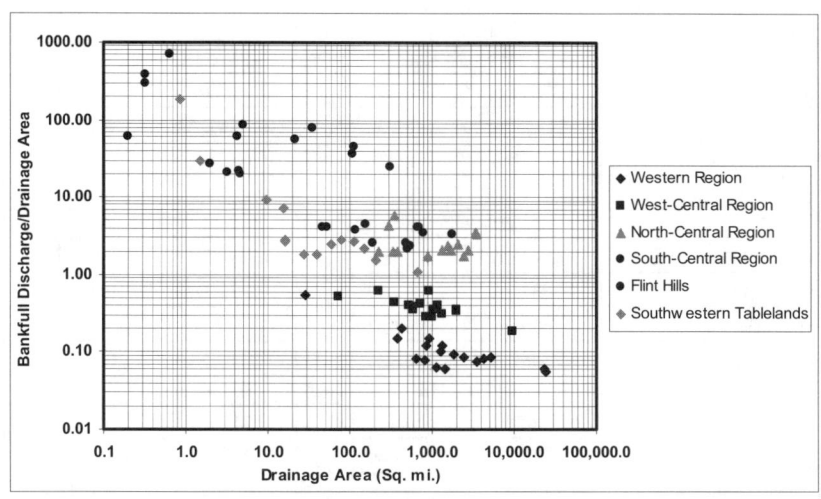

Figure 4: Bankfull Discharge/Drainage Area vs. Drainage Area.

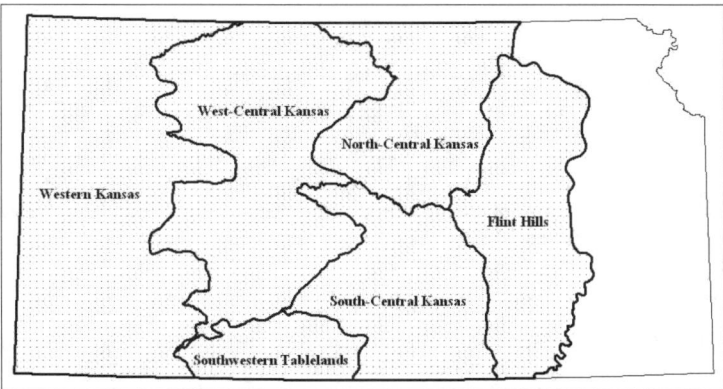

Figure 5: Identified hydrophysiographic provinces for Kansas.

Regression analysis determined the correlation between drainage area and hydraulic parameters at the bankfull stage for the six hydrophysiographic regions (Table 1). The datasets were calculated using the regional curve operations in RIVERMorph (2001-2003).

Table 1. Regional runoff curves for identified hydrophysiographic provinces.

Hydraulic Parameter	Regression Line	R^2	n	Regression Line	R^2	n
	Western Kansas			West-Central Kansas		
Bankfull Discharge	$0.8761DA^{0.7046}$	0.987		$1.558DA^{0.7855}$	0.925	
Stream Width	$1.44536DA^{0.5099}$	0.760	18	$2.5729DA^{0.4136}$	0.516	16
Mean Depth	$0.7229DA^{0.0092}$	0.0007		$0.3063DA^{0.3565}$	0.057	
Cross-Sectional Area	$1.0781DA^{0.5991}$	0.940		$0.775DA^{0.7725}$	0.610	
	North-Central Kansas			South-Central Kansas		
Bankfull Discharge	$3.2544DA^{0.9654}$	0.865		$5.2840DA^{0.9189}$	0.954	
Stream Width	$1.3587DA^{0.6511}$	0.839	14	$10.5616DA^{0.3851}$	0.461	12
Mean Depth	$3.9659DA^{0.0801}$	0.022		$0.3739DA^{0.4092}$	0.348	
Cross-Sectional Area	$8.2508DA^{0.6741}$	0.820		$3.9968DA^{0.7914}$	0.878	
	Flint Hills			Southwestern Tablelands		
Bankfull Discharge	$99.9905DA^{0.7377}$	0.902		$33.7470DA^{0.4122}$	0.932	
Stream Width	$25.0995DA^{.3454}$	0.649	15	$10.9349DA^{0.3054}$	0.581	15
Mean Depth	$1.1226DA^{0.3309}$	0.859		$0.5952DA^{0.1672}$	0.854	
Cross-Sectional Area	$28.2157DA^{0.6762}$	0.963		$6.5289DA^{0.4716}$	0.831	

CONCLUSIONS

Regional runoff curve analysis strongly correlated with most hydraulic parameters at the bankfull stage. Hydrophysiographic provinces matched well with mapped ecoregions (there are disparities, but the fit is generally close). The hydraulic parameters of width and depth did not correlate well with drainage area for several provinces. These correlations would have been closer if regression analysis was performed by stream type in each province. Not enough surveys were completed to perform this operation. Also, eastern Kansas is not complete because not enough surveys have occurred. More surveys are needed to delineate hydrophysiographic provinces and calculate regional runoff curves.

Acknowledgements

I would like to thank Phil Balch for his guidance and help during these two projects. Also, I would like to thank Dr. Tim Keane for his review of data and assistance with field work. Survey crew members who collected data necessary to complete these projects included Rich Beavens, Michael Martin, Kelly Warren, Sam Aberle, Andrew Adams, Katie Kingery-Page, Edward Larson, Shaun Morrell, Brian Hughes, April Felker-Hay, and Adriane Ohlde.

REFERENCES

1. Bathurst, J.C. 1997. Environmental river flow hydraulics. In C.R. Thorne, R.D. Hey, and M.D. Newson. (eds.). *Applied Fluvial Geomorphology for River Engineering and Management*. 69-94. Chichester, England: John Wiley and Sons.

2. Chapman, S.S, J.M. Omernik, J.A. Freeouf, D.G. Huggins, J.R. McCauley, C.C. Freeman, G. Steinauer, R. Angelo, and R.L. Schlepp. 2000. Ecoregions of Nebraska and Kansas (color poster with map, descriptive text, summary tables, and photographs). USGS. Reston, VA.

3. Dunne, T. and L.B. Leopold. 1978. *Water in Environmental Planning*. San Francisco, CA: W.H. Freeman and Co.

4. Emmert, B.E. and K. Hase. 2001. Geomorphic assessment and classification of Kansas riparian systems. EPA No. CD997520-01.

5. Graf, W.L. 1977. The rate law in fluvial geomorphology. *American Journal of Science*. 277: 178-191.

6. Harman, W.A., G.D. Jennings, J.M. Patterson, D.R. Clinton, L.O. Slate, A.G. Jessup, J.R. Everhart, and R.E. Smith. 1999. Bankfull hydraulic geometry relationships for North Carolina streams. In S.Olsen and J.P. Potyondy (eds.). *Wildland Hydrology, Proceedings of AWRA Specialty Conference*. Bozeman, Montana: 401-408.

7. Harman, W.A., D.E. Wise, M.A. Walker, R. Morris, M.A. Cantrell, M. Clemmons, G.D. Jennings, D. Clinton, and J. Patterson. 2000. Bankfull regional curves for North Carolina mountain streams. In D.L. Kans (ed). *Proceedings of the AWRA Conference Water Resources in Extreme Environments*. Anchorage, Alaska: 185-190.

8. Harrelson, C.C, C.L. Rawlins, and M.G. Potyondy. 1994. Stream channel reference sites: an illustrated guide to field technique. GTR: RM-245. USDA Forest Service. Rocky Mountain Forest and Range Experiment Station.

9. Leopold, L.B. and T. Maddock Jr. 1953. The hydraulic geometry of stream channels and some physiographic implications. Geological Survey Professional Paper 252. Washington, D.C.: WPO.

10. Leopold, L.B., M.G. Wolman, and J.P. Miller. 1964. *Fluvial Processes in Geomorphology*. New York, NY: Dover Publications, Inc.

11. Odem, W.I. and J.P. Gilman. 2002. Hydraulic and Hydrologic Relationships in Streams of the Southwest US. Submitted to American Water Resources Association. April.

12. RIVERMorph. 2001-2003. RIVERMorph Stream Assessment and Restoration Software. Ver. 2.1. Louisville, Kentucky. RIVERMorph, LLC.

13. Rosgen, D.L. 1993. *Applied Fluvial Geomorphology, Training Manual*. River Short Course. Wildland Hydrology. Pagosa Springs, Colorado.

14. Rosgen, D.L. 1994. A classification of natural rivers. *Catena* 22: 169-199.

15. Rosgen, D.L. 1996. *Applied River Morphology*. Pagosa Springs, Colorado: Wildland Hydrology Books.

16. Rosgen, D.L. 1998. *The Reference Reach Field Book*. Pagosa Springs, Colorado: Wildland Hydrology Books.

17. Schumm, S.A., M.D. Harvey, and C.C. Watson. 1984. *Incised Channels Morphology, Dynamics and Control*. Littleton, Colorado: Water Resource Publications.

18. Simon, A. 1994. Gradation processes and channel evolution in modified west Tennessee streams: process, response, and form. Geological Survey Professional Paper 1470. Washington, D.C.: WPO.

19. Thorne, C.R. 1997. Channel types and morphological classification. In C.R. Thorne, R.D. Hey, and M.D. Newson. (eds.). *Applied Fluvial Geomorphology for River Engineering and Management*. 175-222. Chichester, England: John Wiley and Sons.

20. Wolman, M.G. 1954. A method of sampling coarse river-bed material. *Transactions of American Geophysical Union* 35: 54-74.

21. Wolman, M.G. and J.P. Miller. 1960. Magnitude and frequency of forces in geomorphic processes. *Journal of Geology* 68: 54-74.

RESTORING MINNESOTA'S AILING RIVERS

L.P. Aadland, Ph.D.

ABSTRACT

The rivers of Minnesota have been damaged by watershed changes, channelization, and fragmentation through dam construction. Watershed changes such as cultivation of prairie and forest, wetland drainage and urbanization have affected hydrology and sediment yield. Channelization directly alters instream habitat but also initiates degradation and aggradation processes that may extend well upstream and downstream of the project area. Dam construction blocks fish migrations and initiates reservoir sedimentation and tailwater degradation that damages habitat. Traditional "hard engineering" approaches have generally ignored fluvial process and consequently have caused numerous maintenance and ecological problems. This paper focuses on restoration of natural channels and fluvial processes, dam removal, and restored fish passage. The approach involves application of reference reach geomorphic data in channel design. Several case examples will be presented.

KEYWORDS. River restoration, dam removal, nature-like fishways

INTRODUCTION

Rivers are a function of their hydrology, fluvial geomorphology, energy pathways, water quality, biology and interactions among these components (Annear et. al 2002). Watershed and land-use changes, dredging and straightening of channels, and fragmentation through construction of dams and other barriers have been primary means that have compromised stream functions (Aadland et al., in press).

Watershed changes such as urbanization and conversion of prairie and forest to row crop agriculture coupled with wetland drainage and ditch construction increase surface runoff, flood peaks, and alter sediment yield (Dunne and Leopold, 2002). This can cause changes in channel dimension and pattern as a response to the altered hydrology (Leopold 1994).

Direct channel alterations such as dredging and straightening can increase slope and shear stress, initiate headcutting and channel degradation (U.S. Army Corps of Engineers 1994), increase inputs of bank materials, and deteriorate and homogenize instream habitat. Fish biomass and diversity is often substantially lower in dredged or straightened rivers. Thousands of miles of stream have been channelized in Minnesota.

Over 2500 dams have been constructed in Minnesota. Dams intercept sediment, causing aggradation within the reservoir and degradation of the channel downstream of the dam (U.S. Army Corps of Engineers 1994). Reservoirs created by dams inundate important high gradient habitat, block migrations of fish and other biota, and dam operations can cause abrupt changes in flow that alter habitat, strand or dewater fish, eggs, and invertebrates. Dams can also alter temperature regimes and nutrient processes.

River Restoration Techniques

Watershed Restoration

Restoration of watersheds is difficult on a large scale and requires changes in attitudes and policies, creation of landowner incentives and elimination of counterincentives, and, in some cases, changes in laws. The Conservation Reserve Program and various initiatives to restore wetlands and perennial vegetative cover are examples of watershed restoration efforts that benefit rivers.

Channel Restoration

Rosgen (1996) outlined an approach to natural channel design that incorporates the use of reference sites, regional data, and norms for channel geometry by stream type to guide determination of appropriate channel dimensions, patterns and profiles. This approach is different from traditional means of stabilizing channels with hard structures and, instead, reinitiates natural fluvial processes. We have used this general approach to re-meander rivers that had been straightened or dredged. For example, the Whitewater River in Southeastern Minnesota was channelized in 1958. Fish biomass was 21 times higher in a meandering reference reach than in the straightened channel. We reconnected 2.5 miles of historic meandering channel and used a relatively unaltered reach as a reference channel to design an additional mile of channel to replace the straightened reach. This project served to stabilize the stream, restore riparian wetlands, instream habitat, and channel forming processes. Fish biomass (catch per unit effort) in the restored reaches were comparable to the reference reach and over 30 times the biomass in the straight pre-project channel (Figure 1).

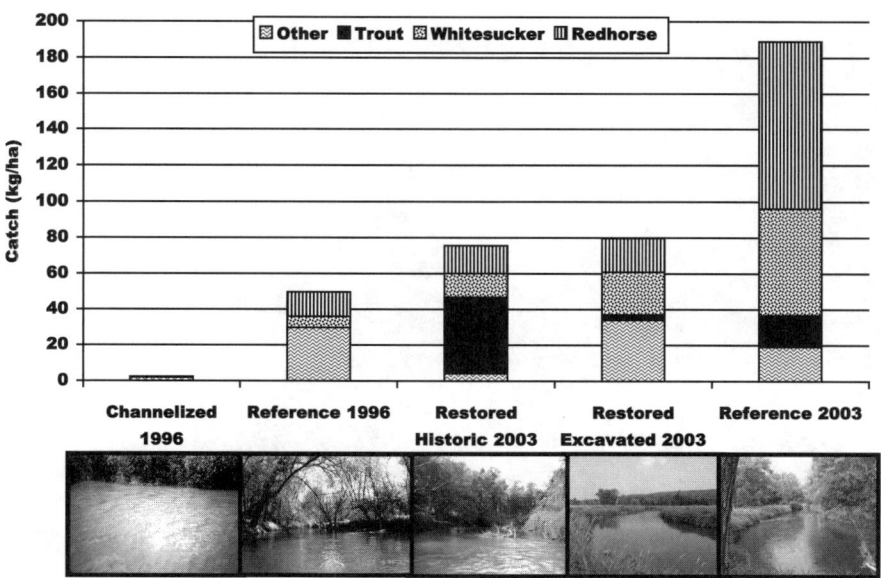

Figure 1. Catch per unit effort of fishes in the channelized (pre-project), reference, and restored reaches of the Whitewater River.

Natural design concepts can also be applied to ditches as a means of improving water quality and aquatic habitat, and reducing maintenance costs. A total of 14,000 feet of a ditched portion of Dalen Coulee in Western Minnesota was restored as part of a flood-damage reduction project we did cooperatively with the Wild Rice Watershed District and Ducks Unlimited. A 160-acre wetland was constructed in the upper portion of the watershed to provide storage and waterfowl habitat.

Figure 2. A restored reach of Dalen Coulee (left), and a constructed wetland in the upper part of the watershed (right).

Ditches that are left without maintenance eventually remeander on their own. Ditches are often designed to contain a 10-year flood while bankfull of natural channels is typically a 1-2 year event. As a result, ditches can be built to have a meandering channel with a 1-2 year capacity channel and a flood plain with a 10-year capacity

Figure 3. A legal ditch that has naturally remeandered to form a stable stream (Lawndale Creek).

Dam Removal

One of the challenges in removing dams is addressing accumulated sediment in the reservoir. When dams are removed, a headcut is typically initiated through this sediment creating an unstable, entrenched and eroding channel that lacks quality fish habitat. We have used natural channel design techniques to restore stable channels in sediment-laden reservoirs. Removal of a milldam on the Pomme De Terre River in Southeastern Minnesota is presented as an example. The dam was partially removed during low water to allow the sediments to consolidate and vegetate. The dam was then replaced with a 2-foot high, 80-foot long rock rapids to provide grade control. The river cut a relatively straight channel through the reservoir sediments. Based on several, naturally stable reference reaches, about a half-mile of meandering channel was excavated and eight rock riffles were constructed for grade control through the former reservoir. While a 350-year flood immediately after construction caused some damages to un-vegetated spoil used to plug the straight channel, damages to the excavated channel were minor and were repaired. The city agreed to allow their property in the old reservoir to naturally vegetate and planted additional trees. Access trails were mowed through the new flood plain. Fishing in the restored channel is reportedly excellent and a number of game and non-game species previously absent in the river upstream of the dam are now found within the former reservoir and upstream.

Figure 4. The Pomme de Terre River before (above) and after (below) dam removal and stream restoration.

Dam Modification

Barrier dams frequently cause extirpation of migratory species in reaches upstream of the dam and can cause basin-wide extirpation of species when critical spawning habitat is lost. While dam removal is the preferred solution to restore these migratory pathways, existing dam functions can preclude this option. Conventional fish ladders generally are not successful in passing many of our warmwater fish species since they do not jump like trout and salmon. Conversion of low-head dams to rapids is one means of restoring fish passage for these species. Lake sturgeon, which historically grew to as large as 400 pounds in the Red River Basin, were extirpated as a result of dam construction and over harvest. Dams blocked access to critical high gradient spawning habitat in the tributaries. As a means of restoring access to critical spawning habitat and eliminating hydraulic undertows associated with low-head dams, we have converted many of these dams to rapids. I developed a design, in collaboration with engineers from the U.S. Army Corps of Engineers, based on observations of natural rapids that incorporates a series of U-shaped boulder weirs similar to those used by Rosgen (1996) and others in river restoration. The weirs create a step-pool configuration that takes advantage of the high burst speed capabilities of fish while providing lower-velocity resting pools between them. At least one weir is used per foot of head loss. The weirs also create converging flow conditions that reduce stress on the riverbanks. Since the hydraulics of these rapids is similar to natural rapids, fish passage is restored and spawning habitat is created within them. We have observed over 30 fish species migrating through these rapids and several species spawning in them. Nature-like fishways that incorporate a step-pool configuration have been widely used in Europe (DVWK 2002) and Canada (Gaboury et. al 1995). I feel that the design presented here has particular advantages in passing the full spectrum of fish species over the full range of flow conditions, and in reducing bank stress in the tailwater. While nature-like fishways do not eliminate many of the ecological problems associated with dams, they are a good means of restoring fish passage and providing spawning habitat.

Figure 5. The Fargo North Dam on the Red River of the North following its conversion to rapids.

Conclusions

The rivers and streams of Minnesota have been severely damaged by past alterations to their watersheds, channels, and connectivity. Use of natural channels as templates for restoration is an intuitively obvious means of restoring fluvial and ecological functions and processes. This approach is different than traditional approaches that considered channel straightening as "channel improvement". The advantages of this self-sustaining approach include reductions in maintenance cost, improved water quality, nutrient uptake, fisheries, biodiversity, aesthetics, and channel stability.

References

Aadland, L.P, T.M. Koel, W.G. Franzin, K.W. Stewart, and P. Nelson. In press. Changes in fish community structure in the Red River of the North. In: *Historical Changes in Fish Assemblages of Large American Rivers*. Bethesda, Maryland. American Fisheries Society

Annear, T. and 15 other authors. 2002. Instream flows for riverine resource stewardship. United States. Instream Flow Council.

Dunne, T. and L. Leopold. 2002. Water in environmental planning. New York, N. Y.: W.H. Freeman and Company.

DVWK. 2002. Fish Passes: Design, Dimensions, and Monitoring. Rome, Italy: Food and Agriculture Organization of the United Nations.

Gaboury M.N., R.W. Newbury, and C.M. Erickson. 1995. Pool and riffle fishways for small dams. Winnepeg, Manitoba: Manitoba Natural Resources Fisheries Branch.

Leopold, L.B. 1994. A view of the river. Cambridge, Massachusetts: Harvard University Press.

Rosgen, D. 1996. Applied River Morphology. Pagosa Springs, CO: Wildland Hydrology.

U.S. Army Corps of Engineers. 1994. Channel stability assessment for flood control projects. Engineer Manual EM1110-2-1418. Washington, D.C.: U.S. Army Corps of Engineers.

Ground Water Pore-Pressure Influences on Stream Restoration

J.A. Magner[1], O. Baird[2]. K.J. Kuehner[3]

Abstract

Ground water exchange with stream channel beds can produce unique restoration challenges. Ground water flow into alluvial substrate stream channels can weaken boundary shear strength of the channel bed. Bed instability limits the quality of aquatic habitat. Channel stability was evaluated on the lower reach of Seven Mile Creek, an 86-km^2 trout stream in south-central Minnesota. Based on assessment a 330-m was restored in a county park using natural channel design techniques. Detailed geomorphic data was collected from the incised stream to flatten channel grade, create pools, and establish a meander bend point bar. A track-hoe was used to excavate bed and bank materials, and place large natural rock and root wades for channel and habitat restoration. For less than $50 per linear meter the restoration activities were generally successful for infrastructure protection based on post storm channel inspection. A before-and-after trout survey indicates that trout populations may have improved with increased percent of pools and enhanced riffle quality. Designed pool depths were not achievable because ground water pore-pressures collapsed pool side-walls when the depth exceeded more than 1.06-m or a side-wall angle greater than 11.4 degrees. Placement of footer rock below cross-vanes and "J"-hook rock vanes was limited to one boulder because sub-bed material collapsed into the excavation within seconds of track-hoe bucket removal. The consideration of ground water pore-pressure should be included in stream restoration designs that interact with ground water discharge zones.

KEYWORDS: River Restoration, Ground water, Fish Habitat

Introduction

The Minnesota Pollution Control Agency (MPCA) is currently developing a fish index of biological integrity (IBI) for streams and rivers across Minnesota (MPCA, 2004). In southern Minnesota, where the study site is located, over 33 stream reaches have been listed for impaired fish biota on the Clean Water Act (CWA) section 303(d) list. Southern Minnesota is intensively managed for row crop production (Payne, 1991) and has been extensively drained via subsurface perforated plastic pipe (Binstock, *in* Magner, et al., 2004). Agricultural drainage in southern Minnesota over the last century has drained wetlands, particularly in the Seven Mile Creek (SMC) watershed (Kuehner, 2004), along with the construction of ditches to improve soil productivity (Leach and Magner, 1992) However, the cumulative influence of large scale drainage, particularly the increase in contributing drainage area, initiated downstream channel instability across southern Minnesota (Magner and Steffen, 2000, and Magner, et al., 2004). When a point source of pollution is not present in a watershed listed for impaired biota, non-point sources become the focus of investigation. Poor fish IBI scores were related to poor fish habitat by Talmadge et al (2002), and Magner, et al. (2003). Sediment deposition in spawning gravels and pools limits the quality of fish habitat (Lisle, 1989, and Lisle and Hilton, 1999) Ditching and stream channelization do not build pools and riffles into the channel bed. Further, if the land use in the watershed is evolving toward more runoff, then the channel sediment

---

[1] Senior Hydrologist, MPCA, [2]Fisheries Biologist, MDNR, [3]Watershed Coordinator, BNCWQB

transport will likely be in disequilibrium (Lane *in* Rosgen, 1996). The presence of pools and riffles in southern Minnesota streams is necessary to achieve a healthy fish community.

Project Background

A joint effort between Nicollet County, Brown-Nicollet-Cottonwood (BNC) Water Quality Board, Minnesota Department of Natural Resources (MDNR) Fisheries, and the MPCA resulted in a 330-m restoration for less than $50 a linear meter. Seven Mile Creek (SMC), an 86-km^2 watershed in southern Minnesota, that drops over 60-m and flows through a Nicollet County Park (Figure 1) and outlets into the Minnesota River. In the lower Minnesota River basin, ground water resurges above the toe-slope of the valley. Selected springs were age-dated using tritium as part of the Minnesota River Assessment Project. The hydraulic residence time of ground water discharging into springs and streams is several decades, with some ground water being pre-1952 pre-atmospheric testing of nuclear weapons, (Magner & Alexander, 1994). Project objectives included the protection of infrastructure and the improvement of trout habitat. Over the last century, 90 % of the historic SMC upland wetlands were drained (Kuehner, 2004), resulting downstream channel disequilibrium; with channel degradation through the bluff and aggradation near the mouth. Because of the extensive drainage in the upper SMC watershed, the lower channel flow ceases in late August except where ground water resurgence provides a steady base-flow. A ground water discharge zone was chosen for restoration using natural channel methods described by Rosgen (1996). Restoration efforts were focused on providing grade control by flattening the channel grade through a reach with active ground water discharge. At locations shown in figure 1, root-wades, cross-vanes and "J"-hook rock-vanes (Rosgen, 1999) were installed to accomplish the above stated objectives.

Methods

In December 2002, a track-hoe with thumb was used to place root-wades and large (0.5-to-1-m) rocks to alter the channel grade and morphology. Prior to the design and placement of wood and rock, several channel cross-sections and a longitudinal-profile were collected along the proposed restoration reach using a Topcon® laser level. Pebble counts were performed using methods developed by Wolman *in* Bevenger and King (1995). The data were used to classify SMC, then size channel facets using regional curves developed by Magner and Steffen (2000). Again in 2003, the above geomorphic steps were repeated to document post construction conditions. During late summer, high evapo-transpiration (ET), and following the 2004 spring thaw (low ET); ground water pore-pressure was measured above, within and below the restored reach using a hydraulic potentiomanometer (Winter et. al., 1988). In early March 2004, before the snow-melt occurred, the relative quality of five constructed pools and a reference pool were defined by measuring pool volume and pool side-wall angle. The reference pool was located downgradient of the restoration reach. In late March, these steps were repeated after the spring flush to evaluate pool volume changes with flow regime. Additionally, the depth of loose sediment over consolidated material was measured in each pool using a graded and beveled pvc tube (Lisle and Hilton, 1999). In September of 2002 and 2003 (pre-and post- restoration) 260-m of SMC was assessed for trout by making three passes with a backpack electofisher. All captured trout were removed from the sample reach until the completion of the sampling.

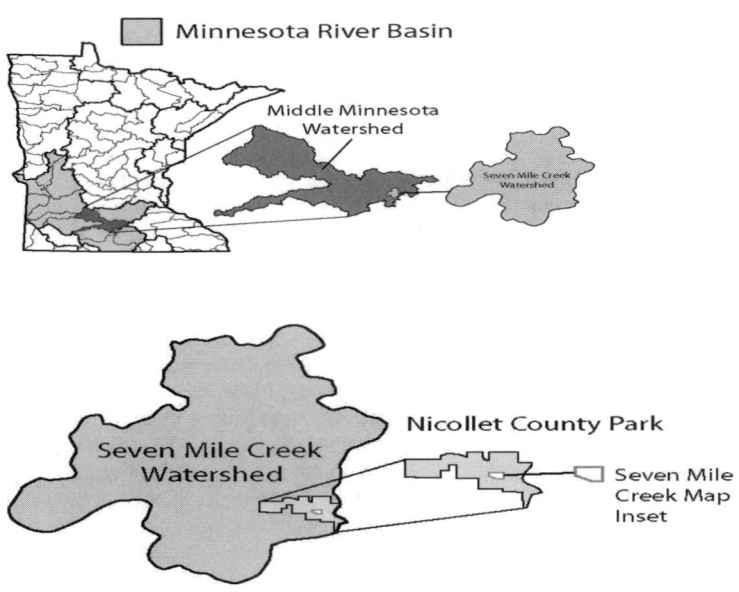

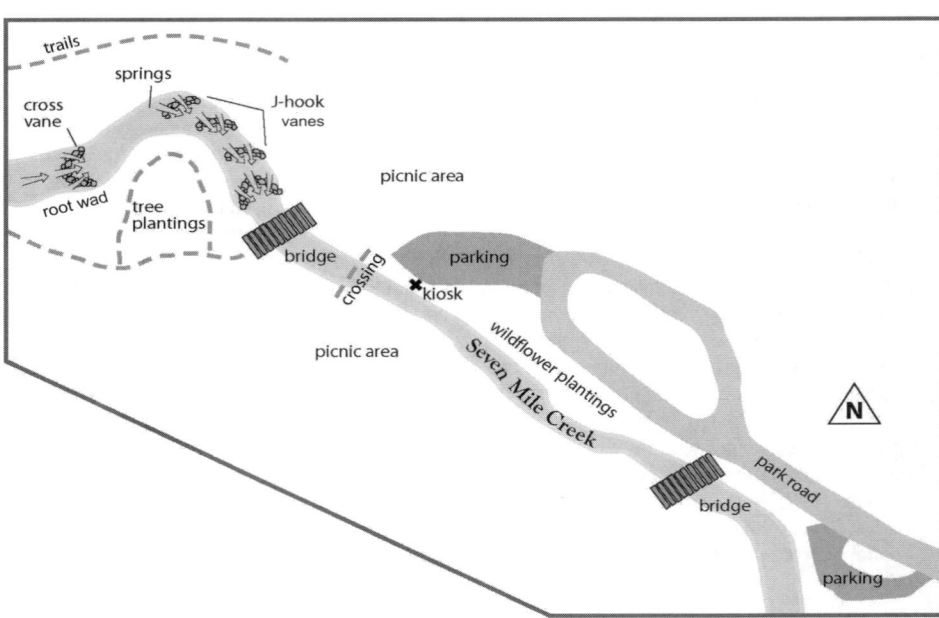

Figure 1. Location of the restoration activities within the Nicollet County Park and the location of the restoration reach within SMC and Minnesota.

Results

Pre-construction data collected in the upper third of the restoration reach showed signs of incising, and was evolving from a "B" type channel toward a "G" type channel. The lower half of the restoration reach located in the valley floor, best fit the B_{3c} Rosgen criteria. The overall average channel slope was 0.015, average riffle slope was 0.038, average pool slope was 0.007, and a maximum pool depth was measured at 0.8-m (relative to the bankfull elevation). Only 16 % of the channel restoration length was identified as pool.

Post-construction data show a shift from primarily riffle/run in the upper third of the restoration to riffle/pool with large gradient changes at vane structures. Re-shaping of the point bar changed the stream type to an E_3 by increasing the floodprone channel width. The average pool slope was similar to the pre-construction average pool slope (0.005). However, over 35 % of the upper third of the restoration was now pool. The lower half of the restoration remained a B_{3c} with a small accessible floodplain, yet riffle quality was improved by decreasing width/depth ratio with the removal of mid-channel bars.

Two cross-vanes and three "J"-hook rock-vanes were placed in the upper third of the restoration using single rock footers placed no deeper than 0.5-m below bed grade. Rosgen's (1999) design criteria (figure 2) called for two footer rocks to be placed below the vane rock 2.5 times the riffle bankfull channel depth or 1.75-m below the existing channel bed. Additionally, after vane construction an attempt was made to dig deeper pools (1.5-m below bankfull), however, within seconds of the track-hoe bucket removal the pool hole collapsed to about 0.8-m. Measurements of ground water pore pressure (vertical hydraulic gradient) were 0.004-to-.025, 0.3-to-0.45, and 0.12-to-0.2 above, within and below the restoration, respectively. Higher vertical gradients were measured in late March 2004. At several locations within the upper third of the restoration ground water seeps were visible during non-storm-event flows.

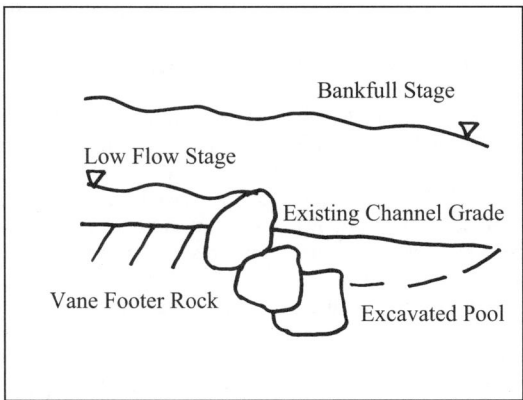

Figure 2. Cut-away profile view of Rosgen's (1996) footer rock placement design.

The maximum pool depths measured in the constructed pools were similar (1 ± 0.06-m) compared with 1.3-m for the reference pool, even after post-storm scouring. The volume of the reference pool was 14-m^3, whereas the five constructed pools ranged from 1.7-to-4.6-m^3. The reference pool side-wall angle was 18 ± 1 degrees, whereas the average constructed pool side-wall angle was 11.4 ± 2 degrees. The pool below the second cross-vane consistently maintained a side-wall angle less than 10 degrees and contained 0.1-to-0.3-m of loose sand over more consolidated material compared to 0.1-m average in the other pools. This pool was created to have a volume similar to the reference pool (15-m^3); however sediment deposition beyond the scour hole decreased the volume of the pool to 4.6-m^3. Pool volumes remained relatively consistent before and after the 2004 soil thaw and snow-melt flush.

Post-construction juvenile trout population numbers were 370 % higher (645 ± 22) compared to the pre-construction numbers (137 ± 17). Biomass of fish increased from 86-kg/ha to 104-kg/ha. In general the number of fish caught in 2003 were smaller with a mean length of 11.4-cm compared to 18.8-cm for fish caught in 2002.

Discussion

Understanding linkages between the stream channel and adjacent aquifer are important for restoration design; particularly where a substantial head exists to drive ground water discharge. Pool volumes, pool side-wall angle and the maximum pool depths for the constructed pools were less than the reference pool. These results are likely related to the consistency of the measured ground water discharge rates and the sandy nature of the sub-pavement bed material. Ground water pore pressure was slightly larger in late March 2004, due to lack of evapotranspiration, nevertheless, the increased pore pressure did not appear to reduce constructed pool volume measured prior to the 2004 snow melt.

Increasing the overall percentage of pool area and enhancing the quality of riffles may have improved the trout population as a result of channel habitat improvement. Flow regime was also a likely contributor to the survival of fingerling trout stocked in June 2003. Cold water temperature and associated stable flow from ground water discharge were dominant during June through September 2003.

Critical concerns about the success of this project could be framed by the quality and longevity of pools and riffles. Based on the measurements of pool volume, pool side-wall angle and maximum pool depth, the five constructed pools were 67-88 % smaller, pool side-wall angles were 36 % flatter, and the maximum pool depth was 20 % less than the reference pool, respectively. The pool behind the second cross-vane was designed (15-m^3) to be larger than the reference pool, but ended up becoming a flat sediment filled dish due to the relatively large ground water pore-pressures. From a construction perspective, dewatering wells could have been used to lower the ground water pore-pressure for the placement of footer rock in vane structures as shown in figure 2. However, the final pool volume would only be as large as the ground water pore-pressure would allow. The reference pool volume was relatively large because the ground water pore-pressures were lower further down gradient than in the upper one-third of the restoration reach. In general, constructed pools were smaller than designed because of the relatively high vertical gradient of discharging ground water through non-cohesive sand. Nevertheless, assuming the sediment balance does not change, pool volume should not decrease based on the current system response to peak flows. An upstream change in sediment size or supply could affect both pools and riffles. Stream discharge has changed over the past century, however re-shaping of the channel banks and point bar should dissipate peak flood flow in the re-constructed floodplain. The removal of mid-channel bars and the associated decrease in width/depth ratio of riffle cross-sections will provide better low-flow velocities and clean substrate for trout fingerling survival.

Conclusions

A deep pool in sandy substrate with discharging ground water is difficult to achieve. The results of this study suggest that constructed pools will likely be no deeper or larger over time than the reference condition. However, increasing the percentage of pools and quality of riffles over the restoration reach may improve trout fingerling survival. The consideration of ground water pore-pressure should be included in stream restoration designs that interact with aquifers.

References

Bevenger, G.S., and R.M. King. 1995. *A pebble count procedure for assessing watershed cumulative effects*. Research Paper 319. Fort Collins CO.: USDA, Forest Service. 17 pp.

Kuehner, K.J. 2004. Hydrologic Changes: A Historical Perspective in Seven Mile Creek Watershed. In *Proc. of Self-Sustaining Solutions for Streams, Wetlands, & Watersheds*, St. Joseph, MI: ASAE.

Lisle, T.E., 1989. Sediment transport and resulting deposition in spawning gravels, north coastal California, *Water Resour. Res., 25:1303-1319.*

Lisle, T.E., and S. Hilton. 1999. Fine bed material in pools of natural gravel bed channels. *Water Resour. Res. 35:1291-1304.*

Leach, J and J.A. Magner. 1992. Wetland drainage impacts within the Minnesota River Basin. *Currents 2: 3-10.*

Magner, J.A., and S.C. Alexander. 1994. The Minnesota River Basin: A Hydrogeologic Overview and Assessment of Selected Spring Resurgence. MRAP, Vol. 2., St. Paul, MN: MPCA.

Magner, J.A., and Steffen L. J., 2000. Stream morphological response to climate and land-use in the Minnesota River basin. In *Proc. of the Joint ASCE Water Resources Engineering, Planning & Management conference,* Reston VA.:ASCE.

Magner, J., M. Feist, and S. Niemela. 2003. The USDA clean sediment TMDL procedure applied in southern Minnesota. *Agricultural Hydrology and Water Quality*, AWRA, Middleburg, VA. ISBN 1-882132-61-0.: AWRA.

Magner, J.A., G.A. Payne, and L.J. Steffen. 2004. Drainage effects on stream nitrate-N and hydrology in south-central Minnesota (USA). *Environmental Monitoring and Assessment, 91:183-198.*

MPCA, 2004. http://www.pca.state.mn.us/water/biomonitoring/bio-streams.html

Rosgen, D.L., and H.L. Silvey. 1996. *Applied River Morphology*. Wildland Hydrology, Fort Collins, CO.

Rosgen, D.L., 1999. *River Restoration and Natural Channel Design Manual*. Wildland Hydrology, Fort Collins, CO.

Payne, G.A., 1991. *Sediment ,Nutrients, and Oxygen Demanding Substances in the Minnesota River: Selected Water Quality Data, 1989-90*. USGS Open-file Report 91-498. Federal Center, Denver, CO.

Talmage, P.J., J.A. Perry, and R.M. Goldstein. 2002. Relation of Instream Habitat and Physical Conditions to Fish Communities of Agricultural Streams in the Northern Midwest. *North American J. of Fisheries Management 22:825-833.*

Winter, T.C., J.W. LaBaugh, and D.O. Rosenberry. 1988. The design and use of a hydraulic potentiomanometer for direct measurement of differences in hydraulic head between ground water and surface water. *Limnol. Oceanogr.33:1209-1214.*

MANAGEMENT CHALLENGES IN STREAM RESTORATIONS OF

URBAN WATERSHEDS

Vishnu V.R. Seri[1]

ABSTRACT

Stream restoration projects in urban watersheds are often confronted with a matrix of unique challenges posed by residents, public agencies, public officials and elected officials. Each of the involved stake holders bring their own unique perspective to these projects, more often making the project managers flounder for direction. The symptoms of an unhealthy and deteriorating stream and its impact on a long term basis are matters of grave concern for the environmentalist even as maintenance staff of public agencies would like to adopt a "quick-fix", economical alternative to address the issues of concern in an expeditious manner. The most important stake holders, viz., the watershed residents are more concerned about their quality of life and aesthetics. Elected officials often favor the most vocal public opinion. To arrive at a please-all solution in such a challenging environment of competing interests and within compressed time schedules stretches the management skills of the project managers. Experience has shown that the project managers need to come up with a matrix of solutions addressing concerns of all stake holders and associating costs and the pros and cons of each such solution. Although this method may not be palatable to all concerned stake holders, the minimum it does is identifying all problems which should be addressed to arrive at a holistic solution. This paper reviews a couple of case studies of stream restoration projects that have been implemented or in the planning stages, summarizes the lessons learnt during these processes and suggests potential remedies which could result in an efficient management of stream restoration projects and provide a better return on investment for tax payer's money.

KEYWORDS: Stream Restoration, Urban Watersheds, Management

INTRODUCTION

Increasing urbanization has resulted in a host of problems in the stream corridors of urban watersheds. Inflows into these streams have increased as a direct result of increase in impervious areas. Although current local regulations attempt to minimize the effects of urbanization, these problems cannot be wished away. Deteriorating physical and environmental conditions within the streams have increased stream restoration projects in urban settings within the past few years. Moreover, stream restoration projects will keep increasing as urban watersheds approach build-out and ageing infrastructure needs to be replaced. Management of stream restoration practices has always been a challenge as opposed to other construction or capital improvement projects. This is primarily because stream restoration projects in urban watersheds are often confronted with a matrix of unique issues posed by residents, public agencies, public officials and elected officials. The available body of knowledge on the management of stream restoration projects, including the comprehensive federal government publication[1], recommends a holistic and comprehensive management by involving all stake holders and forming citizen advisory groups, technical advisory groups etc. to better manage stream restoration projects. Although, the principles enunciated provide good guidelines they do not address most of the issues confronted in the stream restoration projects of urban watersheds. Additionally, engineers and planners are

---

[1] Engineer II, Stormwater Planning Division, DPWES, Fairfax County, 12000 Government Center Pkwy., Suite 449, Fairfax, VA 22035. Vishnu.Seri@FairfaxCounty.gov

being encouraged to adopt "Soft Engineering", which includes bio-engineering practices as opposed to "Hard Engineering", consisting of concrete channels, rip-rap, gabions etc. Currently, bio-engineering practices are difficult to engineer, specify, build and maintain due to the lack of qualified personal and project precedents. To arrive at a please-all, workable and sustainable solution in such a challenging environment of competing interests and within compressed time schedules stretches the management skills of the project managers. Experience has shown that the project managers need to come up with a matrix of solutions addressing concerns of all stake holders and associating costs and the pros and cons of each such solution. Although this method may not be palatable to all concerned stake holders, the minimum it does is identifying all problems which should be addressed to arrive at a holistic solution.

STREAM RESTORATION PROJECT #1

Project Overview:

This project is located within an urbanized, residential part of the county with an approximate density of two houses per acre. The project consists of 1100 linear feet of stream within the county's Park Authority property. At the upstream end of the project, an area of hundred acres drains into the stream with an additional sixty acres draining from two storm drains and two side tributaries. The stream meanders along the backyards of single family homes of 0.5 acre to more than acre. Along the stream and within the Park Authority property are fairly mature trees. This project was initiated based on the problems identified in a watershed planning effort and also from citizen's complaints about stream bank erosion. The park authority of the county in conjunction with the stormwater planning division and Natural Resources Conservation Service (NRCS) collected, analyzed and calculated several Rosgen Classification parameters. This information was subsequently utilized by a consultant to finalize a stream restoration design in 1999.

The Issues:

The stream restoration design of 1999 called for introduction of a sinuosity in the subject stream to restore it to a calculated reference condition. More than eighty trees, some of them mature, had to be removed to implement the proposed stream restoration effort. Although many meetings were held with the local Home Owners' Association (HOA) during the course of the project design, the bone of contention of tree loss was always left to simmer till the very end while all other issues of concern, such as access, easements, construction disturbance, safety etc. were addressed. As most of these trees were within the Park Authority property, the assumption amongst Park Authority staff was that there wouldn't be any road blocks in implementing the proposed project design. However, during an HOA meeting to finalize the proposed project design, sufficient opposition was voiced against the removal of many mature trees, chiefly due to aesthetic and environmental reasons. The county staff was advised to put the project on hold. The project was revived in 2004 to find an amicable solution to the stream restoration effort. A multi-disciplinary scoping team consisting of engineers, landscape architects and ecologists performed a reconnaissance of the stream as per the Unified Stream Assessment / Unified Subwatershed and Site Reconnaissance (USA/USSR)[3] protocol, developed by the Center for Watershed Protection. The unanimous recommendation of the scoping team was that the stream could be rehabilitated with spot treatment at identified trouble spots rather than going through a large scale restoration effort as proposed in the 1999 design plans. Although these recommendations were welcomed by the HOA, a unanimous acceptance was still lacking amongst all stakeholders due to the legacy of issues and personalities involved.

Lessons Learned:

It is abundantly clear that the stream restoration design effort of 1999 was a futile exercise. A close observation of the project history reveals that there was always dissenting voices which were vocal in their opposition to trees being removed, although they were on the park authority property. Input from the HOA was not solicited and encouraged during the design phase based on the assumption that since the park authority was the property owner there wouldn't be any significant opposition. Moreover, the design was a uni-dimensional effort aimed solely at restoring the stream to it's reference state without due consideration to the environmental cost-benefit analysis.

STREAM RESTORATION PROJECT #2

Project Overview:

This project is also located within an urbanized part of the County but with a higher density and an equal split of land use between commercial and residential uses. The stream restoration project included 2200 linear feet of stream. The drainage area at the downstream end of this section is one square mile. Development within the watershed changed the land characteristics by increasing the overall impervious area. Although current regulations require that the effects of urbanization be offset through adequate SWM measures, the majority of the watershed was developed before these regulations came into effect. The stream and its tributaries were formed in response to the lower flow rates and as these rates increased with development, the stream no longer had the capacity to transport the flood flows. In a natural process the stream responded by adjustments in cross sectional areas and slopes along the stream corridor. This resulted in severe stream bank erosion and the concomitant loss of property, infrastructure and habitat. To restore the stability of the stream corridor the County obtained the services of a design consultant who could incorporate geomorphic design techniques and bio engineering stream stabilization practices. Design and construction activities on project were carried out during 2002 and the early part of 2003. However, severe rain fall events of late 2003 coupled with a hurricane led to substantial damage to the restoration activities. Between the County, consultant and the contractor it was decided that the stream needed further rehabilitation to perform to the earlier designed state. These stream rehabilitation measures were performed in the spring of 2004.

The Issues:

It was clear that the rainfall events of 2003 were not the only reasons for the failure of the first round of stream restoration efforts. This became evident during the construction of the rehabilitation efforts of 2004. Additionally the lack of experience in applying geomorphic techniques to the design restoration efforts, a lack of understanding on the part of the inspection team and a lack of flexibility in the construction contract were some of the major reasons for the partial failure of original stream restoration design. Moreover, available payment methods for County prevented the fine tuning of design plans in the field.

Lessons Learned:

Close and continual interaction between design engineers and contractors is essential for the success of stream restoration efforts. This cooperation should offer the flexibility to adapt the design to field encountered situations. Inspection staff should be properly trained in the principles and practice of bio-engineering. Moreover, construction contracts should offer the flexibility to fine tune design efforts and adapt to field learned conditions.

CONCLUSION

Stream restoration projects in urban watershed pose unique challenges as evidenced by the above two case studies. Going through a design phase in the anticipation of community buy-in at a later stage will most likely result in an aborted take-off of stream restoration projects. Moreover, stream restoration efforts should take a holistic approach by factoring in the environmental cost-benefit analysis into the overall design of the project. In addition there needs be to recognition that there has been a paradigm shift in stream restoration efforts from "Hard Engineering", consisting of concrete channels, gabions, rip-rap etc. to those of "Soft Engineering" consisting primarily of bio-engineering practices. Unfortunately, there hasn't been a simultaneous adaptive perspective shift in the construction and inspections practices for such projects. Bio-engineering practices, by their nature cannot be designed and specified as other construction practices. Construction inspectors need to be properly trained in the principles and practices of geomorphic design and bio-engineering. A collaborative effort between the design engineers and contractors with a built in flexibility and based on shared end-vision is likely to result in far higher success rate. Additionally, construction contracts should have the built-in flexibility to be modified for the field learned conditions. Applying age old construction techniques and management principles to these projects is bound to result in continual frustration and failure of stream restoration projects.

REFERENCES

1. Stream Corridor Restoration, Principles, Processes and Practices, the Federal Interagency Stream Restoration Working Group, 1998.

2. Unified Stream Assessment / Unified Subwatershed and Site Reconnaissance, Center for Watershed Protection, 2003.

Stream Restoration and Stabilization
In an Urban System

E. Cummings

A. Ludwig

B.K. Schaffer

D. Schluterman

ABSTRACT

The demand for urban green space is increasing while availability is decreasing. The value of urban green space is that it conserves critical characteristics of the natural landscape within the urban setting. The characteristics are essential for the function of ecological processes necessary for the sustainability of an ecosystem. These goods and services we extract from ecosystems are called ecological services. The goal of this project was to demonstrate the technologies and methods by which ecological services could be restored in a disturbed urban stream system.

Specific objectives included bio-retention of storm water and parking lot runoff, re-establishment of fish pool habitat, implementation of natural bank stabilization, integration of riparian zone buffers, and regulation of stream geometry for maximum in-stream ecological services. We designed off-channel subsurface bio-retention cells to reduce pollution loading to the stream from a nearby parking lot and to maximize retention volume of water in the cells to recharge stream base flow. We integrated the bio-retention cell infiltration zones with the geomorphologic design of the stream channel to provide conditions desirable for fish and other aquatic communities as well as to discourage algae blooms and mosquito infestation. We designed riparian zones with native flora to provide needed habitat for terrestrial animal communities. Finally, we used natural bank stabilization features such as root wads, native tree logs, and boulders to reduce bank erosion and channel entrenchment and to enhance refugia for aquatic organisms.

KEYWORDS. Urban Stream Restoration, Urban Greenway, Subsurface Retention Cell

INTRODUCTION

The City of Rogers created a workforce to develop a trails system based on a Greenway design theme. Urbanization of the watershed and past channel alterations has severely impacted several stream systems in Rogers. Through federal, state, and local agencies, the city acquired funding for a Greenway Demonstration Project on Blossom Way Creek, between Dixieland Road and 26th Street. The goal of the project is to demonstrate methods and technologies for protecting critical ecological services in urban streams. The project's tasks included 1) water quality and quality monitoring, 2) collect and analyze hydrologic and geomorphologic data, 3) develop and implement educational curriculum, 4) develop greenway design, 5) implement greenway design, 6) outreach and technical transfers, and 7) reports.

We were approached by Dr. Marty Matlock, a member of the University of Arkansas faculty to assist in the completion of task four, development of greenway design. The objectives were to decrease erosive forces, restore in-stream habitat, alleviate peak flows, restore terrestrial habitat, and provide an aesthetically pleasing view for the proposed nearby recreational areas. Both branches of the stream had been channelized, creating an increased slope in the reach, which needed to be lowered to decrease the sediment carrying capacity and reduce erosion. Because of the channelization of the stream, the channel had become incised, causing a reduction in ability of high flows to dissipate energy by spilling into the flood plain. Due to heavy urbanization in the watershed, frequency of high flows was increased and time of concentration decreased, causing larger erosive forces onto channel banks and thalweg. Retention of peak flows for attenuation and recharge of base flow of both channels was also needed.

Design Process

We divided the design process into different elements. These included the following elements; channel sinuosity and profile design, large grade control structure design, retention cell design, and in-stream riffle design. We divided the responsibility for the various aspects among the team members and completed the following designs.

<u>Grade Stabilization Structures</u>

During the initial surveys of the project area by Mr. Morgan and his team, it was confirmed that the bridge on 26th street and the increased flows due to urbanization had combined to create a stream channel degradation pattern that had proceeded upstream during storm events, until it was located approximately 200 feet below the confluence of Blossom Branch and the southern tributary in our project area. As Mr. Morgan had discussed in his presentation to the Greenway Committee at the November meeting, the stabilization of the stream-bed was considered to be a top priority of the project . Our design includes a large stabilization structure, commonly called a Newbury Riffle (named for Robert W. Newbury who designed it). This structure is constructed of large rocks that will not move during a large storm event – a 100-year event – that will fall out as the channel degrades, armoring the channel bottom and sides to prevent further degradation. The only maintenance needed over time is to add more rock to fill in what is displaced. The structure currently holding the "head-cut" in place is a large maple tree root which may give way at any time, therefore, making this the most time critical aspect of project construction.

<u>Channel Sinuosity and Profile Design</u>

The channel of Blossom Branch has been straightened and dredged historically to help remove flood water from the area quickly. The classic method of dealing with these streams includes armoring the channel to prevent erosion and building them wide and deep to remove the water quickly (Figures 3.3a & 3.3b).

The problems with this approach are varied. First, the aquatic ecosystem is completely destroyed, and the various fish, macroinvertabrates, insects, and other animals are completely displaced. Second, as the watershed becomes more urbanized with larger peak flows, the water is not retained in the natural stream system and becomes a flood problem for downstream areas. The residents of these areas are then forced to destroy natural habitats and ecosystems to deal

with

these high water problems. Our design is based on managing these increased flows in a manner that enhances the natural aquatic ecosystem and allows the high flow water to reach the flood-plain quickly, so the energy is dissipated rather than causing increased erosion and flooding downstream.

Riparian Zone Design

A riparian buffer strip provides many useful services to aquatic and terrestrial habitat. These services are enumerated in NRCS (2003):

- Create shade to lower water temperatures to improve habitat for aquatic organisms.
- Provide a source of detritus and large woody debris for aquatic and terrestrial organisms.
- Create wildlife habitat and establish wildlife corridors.
- Reduce excess amounts of sediment, organic material, nutrients and pesticides in surface runoff and reduce excess nutrients and other chemicals in shallow ground water flow.
- Provide protection against scour erosion within the floodplain.
- Restore natural riparian plant communities.
- Moderate winter temperatures to reduce freezing of aquatic over-wintering habitats.
- To increase carbon storage in plant biomass and soils.

The number and quality of the services provided is the main reason that this strategy was chosen as a design component.

Subsurface Retention Cell Design

The purpose of subsurface retention cells in our design is two-fold. First, the water is to be stored and released slowly to enhance aquatic habitat during low-flow periods. Second, the storage of parking lot runoff, even for short intervals, will allow some natural biological degradation of pollutants facilitated by contact with substrate surface. Our design is based on use of the abandoned stream channel when the new sinuosity design is implemented. The old channel of Blossom Branch nearest the High School will have the parking lot runoff directed to infiltration zones. The parking lot runoff provides a large source of water from small rainfall events, and it currently enters the channel in the vicinity of the upstream inlet area. The use of subsurface zones prevents the creation of mosquito habitat and the spread of mosquito borne illnesses such as West Nile Virus, which is important for recreational users of the area.

Design Details and Calculations

Grade Stabilization Calculations

We decided on the Newbury Riffle design for our constructed riffle below the confluence. To initiate the design, we began by determining the slope of the upstream and downstream faces of the riffle (Fig. 4.1a). To facilitate fish passage over our riffle, we followed the specifications for walleye spawning riffles used in restorations on Mink Creek in Manitoba and Hamilton Creek (Newbury, 1993). This dictated an upstream face slope of 1 to 4 and a downstream face slope of 1 to 20. Our riffle height is 2 ft which gives us an overall length of 48 ft.

To estimate the volume of rock needed for our large constructed riffle we used the product of the riffle length, height, and bank full channel width (Newbury, 1998). The following volume was calculated:

$$V = \text{length x width x height}$$

where our riffle length is 48 ft, riffle height is 2 ft, and bankfull width is 27 ft. This produces a volume of 2592 ft³ or 96 yd³ of substrate needed. Additional large boulders will be added to tie the structure to the bank.

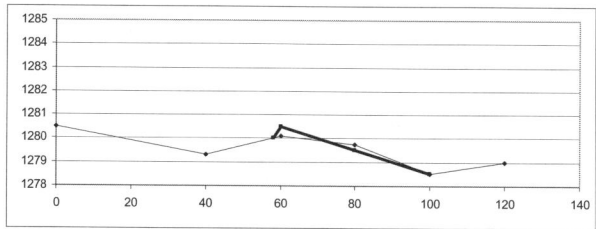

Figure 4.1a Graph of current grade vs. slope of Newbury riffle.

To determine the minimum size of rock needed in the lower constructed riffle, we used the approximation relationship between tractive force and incipient diameter of bed material by the following formula:

$$\tau \ (kg/m^2) = \text{incipient diameter (cm)}$$

To find the tractive force we used the following equation:

$$T = 1000 \times d \times s$$

where d is depth of flow (m), s is slope of water surface, and T is tractive force of flow (kg/m²). We used the FEMA 100 year flood depth as the depth of flow to introduce a sufficient factory of safety. This depth was 7 ft or 2.13 m. We used the slope of the channel through the reach for the slope of water surface. This slope was calculated as 0.005. Using these values, an incipient particle diameter was calculated at 10.65 cm. Using a factor of safety of 2 we get a minimum rock diameter of 21.3 cm (Newbury, 1993).

A separate method was used to confirm the rock diameter. This method used the same relationship between tractive force and incipient diameter, but used bank-full height and slope of downstream face of constructed riffle in place of 100 year flood depth and water surface slope. The equation is as follows:

$$\varphi = 1500 \times D \times S$$

where φ is the diameter of stable rock (cm), D is depth from crest of riffle to floodplain (m), and S is slope of downstream face of riffle. The crest of the riffle is going to be two feet above the channel bottom, and the flood plain depth is 3 feet. Therefore, D is 1 foot or 0.3048 meters. The slope of the downstream face of the riffle is 1 to 20 or 0.05. With these values, the equation gives us a diameter of 22.86 cm (Newbury, 1998).

Construction of the riffle will be done under engineer supervision at all times. Grade stakes will be placed at the start of the riffle, the peak of the riffle, and the end of the riffle. The actual rock size will be variable and will consist of a size range of 12 – 22 inches. Large rocks will be strategically placed on the face of the riffle to establish a ladder for fish passage during low flows.

Channel Sinuosity and Profile Design Details

The constraints and calculations performed concerning the sinuosity and profile were completed prior to our participation

The drawings in Fig. 4.2a show the projected cross-sections that will allow for a low flow channel to maintain equilibrium sediment loads during these conditions and immediate access to the flood-plain during high flow events, while encouraging a healthy riffle-pool-glide flow regime. Figure 4.2b shows the current channel and Figure 4.2c shows the new channel.

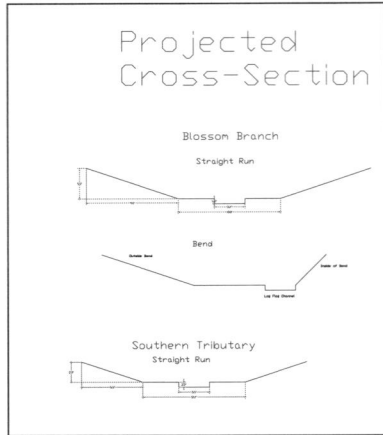

Figure 4.2a Drawings of Cross-Sections for New Channel

Figure 4.2b Picture showing existing channel.

Figure 4.2c New Channel Construction

Subsurface Retention Cell Calculations

To find the losses due to evaporation of the designed aquatic habitat provided by the plunge pool just below the outlet of the retention cell, Dalton's Law was used in the following form:

$$E = C(e_s - e_d)$$

Where:

E = rate of evaporation (mm/day)

C = constant

e_s = saturated vapor pressure at the temperature of the water surface (mm Hg)

e_d = actual vapor pressure of the air (e_s x relative humidity) (mm Hg)

To find C, the following equation was utilized:

$$C = 15+0.93W$$

where W = average wind velocity at height of 0.15m (km/h)

The average temperature during July, the hottest and driest month of the year, was 78 F, the average relative humidity was 60%, the stream water temperature is assumed to be 25 C, and the average wind speed is 10.13 km/h (NOAA 2004). With these parameters, the following calculations were achievable:

$$e_s = 23 \text{ mm Hg}$$

$$e_d = 0.6e_s = 0.6(23) = 13.8 \text{ mm Hg}$$

$$C = 15+0.93W = 15+0.93(10.13) = 24.43$$

Therefore, the loss of depth due to evaporation could be found as:

$$E = C(e_s-e_d) = 24.43(23-13.8) = 224.756 \text{ mm/month} = 7.5 \text{ mm/day}$$

With the designed pool dimensions as 2m x 15m, the flow lost to evaporation could be found as 0.225 m³/day.

The cells will be constructed in the existing channel areas to be abandoned. The length of the Blossom Branch channel is 400ft long and averages 3 feet deep. The top 12 inches of space will be covered by topsoil to allow planting of grass and small shrubs, which results in a cross-sectional area of 21 ft² (Fig 4.3b). Using these dimensions, the total volume of the retention area will be 8400 ft³. This area will be filled with creek gravel, which has a porosity of 50%, resulting in a total water capacity of 4200 ft³. From the picture in Fig. 4.3a, we see that not all the capacity can be used due to the slope of the channel. The estimated maximum water capacity of this cell is 3650 ft³ or 103.4 m³. Using the volume of 103.4 m³ and the desired flow rate of 0.3 m³/day, the retention cell will make up for losses due to evaporation over 345 days.

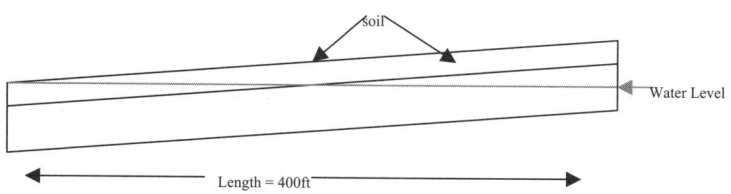

Figure 4.3a Profile View of Retention Cell

To size the outflow channel from the retention cell use Darcy's Law for flow through porous materials:

$$Q = AK(dh/L)$$

where Q represents mass flow rate, A is cross-sectional area, K is hydraulic conductivity, dh is the change in height of porous bed, and L is the length of gravel bed. We calculated a desired Q of 0.3 m^3/day which converts to 0.44 ft^3/h

The height of our bed is 3 ft and the length is approximately 360 ft. Using a typical hydraulic conductivity of gravel beds of 3000 ft/day (SDSU, 2004), calculations for the desired area resulted in an area of 60.82 in^2. This channel will be 8 inches by 8 inches of gravel surrounded by geotextile material to keep soil from plugging channel.

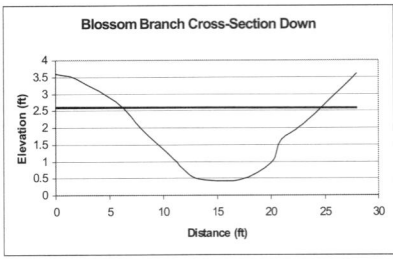 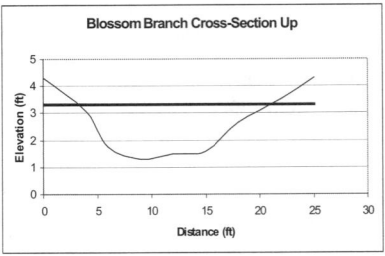

Figure 4.3b Cross-Sections of Blossom Branch Retention Cell Area

The retention cell will be recharged with flow from the high school parking lot. The parking lot consists of 380,000 square feet of asphalt that drains to the retention cell area. Using this area, and assuming that in a relatively small rainfall event in midsummer 50% of the water will be lost due to evapo-transpiration before getting to the cell, we calculate a rainfall event of merely 0.3 inches is needed to recharge the cell.

Riparian Buffer Details

Restrictions on space due to prescribed uses of certain areas lead us to use the minimum required width of 35 ft for the riparian buffer. This means that the riparian buffer will extend to 35 ft from the top of bank on both sides of the stream (NRCS, 2003). To obtain the desired aesthetic appeal, a varying array of trees, shrubs, and wildflowers are being recommended by the City of Rogers Parks and Trails Committee for the detailed design of the riparian buffer, which is not yet finalized. The Arkansas Forest Commission will also make recommendations on the growth pattern and favorable planting procedures of plants to be used.

CONCLUSION

Our design, while based on accepted practices and sound Engineering Principles, is a natural system and contains a number of unknowns. Some of the unknowns we are aware of concern future watershed development. As urbanization progresses in this part of Northwest Arkansas, the concentration time and peak flows on Blossom Branch will change. We have designed our system to be robust, with a large factor of safety where applicable. However, the inability to predict future events and the constraints placed on our design require us to report that some of our structures may not be adequate to perform in extreme events.

We have attempted to design this reach of stream to maximize aquatic and terrestrial habitat in low-flow conditions, as well as maintaining a stable channel in high flow events. Our hope is that a natural riffle-pool-run flow regimen will be restored in low to medium flows, and the flood plain accessibility/riparian structure will reduce flow energy in high flow events. This is a demonstration project that allows us to test various ideas and structure types. The Ecological

Engineering Group at the University of Arkansas, as well as the concerned agencies in the area, will continue to work to monitor and extend the project to include all of the applicable Greenway trail system in the city of Rogers. The lessons learned in this small reach of Blossom Branch may hopefully be implemented in public policy of the various municipalities in the region to prevent further degradation of value of our natural stream systems. This project can indeed demonstrate the coexistence of these systems with the continued growth and development of urban area.

Acknowledgements

Patti Erwin, Arkansas Forestry Commission

David Evans, Arkansas Game and Fish Commission, Stream Team

Mike Henley, Nature Conservancy

Cully Hession, University of Vermont

REFERENCES

1. FISRWG (10/1998). *Stream Corridor Restoration: Principles, Processes, and Practices.* By the Federal Interagency Stream Restoration Working Group (FISRWG) (15 Federal agencies of the US gov't). GPO Item No. 0120-A; SuDocs No. A 57.6/2: EN 3/PT.653. ISBN-0-934213-59-3.

2. Newbury, R.W. M.N. Gaboury. and C. Watson. *Field Manual of Urban Stream Restoration* Illinois Department of Natural Resources. 1998

3. Newbury, R.W. and M.N. Gaboury. *Stream Analysis and Fish Habitat Design – A Field Manual.* Newbery Hydraulics Ltd: British Columbia, Canada. 1993.

4. NOAA. 2004.National Weather Service Forecast Office. Available at www.srh.noaa.gov/tulsa

5. NRCS. 2003. Conservation Practice Standard for Riparian Forest Buffers. NRCS CPS code 391. Natural Resources Conservation Service.

6. Rosgen, Dave. *Applied River Morphology.* Minneapolis, Minnesota. 1996.

7. Schwab, Glenn O. et al. *Soil and Water Conservation Engineering.* John Wiley and Sons, Inc: New York. 1993.

8. SDSU.2004. Applying Darcy's Law to a Confined Aquifer. San Diego State University. Available at www.geology.sdsu.edu/classes/geol351/darcys.htm

Urban Impacts on Stream Morphology in the Ozark Plateaus Region

R.T. Pavlowsky[1]

ABSTRACT

Urbanization typically increases runoff and sediment loads in streams which can cause geomorphic changes in channel size and planform. It is important to understand and predict these channel changes since they are often associated with flooding, sedimentation, water quality, and habitat management problems. This study examines the influence of urbanization on channel geomorphology in watersheds that drain metropolitan Springfield, Missouri, the third largest city in the state located in southwestern Missouri on the Springfield Plateau of the Ozarks Plateaus Region. Regression analysis is used to describe the spatial variations in channel morphology using drainage area, land use, and riparian vegetation variables. Interestingly, channel cross-section dimensions for a given drainage area did not differ between rural and urban streams. But compared to rural streams, urban streams have shorter riffle spacing, shallower pools and larger bed materials. The magnitude of urban influence on channel form in this study is generally within 20% of the rural reference streams, however, studies in other regions often report changes of 200% or more. The limited response of these Ozark streams to urban disturbance can be explained by pre-conditioning of watershed hydrology by historical land disturbances, presence of cohesive banks with natural gravel armoring near the bed, and karst bedrock-control.

KEYWORDS. Channel morphology, urban impacts, Ozark streams

INTRODUCTION

Urbanization can destabilize streams and influence channel form and sediment transport at the watershed-scale by accelerating the hydrologic response through the addition of impervious areas, expansion of the drainage network, and reduction of channel roughness. Channel beds and banks can become unstable in urban watersheds because the channel-forming discharge (i.e. 1- to 2-year flood) typically increases by >3 times predevelopment levels (Dunne and Leopold, 1978). More information on how urban developments impact stream channel form is needed for regions where field data is presently lacking. This paper examines the influence of urbanization on channel morphology in South Dry Sac Watershed located on the Ozark Plateau in southwest Missouri. The watershed drains the northern part of Springfield, the third largest city in the state. Urban growth rates in the area rank as some of the highest in the country. Information on channel geomorphology and the role of human activities in causing stream erosion and/or sedimentation problems in the Ozarks is timely since recent environmental initiatives in storm water control, water quality protection, and channel restoration need to consider the physical behavior of the channel system in order to be effective.

Channel changes associated with a shift from rural (or forested) conditions to urban land use within a watershed has been an important topic of study in geomorphology for almost 40 years (Wolman, 1967; Booth, 1990; Pizzuto et al., 2000). It is well understood that urban hydrology can cause adjustments in channel slope, width and depth, and sediment caliber and that these parameters are commonly used to characterize channel form within a watershed (Rosgen, 1996).

---

[1] Department of Geography, Geology, and Planning, Southwest Missouri State University, Springfield, MO, rtp138f@smsu.edu.

It is thought that the threshold for geomorphic response may be exceeded when urban area reaches 10 to 20 percent of the watershed (Hammer, 1972; Doll et al., 2002).

STUDY AREA

The South Dry Sac watershed (79 km^2) is located in Greene County, Missouri and its main stem flows from east to west into the Little Sac River, then into the Osage River system which flows into the Missouri River. Its drainage area contains 25% urban, 18% forest, and 57% grassland areas. The rural areas are now largely used for beef cattle production and low density residential developments. However, agricultural settlement and land clearing in the watershed began in the 1830s and row-cropping was common up until the 1960s. Urban areas are concentrated in the southwest portion of the watershed where urban settlement began in the 1880s. The river drains relatively uniform, horizontally bedded limestones into which a well-developed karst terrain has formed. Upland soils generally consist of several meters of cherty "red-clay" limestone residuum overlain by one meter or less of silty Pleistocene loess. Alluvial soils typically consist of cohesive silt/clay upper banks and clast-supported, cherty gravel lower banks. Channel bed elevations are at or near bedrock in most of the 2nd order and larger streams in the region. The watershed receives about 100 cm of rainfall annually and produces a mean annual discharge at its mouth of approximately 0.9 m^3/s. The watershed is a primary recharge area for the karstic Fulbright Spring which provides about 20% of Springfield's drinking water supply.

METHODS

Auto-level surveys of the channel cross-section and longitudinal profile were collected at 32 reaches representing a wide range of urban influence and riparian buffer characteristics. For this study, a reach is defined as the length of channel extending over three consecutive riffle crests in a representative section of the stream. The study cross-section located across the middle riffle crest. Width and mean depth measurements were determined for two channel stages in this study. The active (or bankfull) channel stage is identified by the top of lower bench or alternating bar deposits inset within the larger low terrace channel. The width of the active bench or floodplain is generally <5 meters. The active channel banks are typically formed of relatively coarse sediment capped with only a few centimeters of silty material. The low terrace stage is identified as the top of the main channel bank at the point where flood water would be expected to initially spread out onto the much wider valley floor. Slope was measured using a regression line through the three riffle crest elevations on the longitudinal survey. Riffle-spacing and maximum pool depth compared to the downstream riffle is derived from the longitudinal survey. Sinuosity was measured in the field with a 100 meter tape for most streams, with aerial photographs used for larger reaches.

Bed sediment size data were collected in two ways. The size distribution (i.e. D50 and D84) was derived by measuring the intermediate axis of 50 "pebbles" collected from 5 active channel transects with 10 equally spaced "blind touch" samples collected along each transect. The transects were spaced at one active channel width intervals with the middle transect located on the cross-section survey line. The maximum clast size of the reach was determined by the average of the 10 largest bed sediment clasts found in the reach. Hydraulic roughness values (Mannings n) were calculated empirically using the method described in Pizzutto et al. (2000) which requires values for sinuosity, D50, mean pool depth (or max pool depth/2), and mean channel depth. Discharge was calculated for both the active and low terrace channel using the continuity equation. The reoccurrence interval for the flood required to fill each channel to maximum stage was estimated by extrapolation of regional regression equations developed from USGS gage network flood data for Ozark streams in southern Missouri.

The percent urban land use within each sub-watershed above the GPS-referenced sample reaches was determined using Landsat Thematic Mapper and ArcGIS software with spatial data provided by the National Land Cover Dataset from 1987-1993. Subwatersheds containing more than 15% urban area were classified as "urban" and those with less than 15% urban area as "rural."

Riparian buffer composition was calculated as percent of grass or forest buffer along the length of both right and left banks for a distance of 30 active channel widths upstream of the survey transect.

RESULTS

Channel Characteristics

Thirty-two reaches are evaluated in this study for drainage areas ranging from <1 to almost 80 km^2 (Table 1). The sample was evenly divided between rural and urban subwatersheds with 16 in each category. There was a slight tendency to sample more urban first and second order streams due to access limitations and poor channel development in some rural areas. Riparian grassland cover ranged from 20 to 42 percent among the Strahler stream order classes. As expected, reach elevation and slope decrease with watershed size. More significantly, reach sinuosity is relatively low with median values in the 1.05 range with most reaches having values less than 1.10. It is common in this watershed to be able to observe three or more riffle crests from a point standing in the channel. Stream channels in the South Dry Sac watershed are bedrock-controlled and flow relatively straight for a few meander lengths after which they encounter the bluff line and then flow back out across a relatively narrow valley. In some instances, channels flow straight for long distances along the bluff line and appear to be semi-permanently locked in place. Bedrock exposures are common on channel beds in some reaches and gravel deposits are typically <1 meter thick over bedrock in all streams of second or larger order.

Table 1: Sample Reach Characteristics

Stream Order	Ad (km2)	Elevation (ft asl)	Slope (m/m)	Sinuosity (m/m)	Urban Area (%)	Grass Buffer (%)
First (n=7)	0.4	1,251	0.019	1.05	84	30
Second (n=9)	2.0	1,249	0.013	1.05	69	42
Third (n=7)	10.7	1,179	0.012	1.08	16	20
Fourth (n=9)	52.9	1,133	0.004	1.05	9	35

The active channel is significantly smaller than the low terrace channel. The median channel width for the low terrace channel ranges from 2.7 times that of the active channel in first order streams and decreases to 1.6 times in fourth order reaches (Table 2). Median width:depth ratios for the active channel decrease downstream from 27 in the first order streams to 16 in fourth order streams, while no similar trend is shown for the low terrace channel. The capacity of the active channel at it highest stage is usually the 1-year flood or less (Table 2). The capacity of the low terrace channel ranges from the 2- to 5-year flood. These streams appear to reflect moderate fluvial force with low to moderate slopes and stream power values <30 W/m^2 for the active channel and <100 W/m^2 for the low terrace channel.

Given the low sinuosity of these channels in general, evidence of lateral erosion, point bar development, and a distinct "bankfull" or active floodplain is lacking in most reaches. Thirty to 50 year old trees often grow out of the low terrace banks indicating that they have been in a stable position for sometime. Studies of other Ozark streams have shown that both stable reaches and unstable or "disturbance" reaches exist in sequence, but in most cases disturbance reaches remain in place through time and represent only a small portion of total stream length (Jacobson nad Gran, 1999). Disturbance reaches tend to be located where bluff lines interfere with channel flow and in association with sedimentation areas below the confluence of tributaries. Disturbance reaches were not sampled in this study in order to study the more systematic and representative variations in channel morphology related to watershed and riparian buffer factors.

Table 2: Active and Low Terrace Channel Properties (median values)

Stream Order	Width (km2)	W:D (m/m)	Roughness (Mannings n)	Velocity (m/s)	Stream Power (W/m2)	RI (years)
Active Channel						
First	2.4	13.2	0.039	0.98	24.3	0.9
Second	3.4	16.4	0.041	0.93	18.6	0.8
Third	5.9	25.9	0.045	0.81	21.3	0.8
Fourth	13.3	26.7	0.043	0.79	11.3	0.8
Low Terrace Channel						
First	6.4	26.7	0.038	1.16	31.8	4
Second	8.0	17.6	0.037	1.64	59.1	2.2
Third	11.4	20.7	0.040	1.80	70.5	4
Fourth	21.2	16.1	0.039	1.64	28.0	3

Although the streams studied here consist mainly of straight reaches, relatively sinuous channels appear to be clustered in the first, second, and third order streams located in the eastern, upper portions of the basin most distant from the mouth (Figure 1). This area is generally the most intensely farmed in the watershed and stream courses tend to flow through relatively wider valleys with more alluvial fill. Thus the influence of bedrock control on channel form may be reduced in these locations and thus allow a more meandering planform to develop. It is also possible that bank disturbances by riparian forest clearing and grazing may have initiated lateral channel erosion and meandering, but no proof for this process is presented here.

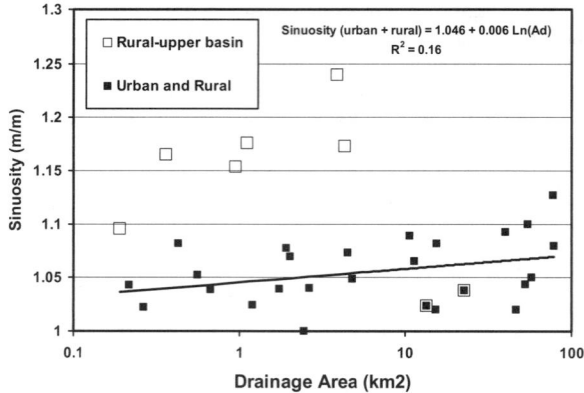

Figure 1: Sinuosity Trends. Double square symbols mark the two 4th order reaches sampled from the upper basin that have low sinuosity values like the rest of the reaches sampled.

Lack of Urban Effect on Channel Width and Depth

Relatively good correlations between drainage area and active channel dimensions are found in this study (Figure 2). However, width and depth relationships did not vary between urban and rural streams in this study. The specific causes of the lack of response of active channel width and depth to urban influence are not tested here, however, several explanations can be offered at this time. First, while the rural subwatersheds in this study may not be affected by urban factors, they have been subjected to land disturbances associated with agriculture since the 1830s. It is common to find 0.5 meters or more of post-settlement alluvium at the top of low terrace deposits. Thus, the "rural" streams studied here are not in a pristine "reference" state and probably reflect a transitional form between disturbed and fully-recovered rather than those streams that are more pristine in nature.

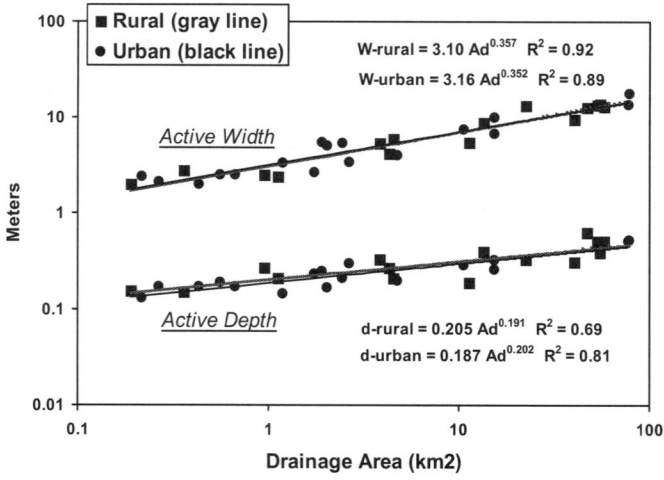

Figure 2: Active Channel Cross-section Trends.

64

Second, the low sinuosity of these streams indicates that lateral channel adjustments are constrained and hence floodplain development is also hindered. Besides the reduction of lateral adjustment rates, floodplain sedimentation rates are also low. Floods in the 2-5 year range are contained in the low terrace channel and not allowed to spread out upon a wider floodplain and dissipate energy or store sediment. Thus, while lateral erosion is limited and the channel may be considered stable, boundary conditions are still erosional and most fine-grained sediment and even fine gravel is transported through the reach to be stored at disturbance zones or exported from the system. More research is needed to document the sediment transport dynamics of these channels. Nevertheless, there is little opportunity for fine-grained sedimentation and active floodplain formation to occur and so these Ozark streams may generally be slow to recover in a geomorphic sense.

And, finally, low terrace bank materials tend to be resistant to erosion in this watershed. Bank deposits in this watershed are largely composed of resistant cohesive banks with gravel basal units that are not subject to failure very often. Sand content is also very low in these bank sediments as compared with other areas in the Ozarks. In addition, the effect of inundation on bank saturation and failure is also limited. With the exception of the lower reaches and several reaches served by spring flow, these are ephemeral streams subject to flashy stage changes during floods. Thus, potential for bank failure due to hydrostatic forces occurring during wetting and drying cycles and pore dewatering in minimized.

Urban Effect on Longitudinal Profile and Bed Sediment Size

In contrast to the channel cross-section, the longitudinal profiles of these streams differ slightly between rural and urban subwatersheds. Riffle spacing and pool depth both tend to decrease in urban streams as compared to their rural counterpart (Figure 3). Given the lack of opportunity to expend energy with lateral channel adjustments as described previously, it is reasonable to expect vertical adjustments in bed form may occur in response to urban factors. While channel incision of <0.5 m depth into stream beds sometimes occurs in these streams, shallow bedrock generally limits depth adjustments. Thus, sedimentary bedforms such as riffles and pools would be the most sensitive to urban effects since they are free to adjust rapidly. Moreover, sediment size tends to increase downstream and in urban streams (Table 3). While the downstream increase in bed sediment size may be influenced by the availability of larger materials from bluff erosion, the confluences of urbanized tributaries are mostly located along the downstream section of the main stem of the South Dry Sac River. Thus, decreasing riffle spacing and pool depth and increasing sediment size is believed to be the result of fluvial response to urban factors in this watershed.

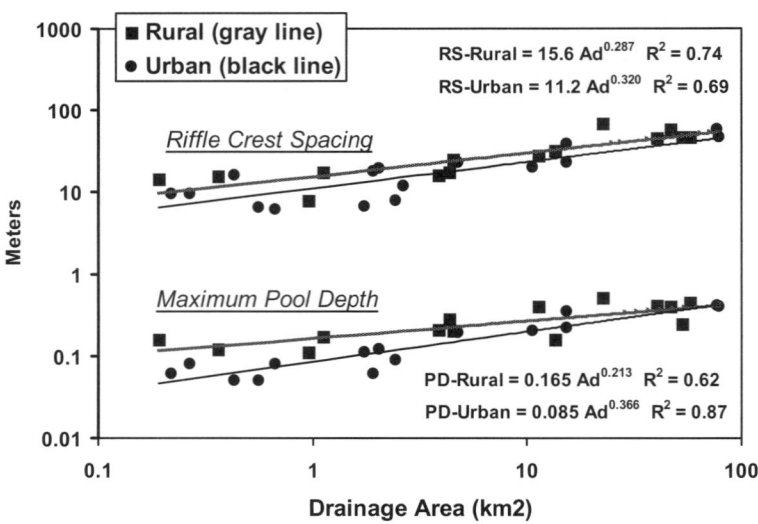

Figure 3: Active Channel Longitudinal Profile Trends

Table 3: Bed Sediment Size by Subwatershed Size and Land Use (median values)

Watershed Type Size	Land Use	n	D50 (cm)	D84 (cm)	Dmax (cm)
<1 km2	Rural	3	1.9	4.2	14.7
	Urban	5	2.0	5.7	20.0
1-10 km2	Rural	4	3.1	6.0	23.1
	Urban	7	3.0	6.0	28.6
>10k km2	Rural	8	2.5	5.0	24.2
	Urban	5	4.5	10.0	24.3

Lack of Riparian Vegetation Influence

This study found that riparian vegetation type did not influence channel form. While the low terrace channel generally is 4 to 5 times larger in cross-section area than the active channel, variations in the trend could not be explained by riparian vegetation or urban factors (Figure 4). Similarly, riparian vegetation was not useful for explaining variations in channel size and shape (Figure 5). The lack of riparian vegetation influence on channel morphology probably reflects three factors. First, the threshold of erosion for channel banks is relatively high as previously discussed so vegetation root influence may only exert a secondary effect. Additionally the influence of riparian vegetation influence may be too variable over time or occur at a scale not linked to geomorphic recovery processes. Second, extensive grassland riparian areas were hard to find in this area since most streams are lined by trees. In addition, tree growth often extended down onto the low terrace banks. And, finally, these streams tend to run dry most of the time due to the effect of karst drainage and "losing" stream hydrology. Thus, cattle do not water in the streams and trample the channel banks in most pastured grassland areas in the watershed.

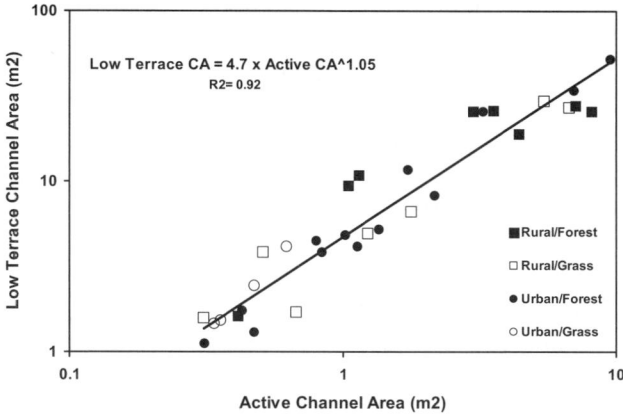

Figure 4: Active and Low Terrace Channel Cross-section Area Trends

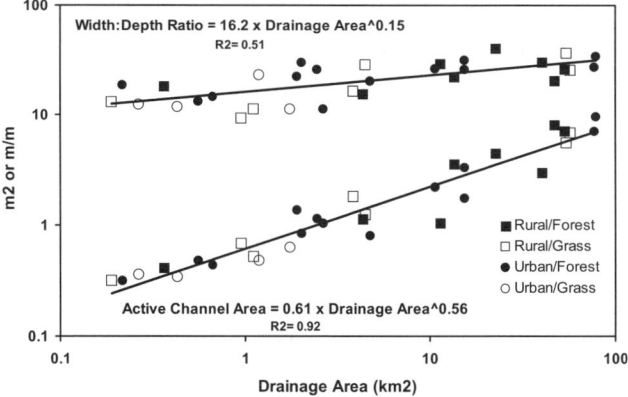

Figure 5: Buffer Influence on Active Channel Size and Shape

CONCLUSIONS

Streams in the South Dry Sac watershed are generally composed of an active or bankfull channel adjusted to the 1-year flood within a larger, low terrace channel which contains the 2- to 5-year flood. The low terrace channel can probably be described as a very narrow meander belt along a relatively straight channel. The low terrace channel tends to be 1.5 to 3 times wider and 4 to 5 times larger in area compared to the active channel. No influence of urbanization on channel width and depth is found in this study. Reasons for this outcome include: (1) lack of an undisturbed "rural" reference for direct comparison, (2) limited potential for lateral fluvial adjustment due to bedrock control and cohesive banks; and (3) slow geomorphic recovery rates of these systems due to low rates of fine-grained sedimentation and floodplain development. However, urban subwatersheds tend to have shorter riffle spacing, shallower pools, and larger bed sediment diameters compared to rural counterparts. Apparently, channel bed form and sediment size properties are more sensitive to urban hydrology and related sediment regime changes than cross-section form in these bedrock-controlled streams draining karst.

While this study emphasizes the lack of detection of a distinct urban influence on channel width and depth, there is clear evidence of stream erosion and sedimentation associated with urban development in the area. Field studies indicate that small drainage-ways and streams not influenced greatly by bedrock-control are subject to bed and bank erosion below urban developments (Martin, 2001). However, this study underscores the importance of understanding the role that geomorphic resistance factors and recovery potential plays in evaluating the physical stability of stream systems at the watershed-scale.

Acknowledgements

This research was supported by a Missouri Cooperative Agricultural Research Grant through the Department of Conservation and Department of Natural Resources. Graduate students John Horton and Aidong Cheng helped complete the fieldwork for this project.

REFERENCES

1. Booth, D.B., 1990. Stream channel incision following drainage-basin urbanization. *Water Resources Bulletin* 26(3):407-417.

2. Doll, B.A., D.E. Wise-Fredrick, C.M. Buckner, S.D. Wilkerson, W.A. Harmon, and R.E. Smith, 2002. Hydraulic geometry relationships for urban streams throughout the Piedmont of North Carolina. *Journal of the American Water Resources Association* 38(3):641-651.

3. Dunne, T., and L.B. Leopold, 1978. *Water in Environmental Planning*. New York, NY: W.H. Freeman and Co.

4. Hammer, T.R., 1972. Stream channel enlargement due to urbanization. *Water Resources Research* 8(6):1530-1540.

5. Jacobson, R.B., and K.B. Gran, 1999. Gravel sediment routing from widespread, low-intensity landscape disturbance, Current River Basin, Missouri. *Earth Surface Processes* 24:897-918.

6. Martin, L.L., 2001. *Geomorphic Adjustments of Ozark Stream Channels to Urbanization, Southwest Missouri*. Unpublished masters thesis. Department of Geography, Geology and Planning, Southwest Missouri State University, Springfield, MO.

7. Pizzuto, J.E., W.C. Hession, and M. McBride, 2000. Comparing gravel-bed rivers in paired urban and rural catchments of southeastern Pennsylvania. *Geology* 28(1):79-82.

8. Rosgen, D., 1996. *Applied River Geomorphology*. Pagosa Springs, CO: Wildland Hydrology.

9. Wolman, M.G., 1967. A cycle of sedimentation and erosion in urban river channels. *Geografiska Annaler* 49A:385-395.

ELM CREEK BANK STABILIZATION PROJECT: SAVING A REGIONAL TRAIL UTILIZING CHANNEL STABILIZATION TECHNIQUES

W.C. Eshenaur, P.E.[1] and J.J. Kusa, P.E.[2]

ABSTRACT

Three Rivers Park District retained SRF Consulting Group, Inc. (SRF) to investigate slope failures and propose slope stabilization solutions for Elm Creek in the vicinity of the Maple Grove segment of the North Hennepin Regional Trail. The goal of this project was to provide immediate slope stabilization to protect the nearby trail, preferably incorporating bioengineering methods. In the affected area, Elm Creek is a highly meandering channel that suffers from increased runoff rates and volumes due to urbanization. Four erosion reaches, each up to 175 feet in length, were identified along Elm Creek. Slope failure heights ranged from two feet to seven feet and the bank erosion was threatening to undercut a portion of the adjacent trail. Stream morphology, extent of bank erosion, soil characteristics, and analysis of development within the watershed were utilized to identify potential bank stabilization solutions. A number of possible bioengineering methods were reviewed for applicability in the project area. Design challenges in the decision process included limited accessibility to the site and stream velocities of up to seven feet per second. Based on the criteria, a bioengineered solution - vegetated geogrids – in conjunction with riprap toe protection, provided immediate bank stabilization to the four separate reaches that were experiencing slope failure. Design and application of vegetated geogrids in an urban setting and analysis of the system after three years of monitoring results are presented in this paper.

KEYWORDS. Minnesota, Elm Creek, SRF, SRF Consulting Group, streambank stabilization, stream, erosion, geogrids, bioengineering, slope failure, riprap, slope stabilization.

INTRODUCTION

Three Rivers Park District, formerly Hennepin Parks, is a natural resources-based park system located within the suburban Minneapolis/St. Paul, Minnesota metropolitan area that manages more than 27,000 acres of park reserves, regional parks, regional trails and special-use facilities. Annually, the Park District serves more than three million park visitors.

Within the park district, an extensive network of trails connect numerous regional parks with local neighborhoods. Many of the trails follow existing creeks and streams which course through the area. The North Hennepin Regional Trail is a maintained paved trailway which connects Coon Rapids Dam Regional Park in Brooklyn Park to Elm Creek Park Reserve in Maple Grove. The corridor features 5.6 miles of paved trails for biking and hiking, and trails for horseback riding (Three Rivers Park District, 2004).

Three Rivers Park District retained SRF to investigate slope failures and propose slope stabilization solutions for Elm Creek in the vicinity of the Maple Grove segment of the North Hennepin Regional Trail (Figure 1). Slope failure heights ranged from two feet to seven feet and the bank erosion was threatening to undercut a portion of the adjacent trail. The goal of this project was to provide immediate slope stabilization to protect the nearby trail, preferably incorporating bioengineering methods.

---

[1] SRF Consulting Group, Inc. weshenaur@srfconsulting.com

[2] SRF Consulting Group, Inc. jkusa@srfconsulting.com

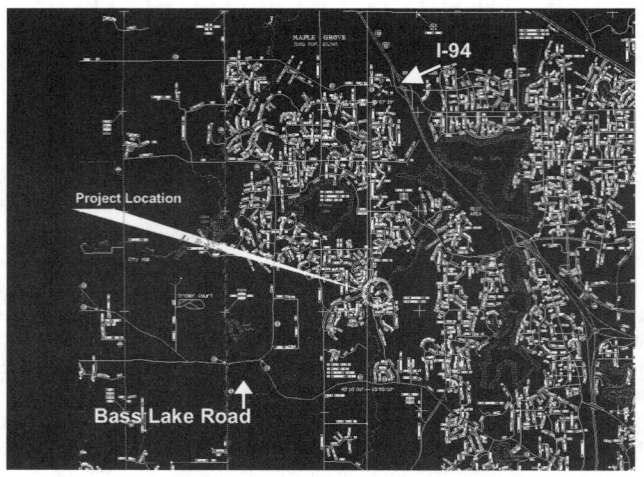

Figure 1. Project Location Map

LITERATURE REVIEW

The type of bank failure identified along Elm Creek is typical of many streambank failures observed by water resource professionals working in urbanized areas nationally. Consequently, a national search of recent projects and guidance documents produced an array of options and opportunities, many of which were applicable to the Elm Creek site. The search relied primarily on published data from various organizations, published design documents, and several site visits to local stabilization projects.

The following organizations were identified for compiling information pertinent to streambank stabilization:

- Minnesota Department of Natural Resources
- North American Lake Management Society
- Dakota County Soil and Water Conservation District
- U.S. Department of Agriculture – National Resource Conservation Service
- Minnesota Pollution Control Agency

Key design documents were also identified and utilized to analyze various bank stabilization techniques. The documents included:

- Annis Water Resources Institute. Grand Valley State University, MI – "Stream Restoration Project Logs"
- U.S. Department of Agriculture and National Resource Conservation Service - "Watershed Technology Electronic Catalog"
- U.S. Department of Agriculture – "Stream Corridor Restoration"
- Marin County Stormwater Pollution Prevention Program
- Minnesota Pollution Control Agency – "Protecting Water Quality in Urban Areas"

70

- Conservation Commission of Missouri. "Tree Revetments For Streambank Stabilization"
- U.S. Fish and Wildlife – "River Restoration Guidelines"
- King County - "Guidelines for Bank Stabilization Projects in the Riverine Environments of King County"
- Maryland Department of the Environment- "Maryland Stormwater Design Manual Volumes I and II"

After analyzing potential solutions, several local streambank stabilization projects of similar scale were visited to determine the effectiveness of potential stabilization techniques. These sites included a designated trout stream, the Vermillion River, located in Farmington, Minnesota and portions of Minnehaha Creek, which flows through the urban heart of Minneapolis, Minnesota.

DISCUSSION

In the affected area, Elm Creek is a very serpentine, very unstable channel that suffers from increased runoff rates and volumes due to urbanization. Previously farmed areas are rapidly being converted to single family and condominium neighborhoods, increasing impervious area in the Elm Creek watershed. The stream discharge time series, as measured by a USGS stream gaging station at Champlin, MN is very flashy - minimum flows of 1.2 cubic feet per second (cfs) and maximum flows of up to 900 cfs (See Figure 2). This type of hydrograph is indicative of a high percentage of impervious area within the drainage basin.

DISCHARGE, CUBIC FEET PER SECOND

Most recent value: 118 10-22-2002 15:00

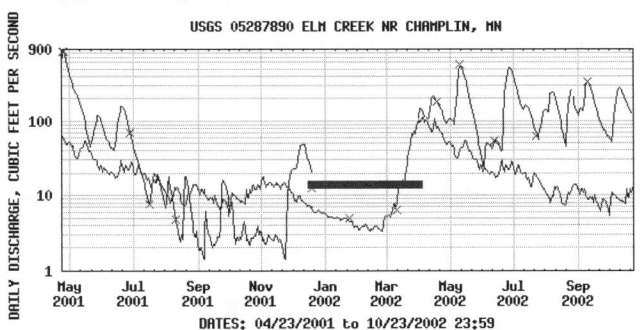

EXPLANATION
—— DAILY MEAN DISCHARGE
—— MEDIAN DAILY STREAMFLOW BASED ON 23 YEARS OF RECORD
× MEASURED DISCHARGE
□ Flow at station affected by ice

Download a presentation-quality graph   Parameter Code 00060; DD 06

Daily mean flow statistics for 10/23 based on 23 years of record in ft³/sec

Current Flow	Minimum	Mean	Maximum	80 percent exceedence	50 percent exceedence	20 percent exceedence
118	1.2	27.6	150	3.84	13.0	33.0

Percent exceedance means that 80, 50, or 20 percent of all daily mean flows for 10/23 have been greater than the value shown.

Figure 2. Elm Creek Discharge Graph

One frequent consequence of increased runoff rates and volumes is a rapid change in channel sinuosity as the stream establishes an equilibrium with new bedload conditions. The change in channel sinuosity is often achieved through erosion of previously stable streambanks within the floodplain. In the case of Elm Creek, a survey conducted by SRF identified four distinct reaches, each up to 175 feet in length in the project area with significant erosion. Slope failure heights ranged from two feet to seven feet. Bank erosion was threatening to undercut a portion of the adjacent trail in three locations. The banks lacked any stabilizing vegetation due to the ongoing erosion of the clay/silty clay soils. Figure 3 shows a typical segment of the first erosion reach.

Figure 3. Bank erosion along Reach 1 of Elm Creek. Existing trail is located adjacent to split rail fence.

Criteria

In order to narrow the range of potential solutions, six criteria were applied to assess the applicability of the various stabilization measures researched. One of the primary criteria considered was aesthetics. A "natural" looking solution would be required to blend with the surrounding trail environment. Durability of the reconstructed banks was also key. The project should ensure bank stability under current and future stream flooding conditions for at least several decades. Site accessibility eliminated several options requiring significant right of way. The adjacent trail is the only access to the sites and the trail is bordered by the stream on one side and by private homes on the other. The situation limited the size of equipment and the allowable disturbance area. As is typical for any project, cost was a factor as well. Three Rivers Park District was interested in minimizing the final cost of the stabilization solution by utilizing readily available materials and applying one type of stabilization technique to all four areas. Stream hydraulics also impacted the choice of stabilization. Based on the flashy nature of the stream, the solution would have to withstand extremely high flow events, yet maintain a "natural" look during low flow periods. Finally, channel sinuosity was taken into account. Although the stream was currently not impacting portions of the third erosion reach as significantly as the other areas, the design took into account continued changes in meander due to expanding development within the upstream drainage area.

Based on these criteria, solutions such as riprap slopes, bendway weirs, and rock gabions were eliminated based on the aesthetic requirement of the site. Other solutions requiring significant excavation into the bank, such as rootwad revetements, were also eliminated due to the limited site accessibility. Based on the research, it was also determined that stand-alone vegetative stabilization techniques were only effective in flows under four feet per second (fps), according to the Minnesota Pollution Control Agency (MPCA) or six fps according to data from the

National Resources Conservation Service (NRCS). Data obtained from a HEC-2 model maintained by the Park District indicated flow velocities of up to seven fps in this area for the 100-year storm event.

<u>Preferred Options</u>

SRF provided Three Rivers Park District with three preferred options, all of this which used a blend of structural and vegetative engineering solutions. The structural solutions included live cribwalls, vegetated gabions, and vegetated geogrids. In addition, it was recommended that dormant posts be utilized in the top portion of the bank to improve long term stability and a boulder vane be installed upstream of all reaches to redirect the thalweg from the stabilized bank. Examples of each stabilization option are shown in Figure 4. The live cribwalls utilize lumber cribs filled with riprap, soil, and live willows to create a structural bank which over time will be overgrown by willows and other vegetation. The vegetated gabions utilize a similar concept, with the exception that metal boxes are filled with riprap, and smaller vegetation is utilized to "mask" the gabions. The vegetated geogrids utilize the structural stability of riprap at the toe of the slope, but rely on a natural geogrid for short term stability and dense willow growth for longer term slope stability.

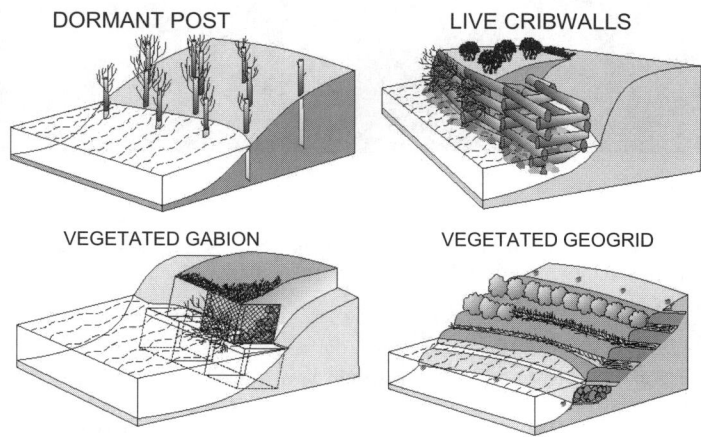

Figure 4. Bank erosion stabilization options.

Three Rivers Park District evaluated the alternatives and decided to utilize the vegetated geogrid option for all four eroded reaches. The availability of willow stakes on a nearby parcel owned by Three Rivers Park District as well as the reduced amount of riprap necessary, were two critical factors for this decision. Contrary to the original design, the Park District decided not to install the boulder vanes in the stream bed. These would have added another factor of safety, but were not critical to solving the immediate bank stabilization problem.

<u>Final Design</u>

The final design of the stabilization project was based on the HEC -2 model data as well as engineering assessment of the site conditions. Section view of the geogrid from the plan set is shown in Figure 5. The riprap base and first tier of riprap were designed to stabilize the stream for the 10-year flow event. It was recognized; however, that the development within the project area impacted both stream flow rates and upstream floodplain limits, which could cause the 10-year flow elevation to change in the near future. The first riprap tier utilized a geosynthetic material which was UV resistant and would withstand high stream velocities. The upper tiers of the geogrid utilized a coir erosion stabilization mat filled with topsoil and were anchored into the existing bank with 18-inch wood stakes. As mentioned before, the coir mats would provide short-term slope stability, on the order of five to seven years. During this period, the vegetation would establish itself and eventually create a naturally stabilized and reinforced slope. The geogrids were layered with topsoil and willow cuttings up to the proposed top-of-bank elevation. The upper slopes were staked with dogwoods and seeded with native grasses.

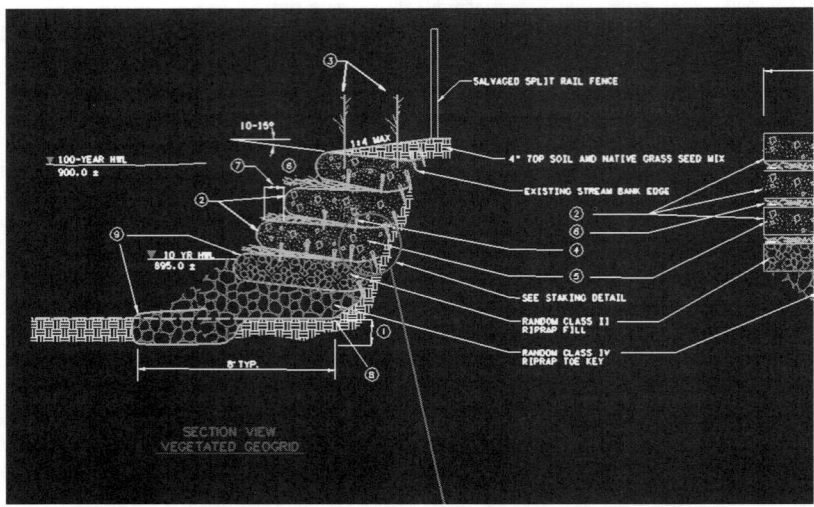

Figure 5. Section View of vegetated geogrid from engineering plan set.

Cross sections were cut through each reach to allow the designer to estimate quantities as well as to provide the contractor with a detail of each construction area, since each reach varied in bank height and slope.

Construction

Although the preferred construction period would have been during the winter months, when stream flow is typically low, the schedule resulted in construction starting in May 2002. Figure 6 shows the construction of the Class IV riprap toe key and geosynthetic wrapped Class II riprap tier. Construction was conducted over approximately a two week period and final seeding was completed in the first week of June. Due to the higher stream flows during the construction period, the contractor utilized a flotation silt curtain to limit sediment deposit within the stream. With the exception of one significant storm during the construction period, which washed out a temporary erosion control measure, no significant issues challenges encountered during construction. This phenomenon can be attributed to Three Rivers Park District, which conducted daily construction management, provided local residents with sufficient information via mailings and one-on-one meetings, and ensured a smooth project delivery.

Figure 6. Construction of riprap key and lower riprap tier.

CONCLUSION

The project can be judged a success based on the goals of the project, to stabilize eroding slopes by implementing a bioengineering solution in four limited access areas. Figure 7 shows the finished bank just one year later. Monitoring of the site has indicated that the willow, dogwoods, and native grasses have all established themselves on the site and no significant erosion of the banks has occurred over the last three years. The banks have quickly achieved a "natural" look, while protecting the adjacent trail.

Figure 7. Reconstructed bank along Reach 1, one year later. This is the same bank area shown in Figure 1.

Lessons Learned

Issues that could have been handled more effectively include sediment and erosion control during the project construction, improved hydraulic information for design, and better construction scheduling. Based on the experience from this project, requiring a heavy duty staked silt fence in shallow areas would eliminate the potential for flotation silt curtain failure during larger storm events. In addition, ensuring adequate installation of all floatation silt fence would help prevent failure occurrences during construction. The data utilized for design of the project was obtained from a HEC-2 model, which was not designed to provide the detailed data that would typically be utilized for design purposes. This would have allowed for refinement of the model and potentially a reduction in the quantity of riprap required for stabilization of the bank up to the 10-year flow elevation. Finally, as mentioned before, the project would have ideally been constructed during low flow periods, thus eliminating the need for a flotation silt curtain and limiting construction delays due to storm events.

Further monitoring of this and other bank stabilization sites within urban areas is recommended, given the lack of quantitative data regarding the use of bioengineering solutions for slope stabilization. Various organizations, mentioned previously, have begun compiling data from these type of projects. This data will assist future designers in choosing appropriate bioengineering techniques for complex and rapidly changing environments. This type of project will hopefully become a long term example of a cost effective application of a bioengineered solution in a limited access urban environment.

Acknowledgements
We would like to thank the following individuals for their participation in the project:
Mike Horn, Three Rivers Park District
Jason Moeckel, Minnesota Department of Natural Resources
Mike McGarvey, SRF Consulting Group, Inc.

REFERENCES

1. Annis Water Resources Institute. Grand Valley State University, MI – Stream Restoration Project Logs - Available at: http://www.gvsu.edu/wri/about/pubs.htm. Accessed June 2004.

2. Conservation Commission of Missouri. March 2004. Tree Revetments For Streambank Stabilization. Available at: http://www.conservation.state.mo.us/fish/streams/revetmen/. Accessed 28 June 2004.

3. Dakota County Soil and Water Conservation District. Available at: http://www.dakotacountyswcd.org/. Accessed 28 June 2004.

4. King County. June 1993. Guidelines for Bank Stabilization Projects in the Riverine Environments of King County. Available at: http://dnr.metrokc.gov/wlr/biostabl/index.htm. Accessed 28 June 2004.

5. Marin County Stormwater Pollution Prevention Program. Available at: http://mcstoppp.org/. Accessed 28 June 2004.

6. Maryland Department of the Environment. October 2000. Maryland Stormwater Design Manual Volumes I and II. Available at: http://www.mde.state.md.us/Programs/WaterPrograms/SedimentandStormwater/stormwater_design. Accessed 28 June 2004.

7. Minnesota Department of Natural Resources. Available at: http://www.dnr.state.mn.us. Accessed 28 June 2004.

8. Minnesota Pollution Control Agency. March 2000. Protecting Water Quality in Urban Areas.

9. North American Lake Management Society. Available at: http://www.nalms.org. Accessed 28 June 2004.

10. Three Rivers Park District. 2004. Park and trail information. Available at: http://www.threeriversparkdistrict.org. Accessed 28 June 2004.

11. U.S. Department of Agriculture, National Resource Conservation Service. Watershed Technology Electronic Catalog. Available at: http://www.wcc.nrcs.usda.gov/wtec/. Accessed 28 June 2004.

12. U.S. Department of Agriculture. Stream Corridor Restoration. Available at: http://www.usda.gov/agency/stream_restoration/newgra.html. Accessed 28 June 2004.

13. U.S. Fish and Wildlife. River Restoration Guidelines. Available at: http://www.r6.fws.gov/pfw/r6pfw2h.htm. Accessed 28 June 2004.

QUALITY ASSESSMENT OF STREAM WATER AND BED SEDIMENTS: A CASE STUDY OF URBANIZATION IMPACTS IN A DEVELOPING COUNTRY

M. N. Tijani[1, 2] and S. Onodera[2].

ABSTRACT

Hydrogeochemical analyses and evaluation of surface water and stream-bed sediment samples from 40 different locations along the stretches of stream networks within Ibadan metropolis, SW Nigeria were undertaken with respect to the influence of urbanization and anthropogenic activities on the quality status of urban drainage system in a developing country. Results show that the surface water samples have generally low TDS with average value of 517mg/l, while average concentrations of the dominant ions are 75mg/l (Na) and 30mg/l (Ca) representing 51% and 18% respectively of the total cations and Cl (av. 86mg/l) and HCO_3 (av. 71mg/l) representing 40% and 21% respectively of the total anions. However, indication of urbanization impacts on the surface water quality is revealed by the NO_3 concentrations with values of 22.8 – 366mg/l, exceeding the WHO limit of 50mg/l in 85% of the sampled locations. The high NO_3 as well as the variability of sodium and chloride concentrations are clear indications of contamination through a number of point-source inputs of domestic sewage, municipal effluents and waste dumps along the stretches of the drainage system in the study area.

The impact of such anthropogenic activities is also revealed by the trace metal enrichment of the stream bottom sediments with enrichment factor (EF) of about 60 (Hg), 3.5 (As), 1.5 (Pb), 0.6 (Cu) and 0.5 (Zn) which imply that anthropogenic sources account for about 90-95% (Hg), 10-80% (As), 10-70% (Pb), 10-55% (Cu) and 10-40% (Zn) of the respective total concentration. Also an assessment of the metal bioavailability revealed that 19.6% (Cu), 10.7% (Pb) and 29.4% (Zn) of the respective metal concentrations are in adsorbed form, the possible remobilization and trophic transfer of which calls for environmental concerns. This study highlighted the need to understand the interactions between the populated urban catchments and the drainage systems and in essence the environmental pollution management of urban drainage systems in developing countries.

KEYWORDS. Stream water, stream sediment, quality assessment, heavy/trace metals, contamination, urbanization impact, Ibadan, Nigeria.

INTRODUCTION

Within the last quarter of the last century, there were much interest on environmental pollution and in particular about geochemical distribution and fate of heavy metals in both water and sediment phases of urban drainage system. Though significant advances had been made in the developed regions of the world, there are still increasing concerns about the impacts of urbanization, agricultural, mining and industrial activities on drainage networks in the developing regions of the world, especially in areas with inadequate land use planning and proper waste disposal and management systems (Ajayi and Mombeshora, 1990 and Mogollon et al., 1996). In such developing countries, contaminations of surface drainage system are mostly related to the consequences of population growth, urbanization, agricultural activities and development of new industrial zones (Olade, 1987 and Paul & Pillai, 1983), while uncontrolled direct dumping of domestic waste and discharge of domestic and industrial sewage water into the urban drainage systems are critical components of trace/heavy metal contamination (Tijani et al.,

---

[1] Biogeochemistry Laboratory, Dept. of Natural Environmental Sciences, Hiroshima University, Kagamiyama 1-7-1, Higashi-Hiroshima, 739-8521 JAPAN. (sonodera@hiroshima-u.ac.jp).

[2] Department of Geology University of Ibadan, Ibadan – Nigeria. (tmoshood@yahoo.com).

2004) especially in areas with lack of strict land-use plan and environmental protection regulations. Though sediments are said to represent the ultimate sinks for heavy metals in the environment (Gibbs, 1977), changing physico-chemical and environ-mental conditions may lead to remobilization and release of sediment-bound metal pollutants into the water column and consequently into the trophic levels of the food chain within an aquatic environment with serious health and environmental consequences.

Based on the above background, geochemical assessment of major and trace element profiles of urban drainage network involving water and bottom sediment samples within Ibadan metropolis, SW-Nigeria, are presented and evaluated in this study with respect to impacts of urban activities on the overall quality of drainage systems within the metropolis. The overall evaluation is expected to gives an insight into vulnerability of urban drainage networks in a typical developing region in response to poor sanitation and waste disposal facilities and other anthropogenic activities within the populated urban catchment of a developing country.

STUDY LOCATION AND GEO-ENVIRONMENTAL SETTING

The study area, Ibadan metropolis and environs form the catchment of the Ogunpa-Ona- Ogbere stream networks with approximately $315km^2$ encompassing urban populated portion of Ibadan metropolis (Figure 1). The study area is characterized by tropical humid climate with two distinct seasons: the wet season which occurs between March and October with an average annual rainfall of about 1,250mm and dry season from November to February characterized by dry, dusty and relatively cold NE-SW trade winds. Ibadan metropolis is characterized by high population density of about 3,250 persons per sq. kilometer in comparison to the national average of 137 persons per sq. kilometer. However, like many urban centers in developing countries, Ibadan metropolis is characterized by poor land-use planning, lack of adequate water supply and sanitary conditions, lack of proper sewage, and waste disposal systems as well as traffic congestion and hold-ups, especially during the rush hours while direct discharge of domestic sewage water and dumping of domestic refuse into the drainage channels are common practices especially within the congested central portion of the city (which constitute the catchment area of the Ogunpa stream).

Geologically, the study area is characterized by Precambrian Basement Complex composed of quartzites, gneisses and migmatite as the major rocks units which are intruded by pegmatites, quartz veins, aplites and dolerite dykes (Olayinka et al., 1999). These form ridges and inselbergs surrounding the adjoining plains and valleys. The overburden is dominated by weathered saprolite units with varied thickness depending on the underlying bedrock while groundwater occurrences are in localized disconnected phreatic weathered regolith aquifers, essentially under unconfined to semi-confined conditions (Tijani, 1994). Morphologically, the unmodified drainage channels form dendritic patterns and run in a southerly direction through much of the city center (Figure 1). Water flows are irregular during the dry seasons, while population pressure had resulted in built-up areas closer to the stream banks and attendant human activities and increasing run-off constituting constant danger in terms of flooding during the peak of the rainy season.

SAMPLING ANALYSES AND DATA EVALUATION

Surface water and stream sediment samples were collected from forty, (40) different locations within the urban drainage networks of Ibadan metropolis, SW.-Nigeria (Figure 1). For the water samples, electrical conductivity (EC), pH and temperature were determined in-situ (in the field) using portable WTW-Conductivity meter (model LF/95) and WTW-pH meter (model pH/91). The collected sediment samples were dried in the laboratory, disaggregated and sieved to obtain the clay fractions (<63μm) which were subsequently subjected to extraction process using a mixture of HNO_3 : HCl (3:1) acid. Major cations (Ca, Mg, Na and K) and trace metals (Pb, Zn, Cu, Cd, As and Hg) were analyzed using Atomic Absorptions Spectrometric (AAS) method, while, analyses of the anions (HCO_3, Cl, SO_4 and NO_3) for the surface water samples were carried out using conventional titrimetric methods. In addition a number of selected stream sediment samples were also subjected to a separate extraction process using 1.0M solution of ammonium acetate in order to determine the adsorbed contents of the respective metals using ICP-AES.

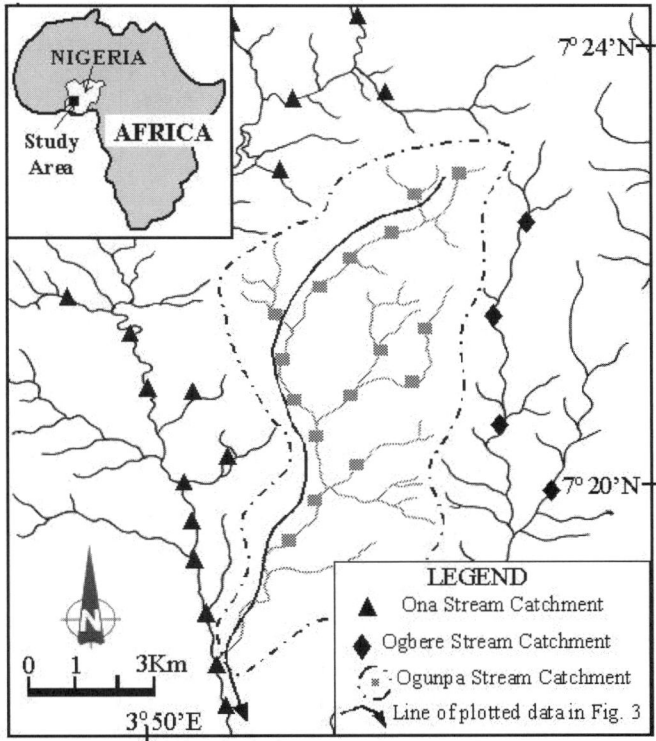

Figure 1. Location Map of the study area showing the sampling points.

Apart from the general evaluation of the quality aspects of the surface water system from the study area, some quantitative indices were used to describe the heavy metal contamination trends:

1. Anthropogenic factor (AF) is a quantification of the degree of contamination relative to the measured background concentration and is expressed as $AF = C_m/B_m$.
2. Enrichment factor (EF) is to quantify the possible contribution of external factors (such as anthropogenic sources) relative to the measured background concentrations and is expressed as $EF = (C_m - B_m)/B_m$.
3. Index of geoaccumulation (I_{geo}) proposed by Mueller, (1979) quantifies the degree of metal contamination in terrestrial, aquatic as well as marine environments and it is expressed as $I_{geo} = \log_2 [(C_m) / (1.5*B_m)]$

where, C_m is the measured concentration in sediment or water, B_m is the background concentration (value) of metal while 1.5 is a factor for possible variation in the background concentration due to lithologic differences. Details of these indices are presented elsewhere (Mueller, 1979; Sutherland, 2000; Manjunatha et al., 2001; Loska *et al.*, 2004; Tijani et al., 2004).

RESULTS, INTERPRETATION AND DISCUSSION

Surface water Quality

The summaries of the results of the hydrochemical analyses of the surface water samples of the urban stream network in Ibadan metropolis (study area) is presented in Table 1 together with the WHO (1993) standards and mean world river (MWR) estimates (Hem, 1985). It can be seen that the TDS for the surface water system is generally low with average value of 516.7mg/l. This is in line with results from similar catchment underlain by low solubility Precambrian Basement Complex rocks. Major ions in the surface water system are within the limits of WHO standards and also comparable to the world river average with the exception of Na, Cl and NO3, which are indications of anthropogenic influence (Figure 2).

Table 1. Summary of the hydrochemical analyses results of the surface water samples from the study area.

Parameter#	Surface water§				*WHO Standard	MWR+
	Min	Max.	Mean	SD		
Temp. °C	25.7	37.4	30.5	2.5	Variable	Variable
pH	5.9	8.9	7.4	0.9	6.5-9.5	Variable
EC(µS/cm)	164	1878	821.0	433.9	400-1480	Variable
TDS	103	1188	516.7	277.9	500-1000	90.0
Ca	2.0	72.6	30.9	18.7	75-200	15.0
Mg	2.0	31.1	14.7	7.0	50-150	4.1
Na	17.8	383.4	92.1	89.4	20-200	6.3
K	7.3	178.5	41.9	42.0	10-12	2.3
Fe	0.03	23.9	1.8	3.8	0.3-1.0	0.5
HCO3	38.0	118.0	67.1	17.7	Variable	58.0
Cl	25.0	150.0	74.8	33.2	250-600	7.8
SO4	10.0	49.0	29.3	12.0	250-400	11.0
NO3	22.8	366.0	104.3	95.1	25-50	1.0
As	0.20	3.30	1.70	0.79	0.01	0.001
Cu	0.001	0.03	0.010	0.006	2.0	0.003
Pb	0.010	0.58	0.089	0.110	0.01	0.003
Hg	0.20	0.40	0.300	0.100	0.001	0.001

\# Zn and Cd are below the detection limits; hence are not reported here.
§ SD = Standard deviation: Min.= Minimum: Max. = Maximum
*WHO Standard, 1993.
+WMR= Mean world river (from Hem, 1985: Martin & Maybeck, 1979).

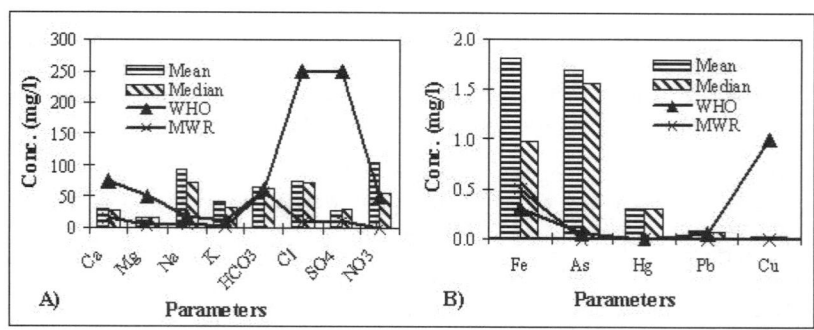

Figure 2. Profiles of (a) major elements and (b) trace elements compared with WHO standards and estimates of mean world river (MWR).

In addition, a number of sample locations with TDS >700mg/l and high NO_3 concentration (80–366mg/l) confirm point-source discharge of untreated domestic/municipal sewage water into the drainage channels as observed during the filed investigation. This is demonstrated in Figure 3 where the peaks of NO_3 do not only coincide with those of TDS/EC, but also have values considerably above 100mg/l within the stretches of urban centers compared to the suburb or peri-urban areas at the upstream and downstream sections. This underscores the influence of uncontrolled anthropogenic discharge of household/municipal effluents into the stream channels within the populated urban center of the study area. For the trace metals, Cu, Pb, Zn and Cd are either below the detection limits and/or occur at low concentrations below the recommended WHO limits for drinking water. However, As and Hg are in higher concentration with average value of 1.70 and 0.35mg/l respectively compared to the recommended WHO limits of 0.01mg/l and 0.001 mg/l respectively for drinking water.

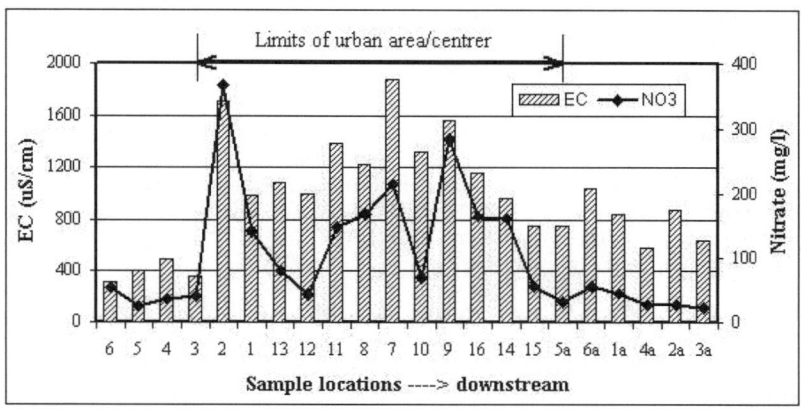

Figure 2. Profiles of electrical conductivity (EC) and NO_3 of the drainage system within the urban centers of the study area.

Metal profiles in stream sediments

Table 2 presents the summary of trace/heavy metal concentrations in the stream sediments together with some environmental quality/contamination indices. Concentration of heavy / trace metals in the stream sediments samples are obviously much higher than those measured in the surface water samples. Fe, Cu, Pb and Zn are most abundant in the analyzed stream sediments with concentration of 3.1– 44.6mg/l Cu (av. 9.7mg/l), 2.5–702.5mg/l Pb (av. 36.5mg/l) and 5.7–115.6 Zn (av. 31.6mg/l) respectively.

Table 1. Summary of trace metal concentrations in the stream sediments and environmental quality indices (average values).

Metals	Stream sediments						AF_{sw}
	Min	Max.	Mean	EF	I_{geo}	$R_{tot/ads}$	
As	0.30	6.80	3.22	2.22	0.77	-	1695
Cd	0.06	0.27	0.12	0.44	-.014	-	-
Cu	3.10	44.60	9.20	-0.36	-1.84	10.0	3.2
Pb	2.50	702.5	36.46	0.35	1.68	27.9	29.7
Zn	5.7	115.6	31.64	-0.53	-2.07	5.7	-
Hg	2.0	11.20	5.75	70.9	6.95	-	300
Fe	56.6	7598	35667	0.49	-0.24	1519	6.0

Min.= Minimum; Max. = Maximum; EF: Enrichment factor; I_{geo}: Geoaccumulation Index
$R_{tot/ads}$: Ratio of total to adsorbed metal concentration
AF_{sw}: Anthropogenic factor for surface water system

81

However, Hg, As and Cd are in relatively lower concentration with average values of 5.8mg/l, 3.2mg/l and 0.12mg/l respectively. The variability of concentrations of these metals within the stretches of drainage channels (like NO_3 in the water column) suggests local anthropogenic input sources through domestic and municipal sewage effluents at various points along the drainage channel. Furthermore, metal partitioning in terms of environmental bio-availability were investigated through comparison of the total metal and adsorbed metal concentration (see Table 2). The ratio of the total to the adsorbed concentration ($R_{tot/ads}$) are 5.8(Zn), 10.1(Cu) and 27.9(Pb) which imply that about 30% (Zn), 20% (Cu) and 12% (Pb) are in adsorbed form as bio-environmental available portion that can be released back into the water phase in response to changes in the physico-chemical conditions and dictated by impacts of human activities. Therefore, while the relatively low concentrations of the trace metals (Cu, Pb, Zn, Cd, As and Hg) in the water column compared to the stream sediments are indications of partition between the water and sediment phase, the proportions of adsorbed concentration in the stream sediments are potential sources of contamination of the water column.

Environmental Quality/Contamination Assessment
Quality status and level of trace metal contaminations in water and sediments in terms of anthropogenic factors (AF), enrichment factor (EF) and geoaccumulation index (I_{geo}) were summarized in Table 2 above. With the exception of Cd and Zn, which are below detection, limits in the surface water system, Cu, Pb, As and Hg have AF values of >1.0 suggesting contamination above the level of mean world river (MWR) and WHO limits, despite the absolute low concentrations in the analyzed surface water system.

For the stream sediments, most of the analyzed trace metals have EF of less 1.0 except for As and Hg with values of 2.3 and 70.9 respectively with respect to the average composition of igneous bedrock (Bowen, 1979), which are more or less an approximation of the basement geology of the study area. However, the estimated Mueller geoaccumulation index (I_{geo}) revealed value of 0.77 for As indicating slight contaminations, while values of 1.68 for Pb and 6.95 for Hg indicate moderate contamination and extreme contamination respectively (see Table 2). While the source of As and Pb may be attributed to anthropogenic activities and dumping of wastes/rubbish on the stream channels, the sources of Hg (as constituent of medicated soaps and cosmetics materials) can be attributed to inputs from discharge of domestic/household waste waters into the stream channels. The environmental implication of stream sediments contamination lies in the fact that the adsorbed portion of these metals (within the sediment phase) could be released back to the water phase while the possible trophic transfer especially at the downstream outside the urban areas where local inhabitants use the water in vegetable nurseries/farming call for concerns.

Multivariate Principal Component Analysis (PCA)
To further characterize the analyzed hydrochemical data and to identify possible controlling factors, principal component analysis (PCA) was employed in this study. Such multivariate statistical evaluation allows the reduction of data and extraction of information that will be helpful for the water quality assessment (Massart & Kaufman, 1983: Simeonov et al, 2003). In this study, Zn and Cd, which are generally below the detection limits, are excluded from the data-set evaluation. For the analyzed surface water system, four (4) main principal component factors were obtained with eigenvalues >1 summing up to 70.6 % of the total variance in the data-set (Table 3).

The first factor, which accounts for 33.9% of the total variance in the analyzed surface water, was correlated with EC, TDS and base (major) cations in both water systems. This factor, tagged "geogenic/weathering" factor is related to chemical weathering dissolution that controls the geogenic / natural chemical inputs to the overall dissolved constituents of the surface water system. The second factor, which weighted on pH, temperature, HCO_3, SO_4 and partly on K could be interpreted as "physico–chemical" control relating to atmospheric dissolution of CO_2 and SO_4 through precipitation and subsequent dissociation to form the respective HCO_3 and SO_4 ions.

$$CO_2 + H_2O = HCO_3^- + H^+ \qquad (1)$$
$$SO_2 + 2H_2O = SO_4^{2-} + 2H^+ \qquad (2)$$

Table 3. R-mode varimax factor scores of hydrochemical data for the surface water system.

Parameter/ Factors	Surface water			
	Factor 1	Factor 2	Factor 3	Factor 4
	0.95 (EC,µS/cm)	0.67 (Temp.°C)	0.61 (Cl)	0.73 (Fe)
	0.95 (TDS)	0.68 (pH)	0.72 (NO_3)	0.49 (As)
Loading	0.76 (Ca)	0.91 (HCO_3)		0.82 (Cu)
values (>0.4)	0.84 (Mg)	0.83 (SO_4)		0.51 (Pb)
and	0.91 (Na)			
Parameters	0.89 (K)			
	0.40 (Cl)			
	0.40 (NO_3)			
Eigenvalue	5.42	1.78	1.46	1.17
% Cum.Var.[+]	33.9	54.2	63.3	70.6

[+] % Cum. Var. = Percentage cumulative variance.
F1 = Chemical weathering/dissolution control; F2 = Physico-chemical control
F3 = Anthropogenic input control; F4 = Toxic/trace metal input control.

However, the third factor, which was loaded on Cl and NO_3, implied both point and non-point anthropogenic contaminations of surface water through inputs of domestic sewage discharges while the fourth factor which was characterized by As, Cu, Pb and Hg suggests an interplay of trace metal inputs through "soil leaching" during run-off and anthropogenic inputs from domestic/municipal effluents.

CONCLUSION

Quality evaluation of surface water and stream sediments from Ibadan metropolis, SW-Nigeria, were discussed as a case study of anthropogenic influence in a typical urban environment of a developing country. From this study, it is clearly evident that population increase and urbanization coupled with lack of proper waste disposal system have considerable influences on the heavy metal contaminations of urban drainage systems in the study area. The study revealed low to moderate contamination of Pb, Hg and As in the surface water compared to the mean composition of world rivers, while the respective concentrations of Pb, Cu, Hg and As in sediment phase revealed an overall enrichment and contamination relative to the background value in the granitic bedrock units in the study catchment. Among the major elements, NO_3 is the most critical in the surface water system, the contamination of which is related to anthropogenic sources mostly from direct discharge of untreated domestic/municipal sewage waters and refuse dumps into the stream channels within Ibadan metropolis.

Evaluation of bioavailability and partitioning of the trace metals revealed that, about 30% of Zn, 20% of Cu, 12% of Pb and less than 1% of Fe were in adsorbed form compared to the respective total metal concentrations. In addition, application of multivariate PCA also revealed interplay of geogenic controls of the weathering/dissolution process and anthropogenic controls on the chemical composition of the surface water system in the study. This study outline the provision of proper waste disposal facilities, formulation of a reasonable waste management policy and urban planning as the critical components of environmental contamination control in a typical urbanized catchment of a developing country.

Acknowledgements
The benefit of JSPS research fellowship by the first author is gratefully acknowledged.

REFERENCES

1. Ajayi S. O. and C. Mombeshora. 1990. Sedimentary trace metals in lakes in Ibadan, Nigeria. *Sci. Total Environ.* 87 & 88: 1–18.

2. Bowen, H.J.M. 1979: *Environmental Geochemistry of Elements.* Academic Press, London: 316p.

3. Gibbs, R. J. 1977. Transport phases of transition metals in the Amazon and Yukon Rivers. *Geol. Soc. Am. Bull.* 88: 829–843.

4. Hem, J. D. 1985. Study and Interpretation of the Chemical Characteristics of Natural Water. USGS Water Supply Paper 2254, 3rd Edition; 263p.

5. Loska, K., D. Wiechula, and I. Korus. 2004. Metal contamination of farming soils affected by industry. *Environment International* 30: 159-165.

6. Manjunatha, B. R., K. Balakrishna, R. Shankar, and T. R. Mahalingam. 2001. Geochemistry and assessment of metal pollution in soils and river components of a monsoon-dominated environment near Karwar, southwest coast of India. *Environ Geol.* 40: 1462–1470.

7. Martin, J.-M. and M. Maybeck. 1979. Elemental mass-balance of material carried by major world rivers. *Mar. Chem.* 7: 173-206.

8. Massart, D. L. and L. Kaufman. 1983. *The interpretation of analytical chemical data by the use of cluster analysis.* John Wiley, NY.

9. Mogollon, J. L., C. Bifano, and B. E. Davies. 1996. Geochemistry and anthropogenic inputs of metals in a tropical lake in Venezuela. *Appl Geochem.* 11: 605–616.

10. Mueller, G. 1979. Schwermetalle in den Sedimenten des Rheins – Veraenderungen seit1971. *Umscha.* 79: 778–783.

11. Olade, M. A. 1987. Heavy metal pollution and the need for monitoring: illustrated for developing countries in West Africa. In *Lead, Mercury, Cadmium and Arsenic in the environment, SCOPE.* T. Hutchinson and K. Meema, eds. John Wiley and Sons; pp. 335 – 341.

12. Olayinka, A. I., A. F. Abimbola, R. A. Isibor, and A. R. Rafiu. 1999. A geoelectrical and hydrogeochemical investigation of groundwater occurrence in Ibadan, SW. Nigeria. *Environ. Geol.* 37(1-2): 31-39.

13. Paul, P. C. and K. C. Pillai. 1983. Trace metals in a tropical river environment – distribution. *Water Air Soil Pollution.* 19: 63–73.

14. Simeonov, V., J. A. Stratis, C. Samara, G. Zachariadis, D. Vousta, A. Anthemidis, M. Sofoniou, and Th. Kouimtzis. 2003. Assessment of the surface water quality in Northern Greece. *Water Research.* 37: 4119-4124.

15. Sutherland, R. A. 2000. Bed sediment-associated trace metals in an urban stream, Oahu, Hiwaii. *Environ Geol.* 39: 611–627.

16. Tijani, M. N. 1994. Hydrogeochemical Assessment of groundwater in Moro area, Kwara State, Nigeria. *Environ. Geol.* 24(3): 194-202.

17. Tijani, M. N., K. Jinno, and Y. Hiroshiro. 2004. Environmental Impacts of Heavy Metal Distribution in Water and stream sediments of Ogunpa River Ibadan, SW Nigeria. *Jour. Mining & Geol.*, (*In press*).

18. WHO. 1993. Guidelines for drinking Water Quality. 2nd Edition, World Health Organization, Geneva.

A Cost Analysis of Stream Compensatory Mitigation Projects in the Southern Appalachian Region

J. Bonham[1] and K. Stephenson[2]

Abstract

Recently the US Army Corps of Engineers (the Corps) has increased the level of compensatory mitigation requirements for streams impacted by surface coal mining in the Appalachian coalfields. Through new permitting requirements (Nationwide Permit 21) the Corps is requiring that applicants submit a compensatory mitigation plan that will offset permanent and temporary losses of aquatic functions and values from fill of intermittent and ephemeral streams. These changes have led to concern by the regulated community over the cost of meeting mitigation requirements. In this study a cost analysis is conducted of projects that could be considered as compensatory mitigation The study estimates total costs for fourteen projects in the areas of channel and riparian zone restoration, abandoned mine land reclamation and acid mine drainage treatment. We conclude that costs are sensitive to project types and size, and a number of regulatory design issues including the regulatory criteria for acceptable mitigation.

Introduction

Section 404 of the Clean Water Act requires that permits be obtained by parties discharging dredge or fill materials into waterways. Under 404, the U.S. Army Corps of Engineers (Corps) administers Nationwide Permit 21 (NWP21) that governs the discharge of fill material into streams from surface mining activities. The Corps developed NWP21 in part to extend the 404 permitting program to the placement of fill from surface mining activities in ephemeral and intermittent streams. Permittees are required to perform "compensatory mitigation" to offset ecological services lost due to such fill activities. Compensatory mitigation occurs via activities designed to restore ecological services in stream channels either on the site of the disturbance itself or at an off-site location. The regulatory objective of 404 programs is to ensure that improvements in aquatic resources from compensatory mitigation offset the reduction in aquatic resources from the impacted areas (achieve no-net-loss).

Several of the Corps districts have, or are developing, in lieu fee programs to secure off-site compensatory mitigation under NWP21 (U.S. Army Corps of Engineers, Louisville District, 2002; U.S. Army Corps of Engineers, Pittsburgh District, 2004; U.S. Army Corps of Engineers). In these programs, the permittee makes a payment to an approved mitigation "sponsor" in lieu of implementing their own mitigation on-site. The sponsor, typically a government agency or a nonprofit organization, takes on the permittee's legal and financial mitigation responsibility and then uses the collected fees to identify, construct, and maintain compensatory mitigation projects. In principle in lieu fee programs set the per-unit fees for stream mitigation by estimating their cost of stream restoration and enhancement projects. This fee, typically expressed as dollars per linear foot, is then multiplied by the amount of stream mitigation (feet of stream) that a permittee is required to restore or enhance to offset the lost aquatic function from the fill.

Mining companies, in complying with these new requirements, have a financial interest in assuring that their compensatory mitigation requirements are met at the lowest possible cost, while also meeting their regulatory obligations to provide ecologically meaningful and

---

[1] Research Associate, Virginia Polytechnic Institute and State University, Blacksburg, VA
[2] Associate Professor, Virginia Polytechnic Institute and State University, Blacksburg, VA
[3] Research Associate, Virginia Polytechnic Institute and State University, Blacksburg, VA
[4] Professor Emeritus, Virginia Polytechnic Institute and State University, Blacksburg, VA; and Resident Scholar, Resources for the Future, Washington, D.C.

successful mitigation. The objective of this paper is to estimate the full cost of providing off-site compensatory mitigation for in lieu fee programs.

Compensatory Stream Mitigation Costs: A Conceptual Overview

An obvious component of the cost of compensatory mitigation is the physical alteration of the landscape associated with a mitigation project. Costs include financial outlays and other opportunity costs of implementing a compensatory mitigation project. In general, four types of activities must be performed for a compensatory mitigation project: 1) pre-construction planning and design, 2) site acquisition, 3) construction and 4) post-construction. Pre-construction costs include the expenses incurred identifying the project site, making a preliminary project assessment, and designing the project. Mitigation guidelines require that project sites be protected in perpetuity. Site acquisition costs refer to the legal protection of the project site, which can be done by, but not limited to, fee simple purchase or conservation easements. Construction costs include labor, materials, capital equipment, and management costs of physically constructing the mitigation project. Post-construction expenses include activities that verify and ensure that the mitigation is achieving ecological success criteria identified by regulators. Post-construction activities would include monitoring the project site, performing remedial action needed to achieve performance objectives, and long-term maintenance of the site. The total cost should also include overhead and management costs of performing and overseeing each of these activities.

The specific types of activities that must be performed (and thus the costs of performing them) could vary with the type of compensatory mitigation project undertaken. Conceptually, stream mitigation can be either in-kind or out-of-kind. Under NWP21, in-kind projects are generally meant to be stream restoration and enhancement activities aimed at improving aquatic habitat. Restoration is defined as the return of a stream to its natural pattern, profile and dimension along with creating aquatic habitat and establishing riparian vegetation and floodplain function (In Lieu Fee Guidelines from Kentucky and West Virginia). Stream enhancement is defined as the establishment of riparian vegetation, the stabilization of eroding stream banks, and the creation of aquatic habitat in-stream (In Lieu Fee Guidelines from Kentucky and West Virginia). A restoration project could be considered an enhancement project with the addition of significant channel modifications. Other types of projects such as abandoned mine land reclamation may improve water quality and hence aquatic resources by reducing sediment or pollutant discharges from upland areas. These projects are called out-of-kind mitigation because the projects do not directly replace the physical characteristics of the stream channel or habitat lost to the fill activity. The types of activities undertaken in an out-of-kind project will differ significantly from in-kind projects, thus influencing the final costs of providing compensatory mitigation.

Stream Mitigation Costs

In lieu fee programs for stream impacts are a relatively recent development within the Corps NWP21 regulatory program. Therefore, very few off-site compensatory mitigation projects have been constructed to offset stream impacts from surface mining activities, and thus little information is available by which in lieu fee sponsors can establish an appropriate fee. This section will provide estimates of the costs of ecological enhancement projects that could be used as compensatory mitigation under NWP21.

The costs of potential off-site compensatory mitigation projects are estimated by compiling the costs of on-going stream improvement projects in rural areas and in similar physiographic regions across Kentucky, North Carolina and Virginia. Projects selected for cost estimation are projects currently used to satisfy compensatory requirements under NWP21 or could conceivably be considered as compensatory mitigation by regulatory authorities. For each identified project, pre-construction planning and design, site acquisition, construction, and post-construction costs are estimated.

Fourteen completed or nearly completed projects were evaluated. Nine of the projects are located in North Carolina, four in Virginia and one in Kentucky. Twelve of the projects are considered in-kind projects centered on stream restoration and enhancement. Two Virginia projects could be representative of out-of-kind mitigation since each involved the amelioration of acid mine drainage from abandoned mine land.

Costs for each project and each cost category were collected through a combination of methods including reviews of official records, interviews with program staff, and professional inferences. Construction costs were the best documented category and collected from official records or reports. Pre-construction costs were reported by North Carolina (N.C. Department of Environmental and Natural Resources 2003). Preconstruction costs for other projects were estimated by agency personnel as a percentage of construction costs. Site acquisition costs were reported for the North Carolina projects and one Virginia project. In other cases, site acquisition costs could not be obtained. In several instances, sites were donated. Since the legal protection of the mitigation sites limits future development and activities, real costs are being incurred even if the sites are donated. To the extent that site acquisition costs are not reported, stream restoration costs will be under estimated. Post construction costs were largely unavailable because either 1) the projects were still in the monitoring phase and the final costs have not yet been realized or 2) the projects did not require post-monitoring activities. However, North Carolina did provide estimates based on costs realized in the first year of monitoring and remediation. Costs for the other projects were estimated based on the North Carolina predictions. Finally, in many instances the extent to which all costs estimates include indirect costs (such as overhead) with each of the four cost categories could not be ascertained, but it is unlikely these costs are fully accounted for in the public cost estimates.

The fourteen projects can be grouped according to size and type. The projects were placed into three groups defined by size, less than 3,001 feet, 3,001 to 10,000 feet, and greater than 10,000 feet. Compensatory mitigation costs (per foot) may be affected by project size because each project contains fixed costs imbedded in each expense component for which economies of scale can be realized. Costs may also be affected by project type. Arguments are sometimes made that out-of-kind mitigation may offer greater potential to improve aquatic resources at a lower cost than strict in-kind mitigation. Unfortunately, projects could not be identified for every combination of project size and type due to the limited number of projects available.

Project costs were estimated in present value terms using a 5% discount rate. The present value of costs was estimated because most of these projects take a number of years to complete. Thus, the timing of cost outlays will vary from project to project. Further, all costs are reported in current dollars. If a project was undertaken before 2000, the costs were adjusted to 2002 dollars using the GDP implicit price deflator.

Results

Combined, the fourteen projects improved 186,191 linear feet of stream at a present value cost of $11,022,674, an average of $59.20 per foot. Mitigation costs show a significant amount of variation across projects with the cost per linear foot for individual projects ranging from a low of $28 to a high of $129.

Broken out by cost category the average unit costs, summarized in Table 1, are $37.40 for construction, $13.63 for pre-construction, $2.70 for site acquisition, and $5.47 for post construction. These costs should be viewed as a lower bound estimate because of the challenges of estimating site acquisition and post construction costs described above.

Pre-construction costs as a percentage of total project cost are fairly consistent across the fourteen projects. As a percentage of total costs, pre-construction expenses in North Carolina and Virginia were 24% and 23%, respectively. The Kentucky pre-construction costs are an underestimate since some design was conducted in-house and was not directly assigned a monetary cost (Sampson). Future projects in Kentucky will be designed externally at an estimated cost of $25 per foot, an estimate made from North Carolina data (Sampson).

Site acquisition costs were on average $2.70 per foot. Site acquisition costs averaged $3.87 per foot for those projects that incurred such costs, and represented 6% of total costs. Because three of the four sites in Virginia were not acquired, the costs of the projects are understated since permanent protection is typically a regulatory requirement under 404 programs. The Kentucky site was donated; however, the landowner incurs an opportunity cost that should be accounted for in a full cost analysis. Two of the North Carolina projects were sited on state-owned land. Full accounting for site acquisition has the potential to significantly increase the cost of compensatory mitigation in the coalfields. Landowners in those areas have expressed reservations about giving up rights to prospective sites because of the opportunity for future mining and other resource extraction (Davis).

Post construction activities cost an average of $5.47 per foot, or 9% of total costs. However, many of the expenses were extrapolated from the nine projects in North Carolina. Whether these post-construction costs would be reflective of post-construction costs that would be incurred under a NWP21 is not known.

Costs vary by both project type (in-kind, out-of-kind) and size. Average costs per linear foot for in-kind projects grouped by size are summarized in Table 2. Total unit costs do appear to be related to the size of the project. Total unit costs are $118.96 per linear foot for small projects, $92.74 for medium projects and $65.22 for large projects. It should be noted, however, that site acquisition costs were not estimated for any of the large projects. Site acquisition costs averaged 5% of total costs for small and medium projects. If this same percentage was applied to the large sites, average total costs of those projects would be $68.48 per foot. After adjusting for site acquisition, the large projects cost 42% less than small projects and 26% less than medium projects. Economies of scale are realized in all phases of the projects with most of the gain in efficiency achieved during post construction—post construction costs of the large projects are 66% lower than those of the small projects. Pre-construction and construction costs for large projects are 50% and 33% lower than those of small projects, respectively.

Results suggest that out-of-kind mitigation has some potential to achieve low cost mitigation relative to in-stream restoration. The average total unit cost for all in-kind projects, shown in Table 3, is $84.09, 79% higher than the $46.98 average cost of the out-of-kind projects. Not surprisingly, the largest difference between the two groups is construction costs, accounting for 65% of the difference between total unit costs for the two groups. The two acid mine drainage projects have an average construction cost of $29.40, whereas the average construction costs of the in-kind projects is $53.70.

It should be noted that such numbers should be interpreted with extreme caution because of the difficulty in comparing the improvements in ecological services from in-kind versus out-of-kind projects. For instance, out-of-kind projects aim to improve the aquatic resources by improving water quality, but estimating the stream length positively impacted by these two projects was difficult. The magnitude of the water quality improvement and how water quality improvements would offset stream fills is unknown. It should also be noted that both of these projects are large projects relative to categories used above. The low linear foot cost of the projects may be a result of achievements of economies of scale.

Conclusion

The cost of the stream restoration and enhancement projects evaluated in this paper are estimated to range from $28 to $129 per foot. Costs can vary significantly by the size of the mitigation project. As project size increases economies of scale can be achieved across all cost categories.

Close adherence to a preference for in-kind restoration (particularly involving channel modifications over stream enhancement) could raise the cost of compensatory mitigation. Evidence suggests that the cost of out-of-kind projects may be significantly lower than in-kind projects. It should be stressed that the number of out-of-kind projects evaluated were limited and more evidence is needed to support this tentative conclusion. In addition, only an amount of the linear feet of stream miles improved by the mitigation project was estimated. These results cannot provide an indication of the qualitative improvement. In essence, the results here treat all stream mitigation improvements (linear feet of improvement) as the same, which may not be the case.

Finally, the mitigation projects evaluated were performed or managed by state government agencies. Given the limitations in public cost accounting, it was difficult to obtain or ascertain the extent to which the cost estimates fully reflect total mitigation project costs. Given that some types of costs (such as overhead or site acquisition) may not be fully attributed to specific projects, the cost estimates reported here should be viewed as a lower bound estimate of costs.

Tables

Table 1. Unit Costs[a] by Expense Category

	Pre-Construction	Site Acquisition	Construction	Post Construction
Average Unit Costs	$13.63	$2.70	$37.40	$5.47
Percent of Total Costs	23%	5%	63%	9%

[a.] Present Value – 5% discount rate

Table 2. In-Kind Project Costs[a] by Project Size

	Pre-Construction	Site Acquisition	Construction	Post Construction	Total Costs
Small (< 3,001 ft.)					
Average Unit Costs	$26.14	$5.65	$68.35	$18.81	$118.96
Percent of Total Costs	22%	5%	57%	16%	100%
Medium (3,001 - 10,000 ft.)					
Average Unit Costs	$21.25	$4.21	$57.28	$10.01	$92.74
Percent of Total Costs	23%	5%	62%	11%	100%
Large (> 10,000 ft.)					
Average Unit Costs	$13.04	-	$45.82	$6.37	$65.22
Percent of Total Costs	20%	-	70%	10%	100%
Aggregate					
Average Unit Costs	$18.31	$2.57	$53.70	$9.52	$84.09
Percent of Total Costs	22%	3%	64%	11%	100%

[a.] Present Value – 5% discount rate

Table 3. Out-of-Kind Project Costs[a]

	Pre-Construction	Site Acquisition	Construction	Post Construction	Total Costs
Black Creek	$9.42	-	$26.93	$5.27	$41.61
Ely/Puckett Creeks	$11.70	$3.29	$29.87	$3.15	$48.00
Average[b] Unit Costs	$11.34	$2.77	$29.40	$3.48	$46.98
Percent of Total Costs	24%	6%	63%	7%	100%

[a.] Present Value – 5% discount rate
[b.] Weighted by project length.

References

Davis, R. Personal Communications. Biologist, Virginia Department of Mines, Minerals and Energy. Big Stone Gap, VA, October 2003 – March 2004.

N.C. Department of Environmental and Natural Resources. *Wetlands Restoration Program 2003 Annual Report.* Raleigh, NC, 2003.

Sampson, W. Personal Communications. September 2003 – April 2004. Fisheries Biologist, Kentucky Department of Fish and Wildlife Resources. Frankfort, KY.

U.S. Army Corps of Engineers – Louisville District. 2002. *Memorandum of Agreement Concerning In Lieu Mitigation Fees Between Kentucky Department of Fish and Wildlife Resources and U.S. Army Corps of Engineers.* Louisville, KY.

U.S. Army Corps of Engineers – Pittsburgh District. February 4, 2004. *Draft Establishment of an In-Lieu Fee Agreement Between the West Virginia Department of Environmental Protection and the U.S. Army Corps of Engineers, Huntington and Pittsburgh Districts.* Public Notice No. 04-M1. Pittsburgh, PA.

U.S. Army Corps of Engineers, et al. November 7, 2000. "Federal Guidance on the Use of In-Lieu-Fee Arrangements for Compensatory Mitigation Under Section 404" *Federal Register,* Vol. 65, No. 216, pp. 66914-17.

STREAM BANKING PROTOCOL: A PROPOSAL FOR MINNESOTA

E. S. Verry, PhD[1]

ABSTRACT

Field methods to classify, delineate, and value wetland acres have been used for wetland banking for decades. These methods judge the value of wetlands destroyed during highway and land development. They also prescribe the number of acres and value of wetlands restored at a wetland bank restoration site. Methods to classify, delineate and value streams destroyed or restored were first authorized by the U S Army Corps of Engineers southwest of St. Louis, Missouri in 1999. The recognition of stream geomorphology principles in the 1990s provides a rigorous professional basis to fulfill a belated promise in the 1972 Clean Water Act: to protect wetlands **and other waters** of the United States.

Many streams and wetlands in Minnesota and the Lake States occur together in the same floodplain. New methods to classify delineate, and value streams and wetlands together are presented in this paper. Rigorous protocols using measured attributes for streams and wetlands in combination promise a fair evaluation, a fair valuation, and a sustainable restoration in a fully functional landscape. Two states are considering combining wetland and stream banking in the same institution. This is a first step. Addressing development pressure with combined stream and wetland restoration in a landscape context is the second step.

KEYWORDS. Wetland Banking, Stream Banking, Stream and Wetland Evaluation Protocols, Stream and Wetland Valuation.

INTRODUCTION

Wetland banking is based on maintaining wetland functions in major and small watersheds. Federal and state administration of development near wetlands has always focused on (1) avoiding wetland impact, (2) restoring wetland function adjacent to an impact area, or (3) restoring wetland function by the contractor at another site. Wetland impact mitigation subsequently progressed to allow contractors to purchase wetland credits from a bank who distributed the contractor's money to others doing wetland restoration or creation (in-lieu-fee mitigation). There is nearly universal acclaim for wetland mitigation success yet serious program reviews show a great distance between concepts, maintenance of ecological functions, bank administration, and net outcomes (Hey and Philippi, 1999; National Academy of Science, 2003). Handling the money has generally been easier than handling the mitigation.

The pressure to find wetland mitigation sites steadily increases with fewer options on the horizon. Wetland mitigation in many Minnesota banks is based on buying a wetter wetland. However, conversion of wet meadows to shallow or deep-water marshes with low head dams has nearly run its course. At the same time new construction (e.g. homes, malls, and highways) is impacting streams as well as wetlands. It is not unusual for a mitigated wetland to cause stream channel incision downstream or a stream dredged for flood passage to unwittingly drain a wetland upstream!

---

[1] Hydrologist, Ellen River Partners, LLC, 21689 Birch Street, Grand Rapids, MN 55744; 218 326 6120, sverry@mchsi.com. Research Hydrologist, Emeritus, USDA Forest Service, North Central Research Station, Grand Rapids, MN.

Often an impacted stream and impacted wetland are together in the same valley. Several states have adopted stream-banking programs following the basic tenants of wetland banking, and in some states (e.g. Kentucky, North Carolina, Georgia, Missouri, and Illinois) wetland and stream banks are or will shortly be combined administratively, while retaining the procedural requirements to avoid, protect, restore, or mitigate wetlands and streams individually.

Applications of landscape, wetland, and stream geomorphology science to land management beg for new options in the administration of wetland and stream banks. Options that allow shared wetland and stream credits include the concepts of wetland acre for wetland acre (value or function) and stream length for stream length (value or function). These concepts must be respected, documented, implemented, monitored, and banked in an open, transparent, readily viewed manner.

Federal and state agencies have progressively shown a desire to move from unit for unit approaches to landscape, watershed, and ecosystem sustainability-based approaches. Even in the face of vague definitions, skepticism, and even derision with these terms, developer, land owner, land user, and agency steward understand the need to manage land change holistically; the need to retain and enhance wetland and stream functions by managing with a sense of valleys rather than a sense of acres alone.

This paper proposes a stream banking protocol based on rigorous field measures of stream condition and compensation ratios and multipliers consistent with the full range of stream restoration. It also suggests a valley framework for combining this new stream assessment protocol with Minnesota's current wetland assessment protocol, MnRAM 3.

STREAM BANKING IN THE UNITED STATES

The Clean Water Act (CWA, 1972) has always addressed streams, rivers, and wetlands; however, most administrative efforts involving restoration, creation, and mitigation banking in the last 30 years have addressed wetlands. This is changing fast.

Even though overall guidance for banking is assigned to the EPA under the CWA, the day-to-day permitting requirements are assigned to the USACE which specifically delegated authority for SOP development to its District offices. In June 1990 the Savannah, GA District of the USACE published a draft Standard Operating Procedure: "Compensatory Mitigation, for Wetlands, Open Water, and Streams". Subsequently, in November, 1990, DOD, EPA, DOI, and DOC published: "Federal Guidance on the Use of In-Lieu-Fee Arrangements for Compensatory Mitigation" under Section 404 CWA and Section 10 RHA (Fed. Reg. 65(216):66914-66917). In 1995 the federal agencies published: "Federal Guidance for the Establishment, Use and Operation of Mitigation Banks" (DOD, EPA, USDA, DOI, DOC, Fed. Reg. November 28, 1985 60(28):58605-58614) that specifically apply to wetlands and **other aquatic resources**.

Based on this guidance, the USACE, St. Louis District, established the Fox River Stream Mitigation Bank in December 1999. Approximately 13,800 feet of Fox Creek and a 100-foot riparian corridor were established with deed restrictions by Fox Creek Mitigation, LLC; 197.2 credits were initially available for sale.

Stream assessment protocols began with the Savannah District USACE SOP for both streams and wetlands in June 2000. Their stream discussions included Rosgen stream type conversions from entrenched systems to floodplain systems and are quite useful for those fully trained in the entire Rosgen course series: I-IV. I consider the Savannah District efforts the First Approximation for stream assessment protocols. The Charleston District in September 2002 published protocols similar to the Savannah District, but encumbered them with summation, division, and power formulas akin to the Brinson method for wetland functions. They are overbearing and difficult to interpret mathematically. I call them the Second Approximation to stream protocols.

The Kentucky Transportation Cabinet (Louisville District, USACE) first utilized the EPA rapid bioassessment protocol for physical habitat and invertebrate assemblages in December 2002. However, monitoring of restored channels showed physical deterioration so now fluvial geomorphology principles are also required. I call the Kentucky version (and a similar Tennessee version) the Third Approximation.

In April 2003, the Wilmington, NC District USACE published a 23 point evaluation sheet that includes Rosgen stream geomorphology variables. I call the Wilmington version the Fourth Approximation. For Minnesota, I suggest a Fifth Approximation using the Wilmington District format, reach level scoring of Rosgen (1994) stream types, monitoring used in Rosgen Level IV evaluations (Rosgen 1996), and revised features of the EPA rapid bioassessment protocol.

A PROPOSED STREAM ASSESSMENT PROTOCOL FOR MINNESOTA

Stream Quality Assessment Worksheets

Table 1 is the first of 3 worksheets. It includes location, proposed channel work, type of state waters affected, land use, and 18 direct measurements of stream geomorphology. Those familiar with the Rosgen stream classification protocol realize considerable field measurement is required to complete Table 1.

Table 2 is the second worksheet with a specific rating protocol for hydrology, stream geomorphology and sediment, riparian characteristics, anthropological alteration, erosion, and biology. Table 3 (sheet 3) is a justification and interpretation for each factor. The biological portion is qualitative although agency knowledge of the fish, invertebrate, and other wildlife is required. An option for Minnesota, like the Wilmington and Louisville Districts is inclusion of the invertebrate and/or fish invertebrate bioassessment protocol from EPA. Like Kentucky, Tennessee, and North Carolina they are presented as options because the time, effort and money for these protocols is much greater than the biology portion in Table 3. In contrast to the physical portion of the EPA rapid bioassessment protocol, these measures are accurate and useful for stream assessment and rating.

The rating values for each of the 21 stream functions or conditions (Table 2) are not equally spaced corresponding to low, medium, and high. Instead they vary and reflect the author's interpretation of value. There are a total of 100 points for rating a stream reach (covering two meander lengths or 20 channel widths in length). Interpretation of the total score is akin to standard academic scores for people. Score ranges at the bottom of Table 2 reflect the author's interpretation: 0 - 24 really poor; 25 -59 poor; 60 – 75 modal; > 75 excellent.

The modal range reflects a large number of streams in Minnesota that have normal, "average", fully functioning stream ecosystems based on their hydrology, geomorphology and sediment, riparian characteristics, anthropologic changes, erosion, and biology. It is possible to get much worse, and difficult to get better; however the >75 range allows a range for truly unique and superbly functioning examples in the state.

The 21 functions and conditions apply equally well to streams of all temperatures, though item 2 gives more points to groundwater sources. State special categories, including trout streams, are referenced in item 17 on worksheet 1 (Table 1).

Mitigation Ratios

Stream mitigation protocols in USACE Districts (and wetland banking protocols) vary with the condition of the impacted stream and the amount of effort expended at a restored mitigation site.

Contrary to wetland banking that also consider mitigation location (in-kind and out-of-kind --- either in or out of the same watershed or county), stream mitigation protocols do not.

Table 1. Proposed Minnesota Stream Quality Assessment Worksheet 1: Summary.

STREAM QUALITY ASSESMENT WORKSHEET

USACE # _____   BWSR#_____   Site # _____(indicate on attached map)

Provide the following information for the stream reach under assessment:

1. Applicant's name: _____   2. Evaluator's name: _____

3. Date of evaluation: _____   4. Time of evaluation: _____

5. Name of stream: _____   6. River basin: _____

7. Approximate drainage area: _____   8. Stream order: _____

9. Length of reach evaluated: _____   10. County: _____

11. Site coordinates (if known): prefer in decimal degrees   12. Subdivision name (if any) _____

Latitude (ex. 34.872312):_____   Longitude (ex. -77.556611) _____

Method location determined (circle):   GPS   Topo sheet   Ortho (Aerial) Photo/GIS   Other GIS   Other_____

13. Location of reach under evaluation (note nearby roads and landmarks and attach map identifying stream(s) location):

14. Proposed channel work (if any):

15. Recent weather conditions:

16. Site conditions at time of visit:

17. Identify any special waterway classifications known: _____Section 10 _____Essential Fisheries Habitat

_____Trout Waters _____Outstanding Resource Waters _____Nutrient Sensitive Waters _____Water Supply Watershed

18. Is there a pond or lake located upstream of the evaluation point? YES NO If yes, estimate the water surface area: _____

19. Does channel appear on USGS quad map? YES NO   20. Does channel appear on USDA Soil Survey?   YES   NO

21. Estimated watershed land use: ____% Residential _____% Commercial _____% Industrial _____% Agricultural

_____% Forested _____% Cleared/Logged _____% Other (_____)

---**MEASURED VALUES**---

22. Bankfull width:_____   23. Floodprone width _____   24. Bankfull depth, mean._____   25. Bankfull depth, max _____

26. Entrenchment ratio ____ 27. W/dmean ratio ____ 28. Channel sinuosity:_____ 29. Channel slope ____30. Valley slope_____

31. Wolman _____% silt & clay _____%sand _____%gravel _____%cobble _____%boulder _____%bedrock.

D50 on riffle_____ D50 total reach _____

32. Stream type (level II) _____, 33.Stream Order_____

34. 8 digit HUC _____, 35. Watershed area _____

36. Pfankuch channel stability score _____

37. Reach condition (Good stable, Fair mod. unstable, Poor unstable)_____

38. Bank Erosion Hazard Index _____

39. Bank Erosion Potential very low, low, moderate, high, very high, extreme _____

Total Score (from reverse): _____

Evaluator's Signature _____   Date _____

94

Table 2. Proposed MN Stream Quality Assessment Worksheet 2: Evaluation and Scoring.

Funct.	#	CHARACTERISTICS	Range	SCORE
HYDROLOGY	1	**Presence of flow/persistent pools in stream** 0 = no flow or sat.; 1 = only pools w/ water; 2 = flow over riffles or steps; 3 = Bkfl chan. 10 - 25% full; 4 = Bkfl chan. >25% full (see weather on page 1)	0 - 4	
	2	**Groundwater discharge** 0 = no discharge; 1 = at least 1seep; 3 = several springs & seeps; 4 = many of each	0 - 4	
STREAM GEOMORPHOLOGY & SEDIMENT	3	**Stream type** D G DA F A B C E 0 0 1 2 4 5 5 5	0 - 5	
	4	**Channel widening** (% of the modal w/d ratio for the given stream type) Stream Type A G F B E C D DA Modal W/D ratio 7 7 20 20 8 24 50 40 0 = > 135%; 2 = 125 – 135%; 5 = 115 – 125%; 6 = <115%	0 - 6	
	5	**Channel sinuosity** (0 for channelized streams; other points are stream type dependent) 0 = channelized; A B D & DA types: 0 = <1.2; 1 =1.2+; C, E, G & F types: 2 = <1.2; 3 = 1.2 - 1.5; 4 = 1.5 -1.8; 5 = >1.8	0 - 5	
	6	**Fresh, in-channel sediment deposition** (visually obvious, new sediment) 0 = fresh, mid- and side-channel islands; or extensive loose sands in an over-wide channel 1 = horizontal point bar accretions >12 inches; 3 = horizontal point bar accretions 6 – 12 in.; 5 = horizontal point bar accretions < 6"	0 - 5	
	7	**Percent fines < 1mm on the riffle** (Wolman pebble count basis) 0 = > 40%; 1 = 20 – 30%; 4 = 10 – 20%; 5 = <10%	0 - 5	
	8	**D50 on the riffle** (Wolman pebble count basis) 1 = silt; 3 = sand; 5 = gravel or cobble; 3 = boulder; 2 = bedrock	0 - 5	
	9	**D50 on total reach (pools and riffles or steps)** (Wolman pebble count) 1 = silt; 2 = sand; 5 = gravel or cobble; 4 = boulder; 3 = bedrock	0 - 5	
RIPARIAN	10	**Floodplain width and wetlands** (ratio of total floodplain to bankfull width) 0 = <1; 1 = 1 -3; 2 = 3 – 5; 3 = >5; +1 floodplain is 1/3 wetland +2 floodplain is 1/3 – 2/3 wetland; +3 floodplain is > 2/3 wetland	0 - 6	
	11	**Riparian vegetation width** (ave. on each side for total corridor; herb., shrub or tree) 0 1 2 3 4 +1 +2 0 <20ft 20 – 40 40 – 60 >60ft ea. side; 70-90%contig. >90%contig.	0 - 6	
	12	**Overhanging cover** (% total of both sides) 0 = <20%; 1 = 20 – 40%; 2 = 40 – 6-%; 3 = 60 – 80%; 4 = >80%	0 - 4	
ANTH-RO	13	**Evidence of past human alteration** (% of total stream corridor) 0 = >50%; 1 = 30 – 50%; 2 =10-30%; 3 = <10% 4 = 0%	0 – 4	
	14	**Evidence of nutrient or chemical discharge** 0 = 4+ outlets; 1 = 3 outlets; 2 = 2 outlets; 3 = 1outlet; 4 = no outlets	0 – 4	
EROSION	15	**Soil trampling and rutting** (% of both sides) 0 = >50%; 1 = 50 -30%; 2 = 30 – 10%; 4 = <2%	0 - 4	
	16	**Slumps** (Presence of slumps as % of high terrace bank in slump condition) 0 = slumps on straight sections or all terrace, high banks slumping; 1 = ½ of terrace banks; 3 = < 10% of terrace banks; 4 = no slumps	0 - 4	
	17	**Terrace banks** (above bankfull elevation; looking at raw dirt bank; % of bank with roots) 0 = no roots; 1 = top 1/3; 2 = top 2/3; 3 = entire bank; 4 = well vegetated	0 - 4	

BIOLOGY	1 8	**Presence of stream invertebrates** (see page 4) 0 = none; 1 = 1or 2 species; 2 = 3 or 4 species; 3 4 or 5 species; 4 = >5 species; EPT present +1	0 - 5	
	1 9	**Presence of fish & mussels** 0 = 0 species; 1 = 1 prey; 2 = 2 - 3prey; 3 = 1 pred.& 1 - 3 prey; 4 = 2 pred & 1 – 4 prey; 5 = 3+ pred.& 3+ prey +1 if mussels present	0 - 6	
	2 0	**Channel habitat complexity** (LWD, mix of fast & slow riffles, deep & shallow pools) 1 fast riffles & shallow pools; 2 fast & slow riffles; 3 & deep & medium pools +1 abund. LWD	0 - 4	
	2 1	**Presence of other wildlife** amphibians +1, reptiles +1, birds +1, mammals +1 if any abundant +1	0 - 5	
		0 – 24 really poor; 25 – 59 poor; 60 – 75 modal; > 75 excellent	**TOTAL SCORE**	

Table 2. cont.

Table 3. Proposed Minnesota Stream Quality Assessment Worksheet 3: Interpretations.

1. The persistence of flow (often perceived as depth of flow) is an over-riding variable for fish habitat assessment. Deep stream water can "correct" for the loss of other habitat factors. Even though the assessment is visual, it is perceived by many as critically important to stream quality. It should be judged in light of the weather and history of weather (e.g. drought periods) described on page 1.

2. Groundwater discharge is a part of flow persistence, but strong groundwater flow usually brings higher base cations (Ca, Mg, K,), near neutral pH, and cold or cooler water temperatures and has long been a critical factor in fisheries evaluations.

3. Stream type as described by Rosgen is not meant to have a value associated with it. In Rosgen's 8 stream types, all 8 are considered stable for the climate and sediment regime they occur in. However, stream restoration will rarely opt to create a D, G, DA, or F channel and hence the wealth of data gained from stream typing is entered here with a value associated with it. The stream type incorporates direct measures of entrenchment and access to the floodplain. Rather than a description of these words without field measurements, the stream type value directly incorporates field measurement.

4. Channel widening is perhaps the single greatest detriment to quality stream habitat because is shallows the system, leaves lots of loosely flowing fine sediments, and renders the hydraulics of the channel incapable of re-establishing a modal width and depth for a given stream type unless massive watershed-wide erosion upstream resets the entire basin.

5. Channel sinuosity is often estimated visually and with great error. The amount of sinuosity and the amount of sinuosity lost by land use changes and its impact on bankfull flow velocities is a function of the stream type. Hence the values are stream type dependent. A loss of sinuosity is a direct loss of total stream habitat. So this value addresses not only quality but quantity of stream habitat.

6. Fresh, in-channel sediment deposition is difficult to see at first. It is widely recognized as a bad thing, and with experience, the observer can tell when a stream has too much sediment moving. The value descriptions are a guide to quantify these visual impressions. It is also a reflection of the Pfanckuch scour and deposition variable that carries a large factor score (24) in that procedure. It is significant because it relates to scour from high banks or low banks (e.g. a pealing back of wetland banks at low heights); so look for both.

7. Percent fines is a direct measure of embeddedness for many state and federal fishery evaluations. However, the embeddednes term is regarded as unreliable when compared between agencies, observers, methods, and stream quality scores. Direct measure of the sediment at riffles, at pools and combined for the total reach allows a similar variable by simply taking the % fines smaller than 1mm. While sand includes particles up to 2 mm and sand is often considered a fine, brook trout routinely spawn in particles as fine a 1 mm. Rosgen generally considers a site embedded when fines exceed 20%.

8. The diameter of the 50th percentile of pebbles on a riffle (the Wolman procedure) is a direct measure of spawning habitat quality.

9. The diameter of the 50th percentile of pebbles for the entire reach (pools and riffles) is required to index the mean sediment size moved by the stream at bankfull flows. The D84th percentile is an estimate of the large particles moved by the river at bankfull flows.

10. Floodplain width and wetlands. The width of the floodplain relative to the width of the stream is an index of how the river and its valley can handle flood flows (eg. spread them out with low velocity, or keep them relatively confined at higher velocities). The presence of wetlands because they are extremely flat can further reduce flood flow velocities. Thus the width is considered first and the relative proportion of floodplain in wetland is added to the width score.

11. Riparian vegetation width refers to the presence of vegetation on both sides of the stream. Many USACE evaluations consider width to 60 feet on each side. The amount of contiguous vegetation is added to the width score. The width score is an average of both sides

12. Overhanging vegetation is best for shade, shadow, and invertebrate flights and has long been a part of fishery habitat evaluations.

13. Evidence of human alteration can be anything including livestock, roads, developments, parks, etc.

14. Many USACE and EPA assessments include a factor for nutrient or chemical discharges into the stream. This is simply evaluated on the number of outfalls in the reach (e.g. tile drains, sewer outfalls, etc.).

15. Soil trampling and rutting is evidence of livestock grazing, erosion and compaction by vehicles: ATV, skidders, farm, hauling vehicles.

16. Slumps can be a major source of sediment to streams and thus they are rated here by their relative occurrence in relation to high terrace banks since these (along with mountain sides) are the major location of slumps. However, in some channel types (G or gully channels) slumps occur with relatively low terrace banks, so if they are occurring along a straight section that indicates strong, confined, bankfull, velocities capable of continually carving out channel sides.

17. Terrace banks are more likely to erode and slump if roots do not bind them throughout their entire height. The bank face considered is that above the bankfull (usually floodplain) elevation. So often, the bankfull, floodplain elevation occurs on the inside of a bend and the high terrace bank on the outside. Consider only the terrace bank above the bankfull elevation.

18. The presence of stream invertebrates is a major index of stream quality; however, invertebrate drift from good sections upstream may mask poor production at the measurement reach. The more species the better, especially if EPT taxa (Ephemeroptera, Plecoptera, and Trichoptera) are present. If so an added point is given.

19. Presence of fish is the proof of the pudding. It is scaled to stream size (the larger the stream the more species). The guides given are in terms of the number of predator fish and prey fish in the reach and are meant in a general way to index fish communities.

20. Channel complexity allows more niches and increases fish, mussel and invertebrate abundance. A mix of habitats assures optimum communities. Channels with only deep narrow riffles and shallow pools rate low. Riffles need both shallow and deep areas, and a mix of deep, moderate and shallow pools provides for all life stages. LWD (e.g. > 8 to 12 inches) in abundance adds to channel complexity.

21.   The presence of other wildlife is often rated high by people referring to riparian areas. It is perceived that a good mix of animal types reflects a healthy riparian area. Hence the scores are additive for each animal category and an extra one for abundance in any form.

I divided the amount of effort expended at a stream mitigation site into three categories: (1) preservation of an existing modal (or better) stream reach, (2) partial restoration of riparian habitat and/or in-channel habitat (work does not change stream dimension, pattern, or profile), and (3) full restoration that includes partial restoration options along with the physical restoration of stream dimension pattern and profile to modal values for a given Rosgen stream type.

Preservation of an existing stream with easements requires more restored stream acres (the stream length times a minimum 100-foot corridor). Full restoration of a stream to a stable dimension, pattern and profile requires construction adjustments to width and depth, sinuosity, channel slope and to the riparian corridor. Full restoration credits for a poor reach at the impacted sites is a 1:1 credit.

Partial restoration enhancements to stream and riparian health occur at less than the reach level so there is no substantial physical adjustment to stream dimension, pattern, and profile. Partial restoration includes riparian vegetation, corridor width, bank protection, bio restoration, and in-stream structures (e.g. J-hook vane, floodplain shelf construction, etc.) without changes in stream geomorphology. The sub-reach scale is defined as less than two meander wave lengths or less than 20 stream widths measured at a riffle (narrow) cross section. Table 4 summarizes the proposed mitigation ratios for all streams in Minnesota. The diminimus stream length considered among the various USACE Districts is 100 to 200 feet. I propose 150 feet for Minnesota.

Table 4. Proposed mitigation ratios for each stream acre (see below) in Minnesota.

Impacted Stream Quality	Kind of restoration at the mitigation site		
	Full Restoration	Partial Restoration	Preservation
Poor <60 score	1	2	3
Good 60 - 75	2	4	6
Excellent > 75	3	6	9

An on-the-ground protocol is needed to delineate stream acres and to incorporate wetland banking mitigation credits used in Minnesota. Wetland banking credits and stream banking credits encompass public value credits associated with riparian lands directly outside the land area used for stream acres. To facilitate a combined approach to stream and wetland banking, the following mapping protocol is suggested Figure 1.

CONCLUSIONS

The stream banking protocols proposed here offer a further refinement of earlier, albeit recent, protocols in several states. The proposed Minnesota protocols offer rigorous, field measured, values directly related to a widely accepted stream classification system. They also include significant portions of stream monitoring measurements such as the Pfankuck (1975) stream stability index, estimates of bank erosion, and practical assessments of stream biology. Their use should avoid the uncertainty associated with early wetland assessment protocols.

While the stream protocol rating and scoring system has been applied to dozens of streams known by the author and reviewers, a broader test of applicability is needed on more stream reaches and for the major stream types in Minnesota.

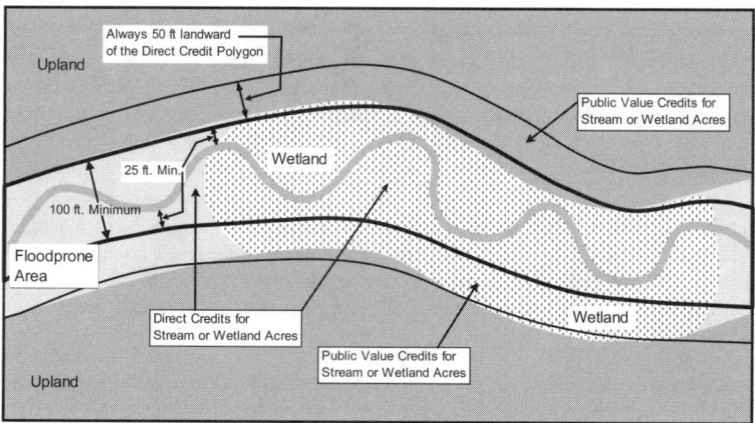

Figure 1. Direct credit stream acres are defined by a smooth polygon encompassing the stream (it is never less than 100 feet wide). The width can expand for larger rivers by always maintaining a minimum of 25 feet on the outside of bends. Public value stream acres are defined by a 50-foot landward polygon on either side of the direct credit polygon. Wetland credits, partial or full, are also applied in the direct and public value credit polygons.

The mitigation ratios proposed, and both the direct credit and public value credit dimensions follow those applied in other states (e.g. 100 ft minimum width, and 50 ft. addition for public value credits). Many details of riparian evaluation, vegetation standards, length and type of monitoring, and workable guides for the application of wetland credits require more than the space available. However, a draft manuscript (Wetland and Stream Banking in Perspective: Considerations for Minnesota), prepared for the Arrowhead Regional Stream Team, discusses these in detail.

Acknowledgements

USDA funding from the Superior National Forest, Duluth, MN to the North Central Research Station, USDA Forest Service, and helpful reviewer comments from members of the Arrowhead Regional Stream Team are gratefully acknowledge.

REFERENCES

1. Hey, D. L. and N. S. Philippi. 1999. *A Case for Wetland Restoration.* John Wiley & Sons, Inc. New York, NY

2. National Academy of Sciences. 2003. *Compensating for Wetland Losses under the Clean Water Act.* National Academy Press, Washington, DC.

3. Pfankuch, D. J. 1975. Stream reach inventory and channel stability evaluation. USDA Forest Service, R1-75-002. Gov. Print. Office. #696-260/200. Washington, D.C. 26p.

4. Rosgen, D. 1996. *Applied River Morphology.* Wildland Hydrology, Pagosa Springs, CO.

5. Rosgen, D. L. 1994. A classification of natural rivers. Catena 22:169-199.

6. USACE. Year. Stream and wetland protocols are available online at the respective USACE district offices.

Learning from Nature's Stability: Building a Multi-Purpose Data Base Applicable to Stream Assessment, Restoration and Education

T. D. Keane,[1] PhD

ABSTRACT

Terms such as 'self-sustaining,' 'sustainable,' 'regenerative,' and 'self-maintaining' are increasingly employed in the discourse of environmental engineering, landscape architecture, environmental planning and design, and in traditional architecture. What do these terms mean and where do we look for definition, for depth of understanding and for consistency of application among the allied disciplines dealing with environmental integrity?

We must look to natural systems for models of sustainability, for efficiency of energy use, for balance between biomass production and symbiotic interaction and for recycling of all materials. Thus, if we are to learn to manage, plan and design sustainably, we must learn from nature; we must learn from natural stable systems.

To examine and document stable, self-sustaining, fluvial systems in **Kansas** we have sought and studied **reference reach streams.** This EPA funded study has employed **Rosgen** (1996) **level III assessment and classification** with modifications and additions deemed appropriate for the hydrophysiographic provinces of the Great Plains and Central Lowlands.

This work summarizes data collected to date, its analysis, preliminary findings, and most importantly, its intended applications. If future works of **stream restoration** are to be sustainable, they must be founded upon a sound understanding of **natural fluvial process** within given hydrophysiographic parameters. Presented here are examples of our data and detailed description of its collection, record and analysis. Elaboration on the multiple ways in which this data will be applied will be offered. Applications include: **education** of students and professionals, **stabilization** and **restoration design** for impaired streams, **new channel design** (ie: stormwater conveyance) and **natural channel design restoration.** Presented here is a brief description of what we have learned in the field, how we will apply this learning, and how others might benefit from this extensive data collection process.

Rosgen, D. 1996. Applied River Morphology. Wildland Hydrology. Pagosa Springs, CO

KEYWORDS: Kansas, reference reach streams, Rosgen level III assessment and classification, stream restoration, natural fluvial process, streambank stabilization, restoration design, new channel design, natural channel design restoration.

INTRODUCTION

"The basic theme can then be restated, as for every problem, that it is necessary to understand nature as an interacting process that represents a relative value system, and that can be interpreted as proffering opportunities for human use – but also revealing constraints, and even prohibitions to certain of these."   McHarg, 1992, p. 127.

---

[1] Associate Professor, Landscape Architecture/Regional & Community Planning, Kansas State University, Manhattan, Kansas 66506

For too long we have operated from a perspective of exploitation and control. Legal scholar A. Dan Tarlock (2000), who has written extensively on watershed and river management issues, suggests: "The idea that the benefits of improving nature always exceeds the costs is difficult to reverse because it is so deeply embedded in the law and philosophy of watershed use. We have been conditioned to appreciate the value of altered and managed riverine landscapes." (Tarlock, 2000, p. 3) Tarlock (2000) goes on to say that most of our rivers "and their adjacent corridors have been perceived as under-used natural resources that should be extensively developed or used for waste disposal. Thus, rivers have often been conceptually and functionally "detached" from their surrounding landscape. . . Both science and law have contributed to the "detachment" of rivers from their surrounding ecosystems." (Tarlock, 2000, p. 3) This detachment allows the perspective of rivers and their watersheds as commodities to be developed or "improved" to meet human desires. We need only look at most of our waterways in the Midwest to see the results of such a perspective.

I have long believed that we need to address environmental problems from a different perspective—one of understanding (as best we can) the components, patterns, and processes of the natural systems in which we work. This entails an attempt to see and understand interrelationships between abiotic and biotic components of ecological communities. If such understanding provides the guidance, the basis for our designs or solutions to environmental problems then the spaces, structures, facilities, systems we build will have a better chance of being sustainable. Such settings will fit their surroundings, be less demanding of maintenance or repair, will look "right," and in the long run will be less threatening and detrimental to all of the inhabitants of the ecosystems in which we find our work.

The purpose of this paper is to explain and illustrate our collection of geomorphic data from Kansas reference reach streams. This effort is aimed at enhancing our understanding of how water interacts with unique combinations of lithology, structure, soils, vegetation, micro-climate, and land use in its drive to achieve a "quasi-equilibrium" or "stable state" fluvial system. Each watershed and its drainage network is in a sense individual, while at the same time, resembling all others in its governance by consistent physical processes. Science, specifically fluvial geomorphology, allows us to quantify these similarities within certain ranges of natural variability.

This paper begins by presenting some key or foundational concepts of fluvial geomorphology. It then offers explanation of our selection of the Rosgen (1996) assessment and classification system, of what our data entails and the methods employed in its collection. Finally, I put forward several applications of our growing data base.

BASIC CONCEPTS

Every hydrophysiographic province, because of homogeneity of lithology, structure, climate and land use, has a characteristic or representative stream type or set of stream types that show a balance or stability. This natural stability is achieved through the balance of fluvial system variables: channel width, depth and slope; velocity, discharge and resistance; the size of the sediment and the sediment distribution. (Leopold, Wolman & Miller, 1964)

The premise here is that this balance or "stable state" is correlated with the highest, long-term, physical, biotic and aesthetic potential for a given hydrophysiographic province. In other words, geomorphic, biologic and aesthetic factors are inextricably linked. Our goal has been to locate and measure the physical parameters of stable, reference reach streams in each of the hydrophysiographic provinces of Kansas. These streams represent the balance and sustainability of physical, chemical, biological and aesthetic factors that we must mimic in our stream restoration efforts.

Basic concepts of fluvial geomorphology relevant to the discussion here include: channel stability, reference reach, hydrophysiographic provinces (regions) and bankfull or effective discharge. Channel stability has been defined in many different ways but the following definition from Rosgen (1996) seems most cogent.

100

"Stream stability is morphologically defined as the ability of the stream to maintain, over time [and in the current climate], its dimension, pattern and profile in such a manner that it is neither aggrading nor degrading and is able to transport without adverse consequence the flows and detritus of its [attendant] watershed." (Rosgen, 1996, pp. 6-1-6-2)

Reference reaches are representative, typical, lengths of stream channels identified within the basin or hydrophysiographic province under study. Reference reaches are not always "stable," rather they are representative of the stream type(s)within the drainage basin being studied.

Hydrophysiographic provinces or regions are areas of homogeneous climate, geology, soils and vegetative communities. Such regions have a consistent lithology and structure, a semi-constant climate (which drives most of the geomorphic change processes) and a regular pattern of relief. In short, hydrophysiographic provinces are discernable areas of homogeneity concerning landform, underlying geology and soils, change process(es), hydrology and biotic communities.

Bankfull discharge is the volume of flowing water that, over time, is responsible for channel formation and maintenance. The recurrence intervals of this flow volume averages 1.5 years.

"The bankfull stage corresponds to the discharge at which channel maintenance is most effective, that is, the discharge at which moving sediment, forming or removing bars, forming or changing bends and meanders, and generally doing work that results in the average morphologic characteristics of channels." (Dunne & Leopold, 1978, pp. 608-609)

Bankfull discharge is often referred to as "effective" or "channel forming" discharge and is usually synonymous with USGS "normal high flow." Each of the concepts or factors described above are prominent in the Rosgen (1996) stream classification system.

ROSGEN (1996) STREAM CLASSIFICATION SYSTEM

Rosgen's (1994, 1996) system of stream classification is more than mere pattern recognition of forms. Rather, this system allows classification, assessment of stream "state" or condition, prediction of trends, and verification of conditions. This classification system is based upon process, as well as form, and it is the most widely accepted and employed system in the United States.

Specific objectives of the Rosgen (1996) classification system include:

1.   The ability to "predict a river's behavior from its appearance."

2.   The development of "specific hydraulic and sediment relationships for a given stream type and its state."

3.   Provision of "a mechanism to extrapolate site-specific data to stream reaches having similar characteristics

4.   Provision of "a consistent frame of reference for communicating stream morphology and conditions among a variety of disciplines and interested parties."

Most important to our data collection efforts is the fact that the Rosgen (1996) classification system is based upon field-collected data from hundreds of rivers. The system employs consistent, objective, quantitative, and replicable measures. Most of the other classification systems currently employed are qualitative and potentially inconsistent. The inconsistency springs from multiple observers whose interpretation of qualitative criteria may vary significantly.

In my study and examination of assessment procedures and classification systems, I have found none that are more well-founded, consistent and broadly applicable than Rosgen's system. Drawing upon the definition of stability offered earlier, the Rosgen system of classification employs delineative criteria based upon stream channel dimension, pattern and profile.

Dimension refers to the channel width, depth and cross-sectional area at the bankfull stage. Within a given hydrophysiographic province there is usually a strong correlation between drainage area and channel dimension. Measurements of channel dimensions are used to determine width/depth ratio, an important factor in Rosgen (1996) Level II classification which is discussed shortly.

Channel patterns have been qualitatively described as straight, meandering or braided forms (Leopold, Wolman, Miller, 1964). Shumm (1997, p. 113) notes that "any division between straight and meandering channels is arbitrary" thus the importance of quantitatively measuring channel pattern parameters. Channel forms "exhibit specific geometric relationships that may be quantitatively defined through measurements of meander wavelength, radius of curvature, amplitude, and belt width." (Rosgen, 1996, p. 2-5) Sinuosity, which is stream length divided by valley length is also an important pattern criterion in stream classification. Sinuosity is an indicator how a river has adjusted its slope to that of its valley.

Stream channel profile is an important consideration not only for the overall gradient of the stream but also for determining bed forms and spacing sequences. Steeper channels tend to be of step-pool form or are rapids-dominated, while lower gradient channels may show bed features such as riffles, runs, pools and glides. Channel profile, bed form(s) and spacing are related to bed material and are important parameters of stream classification.

Level I and II delineative criteria

Rosgen's (1996) classification system is hierarchical and consists of four levels with each level building on the former. As Figure 1 shows, Level I classification is focused upon "geomorphic characterization" of valley morphology and its influence on watershed and drainage network patterns of occurrence. Level I classification can often be done using remote-sensed data combined with field familiarity of basin geology and climate.

Level II stream delineation is based upon field collected data of channel cross-section, a longitudinal profile and plan form or pattern measurements from a "reference reach." The five primary delineative criteria of Level II classification are:

1.   Entrenchment Ratio (width of flood-prone area/bankfull width) describes the degree of vertical containment of the channel.

2.   Width/Depth ratio (bankfull width/mean bankfull depth) indicates the shape of the channel cross-section.

3.   Slope: water surface slope averaged for 20-30 bankfull channel widths.

4.   Sinuosity (stream length/valley length or approximated by valley slope/channel slope) indicates pattern and gradient adjustment.

5.   Dominant Channel Materials: channel material size distribution analysis with the D50 indicating the most prevalent particle size.

These five criteria were selected from data analysis of hundreds of rivers as those that best clustered to allow classification. Level II classification allows the morphological description of stream types A1-A6 to G1-G6 (see Figure 2).

Level III Classification Criteria

We have employed Rosgen (1996) Level III classification in our data collection from Kansas reference reach streams over the past 3 years. As Figure 1 indicates, Level III analysis "results in a description of stream condition as it relates to stream stability, potential and function." (Rosgen, 1996, p. 6-3) Level III classification employs additional parameters to provide a more refined view of the state of the stream reach. The ten additional parameters used in Level III field inventory and assessment are:

1.   riparian vegetation

2.   flow regime

3.   stream size and order

4.   organic debris and/or channel blockage

5.   depositional patterns

6.   meander patterns

7.   streambank erosion potential

8.   aggradation/degradation potential

9.   channel stability rating

10.   altered channel materials and dimensions   (Rosgen, 1996)

In addition to the above parameters, we use the "Stream Visual Assessment Protocol" (NRCS) and a detailed description of riparian vegetation communities. These additions allow what we believe is a thorough assessment and record of the condition of stable reference reach streams in Kansas. In future work, we hope to move to Level IV classification with emphasis on sediment and streamflow measurements.

METHODS: Field Measures

Here I briefly list the various measures we employ in Level III assessment of Kansas reference reach streams and the reasons or purposes of their use.

Longitudinal Profile: measures the thalweg, edge of water, low bank (where obvious) and bankfull stage indicators. The profile provides a measure of water surface slope for the reach and a detailed picture of bed form and diversity. An example of a longitudinal profile is illustrated in Figure 3.

Cross-Sections: taken at riffle, run, pool and glide to establish channel shape, dimension and cross-sectional area. These measures allow calculation of width/depth ratio and entrenchment ratio. The cross-section at a representative riffle is used to measure velocity and calculate discharge. Figure 4 illustrates a riffle cross-section from one of our reference reach streams.

Sediment Sample: of the surveyed reach using the Wolman (1954) "pebble count" procedure. Essentially, this is a stratified random sample and record of channel materials from the bed and banks up to the bankfull stage. Figure 5 illustrates an example of a "reach pebble count." In addition to the "reach count" we conduct "pebble counts" of the active bed for every cross-section surveyed. Sediment samples allow stream classification and "state" or "condition" assessment.

Riparian Vegetation Community(ies) Survey: a careful observation and record of species, their position(s) of dominance and/or importance as well as notation regarding species and age-class diversity Also noted here are current or recent disturbances such as timbering, grazing or clearing. Vegetation surveys are categorized to look at overstory vegetation, understory vegetation and, important in Kansas streams, bank stabilizing vegetation.

BEHI & NBS, Pfankuch and SVAP: are assessments of channel stability and function. Bank Erosion Hazard Index (BEHI) and Near Bank Stress (NBS) are assessed for each bank type occurring on the surveyed reach following Rosgen (2001) procedures. The Pfankuch Channel Stability Evaluation was developed by USFS Hydrologist Dale Pfankuch and is described in his 1995 Forest Service publication. This evaluation involves visual assessment of fifteen different channel stability indicators, each of which is scored and allows a composite rating of channel stability. The Stream Visual Assessment Protocol (SVAP) developed by NRCS is described as an "easy-to-use assessment protocol to evaluate the condition of aquatic ecosystems associated with streams" (NRCS, 1998, p. 1) SVAP also results in a composite score of stream system stability and condition.

Pattern: "plan-view" measures (Figure 6) of reference reach streams include sinuosity, meander pattern, meander length, radius of curvature, belt width and meander width ratio (belt width/bankfull channel width). Some of these measurements may be made from recent aerial photographs but field measures (especially of sinuosity) are often more accurate. Pattern measures allow stream classification and a view of how the stream has adjusted its gradient to the valley morphology.

Bar Sample: a specific sample of point bar materials that allows stream entrainment capacity calculations at different, significant discharge levels. The bar sample procedure results in a prediction of the size of sediment mobilized as bedload during a bankfull flow event. This procedure is further described in Rosgen (1996).

Such is the nature of methods employed and data collected from Kansas reference reach streams. The presentation and analysis of this data will provide a basis for varied and significant application. Some of these applications are briefly addressed in the following section.

APPLICATIONS

The application of the information I have been describing is indeed varied and important to the establishment or re-establishment of self-sustaining riparian-fluvial systems in Kansas and throughout the Central Lowlands and Great Plains. A few general areas of application are offered here:

Education: of students, landowners, concerned citizens, legislators, policy and decision makers, and all other "stake holders" as to the condition, function and trend of the streams of the hydrophysiographic provinces of Kansas and the Midwest. This education can and should occur in multiple venues including classrooms, field studies, laboratory, professional workshops, and planning meetings at municipal, county and district levels. My observation has been that few stakeholders or decision makers are well-informed as to the way water interacts with the unique parameters of their hydrophysiographic province. Without such understanding, the development of sustainable design, planning and management guides will be unlikely.

Research: into channel stability parameters and prediction for Midwestern stream types; sediment transport; Total Maximum Daily Loan (TMDL) suggestions for sediment, based on reference reach (baseline) conditions; evolutionary trends for Midwestern streams; the impacts of various land use and management schemes on stream function and habitat (especially for threatened and endangered species) to name but a few. The physical conditions of streams presented by our data are directly linked to chemical, biological and aesthetic parameters of Kansas riparian systems.

Stabilization: many (if not most) of the streambanks in the Midwest agricultural region are in need of stabilization. As agricultural practices have improved, such as residue management and reduced tillage, it appears that streambanks are becoming the principle source of sediment pollution. Data such as we have collected should inform streambank stabilization efforts aimed at preservation of agricultural lands, urban structures, transportation routes or structures, suburban developments, and the list could go on and on.

Design: of new channels for purposes of stormwater conveyance, flood diversion, irrigation or area drainage. Past design of channels for these purposes has been guided by the goal of rapid removal of "excess" water rather than creation of sustainable, stable channel form and function. Reference reach data, such as we have been collecting, may inform the design of new channels which meet more objectives than the mere removal of water from a given site as rapidly as possible. These "new" streams, in addition to providing stormwater management, would also provide increased habitat, aesthetic value and enhanced property value opportunities.

Stream Restoration: to be truly sustainable, restoration designs must be based upon reference reach data. The data we have been collecting will serve as the blueprint for "natural channel design stream restoration." (Rosgen, 1998) Once a stream channel exceeds a geomorphic threshold of stability, it will likely face a long (and usually costly) period of adjustment before reestablishment of equilibrium conditions. If we cannot afford to wait for these changes to occur naturally, we will need blueprints, concepts, and templates with which to design stable channels that restore sustainable form and function.

CONCLUSIONS

Presented here is a description of data collected on Kansas, stable reference reach streams over a 3 year period. Methods used to collect this data are explained and examples of its record are provided. Applications of the extensive data base are suggested and briefly described. At the time of this writing, our data collection efforts are on-going with much analysis waiting, but this neither negates the immediate application nor the significance of our data base. Many of the applications suggested above can be served by this information and subsequent analysis and conclusions will only enrich the usefulness of this data.

It is my hope that the multiple applications and value of the data described here will inspire others in other regions to learn these methods and develop their own regional data bases. Only when we have specific data bases for each hydrophysiographic province, when we have regional curves which correlate basin area with discharge and channel dimension will we have the true basis of sustainable stream management and restoration design. Our data from Kansas reference reach streams will be made available to all interested parties and hopefully will add to the growing understanding of fluvial geomorphology. My greatest hope is that the information we have collected from the field and the understanding it allows will be employed to better the condition and enhance the sustainability of the streams of Kansas and the Midwest United States.

Acknowledgements

I would like to acknowledge Adrianne Ohlde, April Felker-Hay, and Brian Hughes who assisted with field collection of data. We owe thanks to the many landowners who have granted access to the reference reach streams studied. Finally, my thanks to Phil Balch and Brock Emmert of the Watershed Institute, Topeka, Kansas, who asked me to join this data collection effort.

REFERENCES

Dunne, T. and L.B. Leopold. 1978. *Water in Environmental Planning.* San Francisco, CA: W.H. Freeman and Co.

Leopold, L.B., M.G. Wolman and J.P. Miller.. 1964. *Fluvial Process in Geomorphology.* San Francisco, CA: W.H. Freeman and Co.

McHarg, I. 1992. *Design With Nature.* New York, NY; John Wiley and Sons.

NRCS. 1998. Stream visual assessment protocol. NWCC Technical Note 99-1, SVAP. Washington DC: GPO.

Pfankuch, D.J. 1975. Stream reach inventory and channel stability evaluation. USDA Forest Service, R1-75-002m Washington, DC: GPO.

Rosgen, D.L. 1994. A classification of natural rivers. *Catena* 22: 169-199. The Netherlands: Elsevier Science.

Rosgen, D.L. 1996. *Applied River Morphology.* Pagosa Springs, CO: Wildland Hydrology.

Rosgen, D.L. 2001. A practical method of computing streambank erosion rate. *Level III course materials.* Pagosa Springs, CO; Wildland Hydrology.

Schumm, S.A. 1997. *The Fluvial System.* New York, NY: John Wiley and Sons.

Tarlock, A.D. 2000. Putting rivers back in the landscape: The revival of watershed management in the United States. *Hastings West-Northwest Journal of Environmental Law and Policy.* Winter/Spring, 2000. Retrieved via Lexis Nexis Academic 1/5/2004.

Wolman, M.TG. 1954. A method of sampling coarse river-bed material. *Trans. Am. Geophysical Union* 35: 951-956.

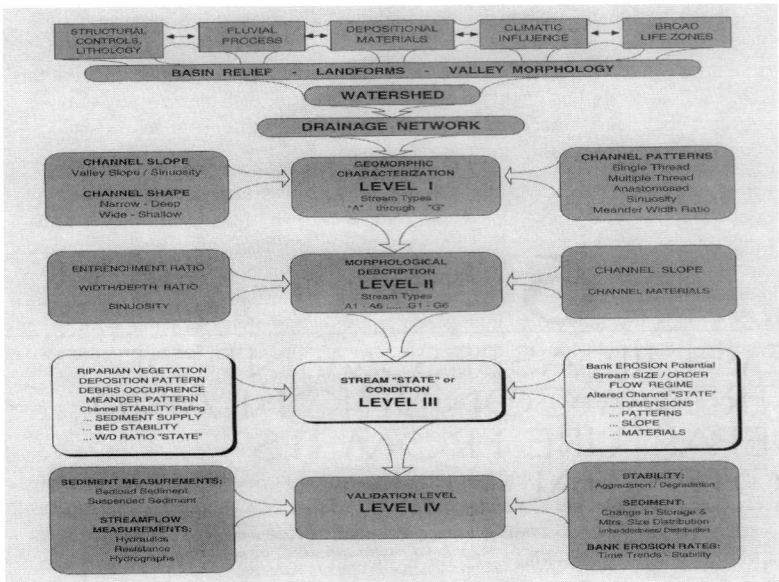

Figure 1. Rosgen stream classification system. (Rosgen, 1996)

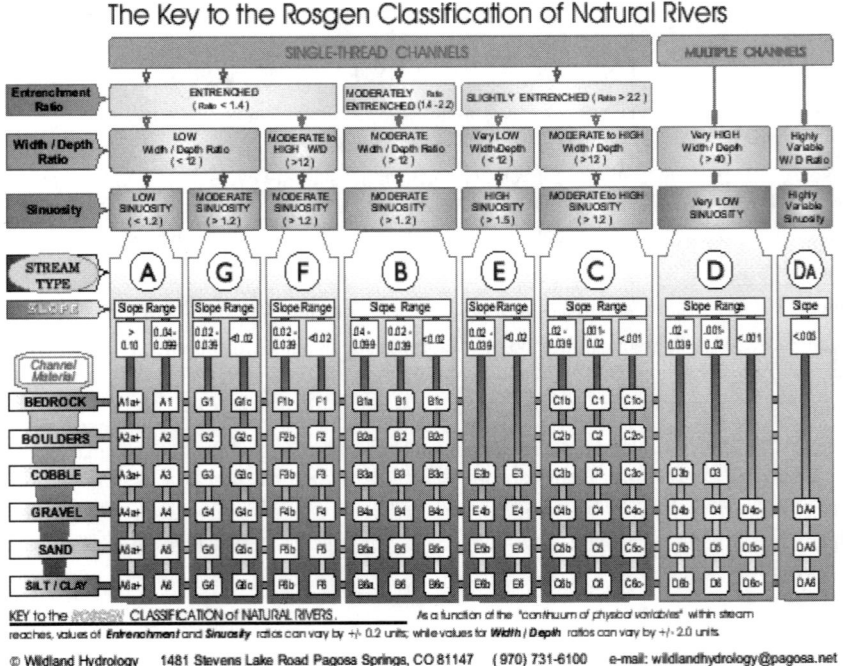

Figure 2. Classification key. (Rosgen, 1996)

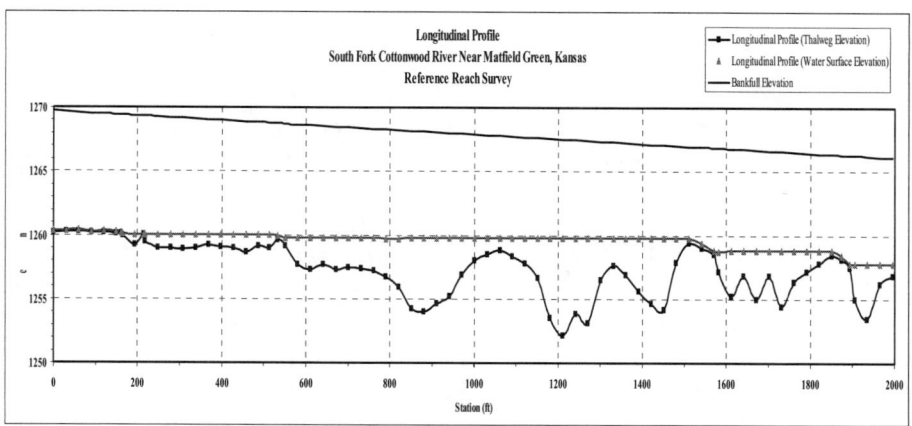

Figure 3. Longitudinal profile, South Fork Cottonwood River, Kansas.

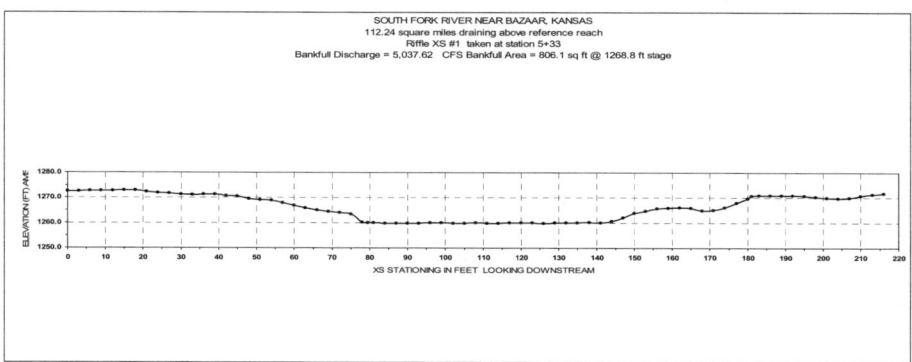

Figure 4. Cross-section, South Fork Cottonwood River, Kansas.

107

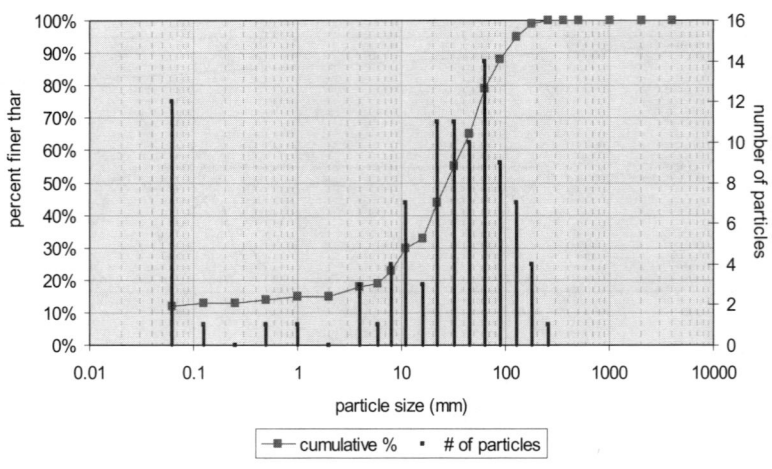

Figure 5. Reach pebble count. South Fork Cottonwood River, Kansas.

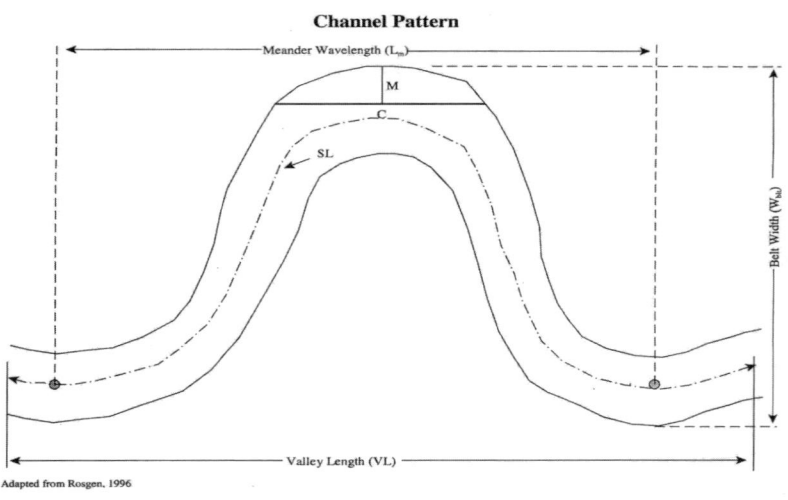

Figure 6. Channel pattern parameters. (Rosgen, 1996)

Soil Erosion Control from Construction Sites: An Important Aspect of Watershed Conservation

M. Pudasaini[1], S. Shrestha[1] and S. Riley[1]

ABSTRACT

Amount of sediment transported from a watershed is an important indicator to determine the extent of the watershed health. It is now well-conceived fact that the rate of soil erosion is an important variable for the sustainable watershed management and planning. This has escorted the development of numerous erosion prediction models in hillslope and/or watershed scale. Revised Universal Soil Loss Equation (RUSLE) is one of the easy to use yet efficient models developed after long time experience and experiments.

Although the global figure of soil erosion from the construction site is small as compared to that from agriculture activities, intensity of erosion is 10 to 20 times greater in the construction sites ranging from 20-200 ton/acre/year. So, the watersheds with massive construction activities are becoming vulnerable to soil erosion. Controlling such massive erosion from the construction sites has become a serious challenge to the conservation planners.

Data from six different erosion plot experiments with possible land use conditions in the construction sites were used to calibrate parameters of RUSLE in this study. Large-scale rainfall simulators developed in University of Western Sydney, Australia were used for 24 different events of rainfall simulation during the study. Calibrated parameters were used to evaluate the efficiency of some specific support practices to control soil erosion from construction sites. Study showed that mechanical methods of erosion control such as short grass strips, gravel bag and silt fencing are the efficient erosion control measures for the construction sites reducing soil erosion by 45%, 63% and 85% respectively.

INTRODUCTION

Exponentially increased human population over the last three quarters of the century is leading to rapid exploitation of natural resources beyond its renewal capacity (Hudson, 1995). Soil is not an exception of this. Natural rate of soil erosion is increased exorbitantly due to the increase in agricultural activities. Mine exploration and massive construction activities heavily disturbed the natural state of the soil (Morgan, 1995). Use of chemical fertilizers increased the erosive potential of the soil This further increased the rate of soil erosion. As a non-renewable resource at human time scale, the masking effect of improved technology to enhance productivity became questionable as it lead to the negative effects of permanent desertification. Only 11% of the global land considered as productive should feed 7600 million people by 2020 (Eswaran et al., 2001). A large portion of the sediment particles carried by natural river system is discharged at the ocean causing the seashore land to be more vulnerable to impoundment. So the land degradation has been a serious environmental issue and will remain as a problem in the 21st century.

More than 80% of land degradation is due to soil erosion out of which 56% is due to the water induced soil erosion (Oldeman, 1992). UN Environmental Program reports that crop productivity on about 20 million hectares each year is reduced to zero or becomes uneconomical because of

---

[1] School of Engineering and Industrial Design, University of Western Sydney, Australia, URL: www.uws.edu.au

soil erosion or soil induced degradation (UNEP, 1991). Human induced degradation has affected 24 percent of inhabited land area (Oldeman, 1992).

Soil erosion has both onsite and offsite effects. Reduction in the productivity of the agricultural land by removal of fertile top soil and depletion of root depth of crops, formation of undulations in land surface due to the formation of rill and gully, inundation of vegetation from the land surface, scouring of the foundations of hydraulic structure, road and other infrastructures are some examples of the immediate onsite effects caused by the soil erosion (Morgan, 1995). Offsite effects include unnecessary sediment deposition over the cropland, extreme pollution of water bodies, sedimentation in the storage reservoirs, channel bed and road drainage structure (Morgan, 1995). Construction sites are significant source of sediment and other non point source pollution. Soil erosion from construction sites, without proper soil erosion and sediment control practices can average between 20-200 tones/acre/year, which is 10 to 20 times greater than typical losses from the agricultural land (NRCS, 1999). Soil erosion from construction sites thus is more severe in terms of intensity than from agricultural land and degrades land more rapidly. Chemicals used in the construction sites and mined land and carried out with eroded soil and water may be poisonous to aquatic life, human and vegetation. Thus soil erosion has multifold effects in natural environment. This is a serious problem for sustainable future of every watershed and whole the environment.

Universal Soil Loss Equation, USLE (Wischmeier and Smith, 1978), later revised as Revised USLE, RUSLE (Renard et al., 1996) is one of the empirical models developed in the USA with more than 10,000 plot years of research data and experience of soil scientists to estimate soil erosion in annual basis from the hillslope. The model was selected for the study as it is easy to use yet giving acceptable results from the construction sites. Parameter estimation is relatively easy and well-explained documentation is available (Renard et al., 1996). Calibration and validation of some parameters were made using data from erosion plot experiments carried out in NSW. Calibrated parameters were used to evaluate the efficiencies of some mechanical techniques of erosion control from the construction sites.

MATERIALS AND METHOD

Calibration and validation of RUSLE was carried out using the data from the erosion plot experiments carried out at Penrith campus of the university and at the Gosford, NSW. The rainfall intensities used in this experiment were to be representative of the low (one year average recurrence interval (ARI)) to medium (five to 10 year ARI) with 30-minute rainfall duration. The intensities were obtained using IFD curves of selected sites published by the Bureau of Meteorology (IEAust, 1987). It was expected that the 30-minute duration would provide enough time to capture all components of hydrographs. Times of concentration for the plots for Penrith area was estimated as 10 minutes and that for Somersby was estimated as 15 minutes.

Erosion plots of 80m length were constructed to represent general length of the land employed in the housing project construction sites in NSW. The width of the plots were kept to 5m to avoid boundary effects. Slopes of the sites were varied between 7% or 8%. Three different treatments: Rotary hoed, rolled smooth and topsoil restored were employed to represent general construction site land use condition. Sediment collection troughs were installed to collect sediment and runoff water from the experiment. Two standardized RBC flumes (B_c=75mm and B_c=150mm) were prefabricated and installed at the plot outlet downstream of the collection trough to measure runoff from each of the plots. These flumes were laboratory tested before commencement of the experiments to ensure verification of calibration results provided by Bos (1991). Every care has been taken while constructing the plots to avoid the influence of external factors during the simulation and the erosion processes. Access roads and boundaries were constructed around the plots.

Large-scale pressure sprays rainfall simulators were used in this study. Recommended pressure and nozzle combinations from the calibration study of the rainfall simulator carried out by Farre (2001) were used for the rainfall simulation. Runoff was measured using the RBC flumes at the outlet of collector apron. Discharge was measured in 30 second or one minute interval, manually by reading the stilling well. Time of the beginning of storm and commencement of runoff was also recorded. Runoff hydrograph is plotted for each run of experiment for each plot. Soil samples were collected from three different locations of each plot before and after experimental runs to enable estimation of the change in soil moisture. The samples were stored inside the core casings sealed with core caps and weighted. The cores were then oven dried at 104°C for 24 hours until complete dryness was achieved. Then bulk density and moisture content of soil were estimated.

NSW Department of Infrastructure, Planning and Natural Resources (DIPNR) carried out the detailed soil analysis and the results provided by the department were used in the model. Model parameters such as organic mater content, percentage clay, percentage silt, percentage sand and percentage rock fragment were also obtained from the detail soil analysis conducted by DIPNR. Soil permeability and soil structure class parameters were estimated using the guideline given in SOILOSS (Rosewell, 1993). All these soil parameters were used to calculate soil erodibility factor K of RUSLE model.

Runoff water was carefully collected at one-minutes interval in clean and washed polyethylene bottles of 500ml capacity. The collection point was the outlet of the apron. Sediment samples thus collected were filtered. The results were used to generate sedigraph (a plot of sediment concentration vs time), which gives the picture of sedimentation concentration over the experimental run. Sedigraphs were used along with the hydrograph to estimate total soil loss exit from the outlet of the collection trough. Considerable amount of sediment was deposited in the collection trough, which was not counted in the sedigraph. Volumetric measurement of the sediment deposited between the outlet of the apron (where sample for sedigraph was taken) and outlet of the experimental plot was carried out after runoff ceased. The total quantity of soil eroded during experimental run was computed as the sum of these two components: from the sedigraph and hydrograph and from the collection trough.

RUSLE estimates the average annual soil loss from the entire hillslope. The equation is given as:

$$A = R \times K \times LS \times C \times P \tag{1}$$

where:

A = Average annual soil loss predicted (t/ha), R = Rainfall runoff erosivity factor (MJ mm/(ha h)), K = Soil erodibility factor, (ton ha h/(MJ ha mm)), L = Slope length factor, S = Slope steepness factor, C = Cover management factor and P = Support practice factor

Rainfall runoff erosivity factor R, in RUSLE represents the impact of rainfall energy in erosion. Rainfall energy per unit depth of rainfall e_k (MJ/ha/mm) was determined using the equation suggested by Rosewell and Turner (1994) given by:

$$e_k = 0.29\left(1 - 0.596e^{(-0.04i_k)}\right) \tag{2}$$

where:

i_k = Intensity of rainfall (mm/h)

The total rainfall energy of single storm is computed as

$$E = \sum_{k=1}^{p} e_k d_k \tag{3}$$

where:

d_k = Depth of rainfall for k^{th} interval of the storm (mm), p = Total no of intervals in the storm

Finally R is estimated by the relation

$$R = E \times I_{30} \tag{4}$$

where:

I_{30} = Maximum 30-minute intensity (mm/h)

As rainfall simulators were used in the experiment with constant intensity throughout the experimental period, I_{30} is equal to the average intensity. Each of the simulation was carried out for 30 minutes resulting in the depth of the rainfall as half of intensity. R factor in this case, hence reduces to:

$$R = 0.145(1 - 0.59e^{(-0.04i)})i^2 \qquad (5)$$

where:

i = Measured average rainfall intensity (mm/h)

Eq. 5 was used to estimate R factor value from each of the experimental runs in this study.

Soil erodibility factor, K, is estimated from the approximated equation of the nomograph developed by USDA-ARS (Renard et al., 1996) as total percentage of silt and fine sand did not exceed 70% and given as:

$$K = 2.77 \times 10^{-7} \left(12 - OM\right) M^{1.14} + 4.28 \times 10^{-3} \left(s - 2\right) + 3.29 \times 10^{-3} \left(p - 3\right) \qquad (6)$$

where:

K = Soil erodibility factor (ton. ha. h/MJ/ha/mm), s = Soil structure class (1-4), p = Soil permeability class (1-6), OM = Percentage organic matter content and M = Product of primary particle size fraction given as (Rosewell, 1993)

$$M = (si + 0.7Fs)(si + Fs + Cs) \qquad (7)$$

where:

si = Percentage silt, Fs = Percentage fine sand Cs = Percentage coarse sand

As the research plots used in the experiment satisfied the requirements of USLE plot, the equation developed by USDA, ARS to estimate LS factor were used. The equations used to estimate L factor for hillslope of length λ were:

$$L = \left(\frac{\lambda}{22.1}\right)^m \qquad (8)$$

where:

m = Variable slope length exponent.

The value of slope length exponent depends upon the ratio of rill to interrill erosion. If β is the ratio of rill erosion to interrill erosion then m is given as

$$m = \frac{\beta}{(1 + \beta)} \qquad (9)$$

For moderately susceptible soil in both rill and interrill erosion, McCool et al. (1989) suggest the equation:

$$\beta = \frac{11.1607 Sin\psi}{3.0\left(Sin\psi\right)^{0.8} + 0.56} \qquad (10)$$

where:

ψ = Slope angle (degrees)

Renard et al. (1996) suggested to double the value of β obtained from the above equation to compute the slope length exponent, m, if the soil is highly susceptible to rill erosion like in case of freshly prepared steep construction slopes. As all the experiments in this study were carried out on bare soil to represent urban construction sites, value of β was doubled before applying it to compute exponent m.

S factor is the slope steepness factor and represents the effect of slope in erosion given by the relation (McCool et al., 1987).

$$S = \begin{cases} 10.8 Sin\psi + 0.03 & for \quad slope < 9\% \\ 16.8 Sin\psi - 0.50 & for \quad slope \geq 9\% \end{cases} \qquad (11)$$

where:

ψ = Slope angle (degrees)

This equation was used to compute S factor and value of L and S were combined to get the value of LS factor.

Value of measured soil loss A, and estimated values of R, K and LS from the above equations were used to estimate combined value of cover management factor C and support practice factor P, known as CP and given by the relation:

$$CP = \frac{A}{(RKLS)} \qquad (12)$$

Half of the data from the experiment from each plot were randomly selected for the calibration of CP and remaining half were used to validate the calibrated value. After validation, calibrated CP values from each of land use condition were used to determine the efficiency of three different support practices.

Error estimation

A new approach of error estimation is introduced to evaluate the efficiency of the model against the measured soil erosion. A perpendicular distance (p) dropped from plotted point of measured versus predicted soil loss (x, y) to a line passing from the origin with slope angle of 45° (OP, Figure 1.) is regarded as error of prediction. This definition is justified because for perfect prediction this distance should be zero. The perpendicular distance (p) thus obtained is normalized with corresponding perpendicular distance from the origin (r). The standard error of prediction (SE) is defined as square root of average of sum of the square of the ratio (p) to (r). Equation 4 gives standard error for (n) data. Value of (SE) ranges from 0 to 1, giving the lesser the value the better the prediction.

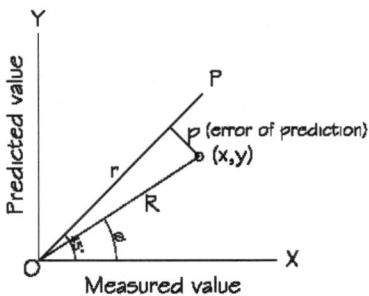

Figure 1. **Geometry of error estimation**

$$SE = \sqrt{\frac{\sum \left(\frac{p}{r}\right)^2_i}{n}} \qquad (13)$$

and model efficiency (η) is given by the equation:

$\eta = 1 - SE$ (14)

which gives the applicability of the model to predict soil erosion

RESULTS AND DISCUSSIONS

Toy and Foster (1998) recommend that for the scalped surface, value of C is 0.15. As no specific support practices were adopted in plot one of each site and land surface was rough to disturb the flow, P factor in this case can be considered to be in the range of 0.7-0.8. This gives CP ranging 0.11 to 0.12. The calibrated value of CP for Penrith soil is 0.13 while that for Somersby is 0.07. Similarly, CP value for plot three (topsoil restored, representing partially filled construction site) was estimated as 0.23 and 0.32 for Penrith and Somersby respectively. Toy and Foster (1998) suggest C value for complete fill condition ranging from 0.85 to 1.00. For partially filled condition, value of C can be taken to be 0.4 to 0.5. Similar support practice value as in plot one can be considered in this case, ranging P value from 0.70 to 0.80. This gives the range of CP from 0.28 to 0.40. P value suggested in different literature (Renard et al., 1996; Rosewell, 1993; Morgan, 1995) for scraped and rolled smooth condition as in plot two is 1.0. But C value for the scraped surface with undisturbed soil and compacted with roller can be taken ranging from 0.15 to 0.20, giving the range of CP from 0.15 to 0.20. Estimated CP for this condition ranges from 0.12 to 0.26. This shows that calibrated values of CP seem reasonable when compared with the values published in different literature.

Table 1. Measured and Predicted soil loss with different support practices

Site	Land use condition	Rainfall intensity (mm/h)	Measured soil loss (t/ha)	Calibrated CP	Predicted soil loss (t/ha)	Predicted soil loss (t/ha) with		
						Short grass strip (P=0.55)	Gravel bags (P=0.37)	Silt fences (P=0.15)
Penrith	Rotary hoed	62	3.45	0.13	3.42	1.88	1.27	0.51
		72	4.67		4.70	2.59	1.74	0.71
	Rolled smooth	28	1.23	0.26	0.81	0.44	0.30	0.12
		67	2.72		5.62	3.09	2.08	0.84
	Topsoil restored	46	3.09	0.32	3.20	1.76	1.19	0.48
		57	5.20		5.02	2.76	1.86	0.75
Somersby	Rotary hoed	76	3.14	0.07	2.78	1.53	1.03	0.42
		74	2.30		2.63	1.45	0.97	0.39
	Rolled smooth	87	6.19	0.12	6.18	3.40	2.29	0.93
	Topsoil restored	44	2.91	0.23	2.94	1.62	1.09	0.44
		70	7.96		7.87	4.33	2.91	1.18

The validation of the calibrated CP values was carried out by using CP values as input to the model. Predicted soil loss from the model was compared with corresponding value of measured soil loss. Figure 2 gives the scatter plot of predicted vs measured soil loss. Standard Error (SE) of prediction using Eq. 13 is estimated to be 0.19 giving the model efficiency of 81%.

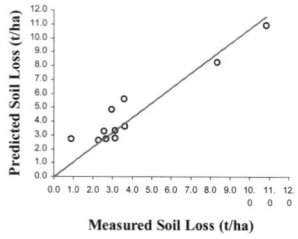

Figure 2. Measured vs Predicted soil loss

There were less than 15% differences in most of the soil loss prediction results in comparison to measured values. Model over predicted soil loss some of the cases. The highest over-prediction was observed in run one of plot three for Somersby site, resulting in 204.5% more soil loss from the measured value. Very low runoff of 0.2mm was recorded in this experiment, although the rainfall intensity applied was 43mm/h, giving R factor as 238 MJ mm/(h ha). This implies that although there was a significant splash erosion causing soil detachment (R=238 MJ mm/(ha h), higher infiltration (42mm/h) caused very low flow, reducing the measured soil loss. RUSLE is not a spatial model and does not estimate soil deposition separately. All the detachment were accounted in the soil loss prediction by the model. But due to the low surface flow, significant quantity of detached particles were deposited within the plot and not accounted in the measured soil loss. This is a valid argument as model over predicted soil erosion in most of the cases. Some deposition is always expected within the plot irrespective of the applied rainfall intensity and runoff produced in the hydrograph recession when the rainfall ceases. This is because of the decrease in flow. Majority of the results support the validity of model leading to usefulness of the calibrated parameters.

Cover management factor C and support practice factor P are the two factors that can be modified by adopting suitable land use and support practices if the soil loss predicted is above the tolerance. But generally in the construction sites, changing the value of cover management factor such as increasing ground cover, mulching or increasing surface roughness are impractical. Then choosing suitable support practices such as buffer strips, establishing barriers etc, will be the best alternatives for erosion control. In such cases, reasonable value of cover management factor C and support practice factor P need to be determined from the recommended CP value. Desirable support practices can be adopted based on the P factor value computed.

Sometimes if single support practice is insufficient to reduce soil erosion to tolerable limit, more than one practice can be implemented. In such cases P factor values from each support practices can be multiplied to get final effective P factor (Toy and Foster, 1998). Toy and Foster (1998) recommended for the slope condition ranging from 5-10%, P value for Short grass strip, gravel bag and silt fences can be taken as 0.55, 0.37 and 0.15 respectively. Using these values with the calibrated CP values it has been found that soil erosion can be reduced by about 45% from each site if short grass strips are employed as support practice. Gravel bags are more effective in soil erosion control, reducing the value by over 60%. Results also show that the silt fence is the most effective support practice and can reduce soil erosion by over 85%. It is important to note that the efficiency of practice factor will vary with change in land slope.

CONCLUSION

The general conclusion of this study is that RUSLE can be a reasonably accurate model to predict soil erosion from construction sites in NSW for single or multiple storm rainfall events of any particular intensity. This is from the evidence that model efficiency of this study is 81%. The parameters will be useful in estimating soil erosion from construction sites of NSW. Suitable management practices can be adopted to reduce soil erosion to a tolerable limit. Model can also be used to observe sensitivity/effectiveness of particular support practice in erosion control. It has been identified that support practices such as short grass strip, gravel bag and silt fence are very effective erosion control devices reducing soil erosion in the range of 45% to 85%. These support practices can be regarded as best management practices (BMPs) as they are efficient and environment friendly. This will help to adopt suitable management practices to reduce soil erosion from the construction activities and help to ensure sustainable watershed management.

REFERENCES

1. Bos, M. G. 1991. Flow measuring flumes for open channel systems *American Society of Agricultural Engineers*.

2. Elwell, H. A. 1981. A soil loss estimation technique for southern Africa In *Soil Conservation:Problems and Prospects*. (Ed, Morgan, R. P. C., Wiley, Chichester, Sussex), pp. 281-292.

3. Eswaran, H., R. Lal and P. F. Reich 2001. Land degradation: An overview In *2nd International Conference on Land Degradation and Desertification*(Eds, Bridges, E. M., I.D.Hannam, L.R.Oldeman, F. W. T. P. d. Vries, S. J. Scherr and S. Sompatpanit) Oxford Press, New Delhi, India, Khon Kaen, Thailand.

4. Farre, G. 2001. Calibration of Rainfall Simulator, Research Center for Sustainability in Ecological Engineering and Water Resources Technology, University of Western Sydney, Sydney.

5. Hudson, N. 1995. *Soil Conservation*. B T Batsford Limited, London, pp. 126-153.

6. IEAust 1987. Australian Rainfall and Runoff: A Guide to Flood Estimation, Institution of Engineers, Australia, Barton, ACT, Vol. 1.

7. McCool, D. K., L. C. Brown, G. R. Foster and et al. 1987. Revised slope steepness factor for the Universal Soil Loss Equation *Trans. ASAE,* 30, 1387-1396.

8. McCool, D. K., G. R. Foster, C. K. Mutchler and L. D. Mayer 1989. Revised Slope length factor for Universal Soil Loss Equation. *Trans. ASAE,* 32, 1571-1576.

9. Morgan, R. P. C. 1995. *Soil Erosion and Conservation*. Longman Group Limited, London.

10. NRCS 1999. Construction Site Soil Erosion and Sediment Control, http://www.il.nrcs.usda.gov/engineer/SoilEro.html

11. Oldeman, L. R. 1992. Global extent of soil Degradation, International Soil Reference and Information Centre, Wageningen, The Netherlands pp. 19-36.

12. Renard, K. G., G. R. Foster, G. A. Weesies, D. K. McCool and D. C. Yoder 1996. Predicting Soil Erosion by Water: A guide to conservation planning with the Revised Universal Soil Loss Equation, US Department of Agriculture, Agricultural Research Services, Agricultural Handbook 703.

13. Rosewell, C. J. 1993. *SOILOSS: A program to assist in the selection of management practices to reduce erosion,* The soil conservation service, Department of Conservation and Land Management.

14. Rosewell, C. J. and J. B. Turner 1992. *Rainfall Erosivity in New South Weles,* Department of conservation and Land Management, National Soil Conservation Program.

15. Toy, T. J. and G. R. Foster 1998. Guidelines for the Use of the Revised Universal Soil Loss Equation (RUSLE) Version 1.06 on MIned Lands, Construction Sites, and Reclaimed Lands.

16. UNEP 1991. Status of desertification and implementation of UN plan of action to combat desertification UNEP, Nairobi, Kanya, pp. 344.

17. Wischmeier, W. H. and D. D. Smith 1978. Predicting rainfall erosion lossess-a guide to conservation planning, US Department of Agriculture, Agricultural Handbook 537.

Application of Water Quality Modeling to Regulating Land Development in Protected Drinking-Water-Source Areas: A Case Study on the Kao-Ping River Watershed, Taiwan

Chun-hsu Lin[1] and Te-thui Huang[2]

ABSTRACT

The purpose of our project is to build a management system for protected drinking-water-source areas where land development is restricted for drinking safety. Under this system, a protected area is further divided into two different zones. In the first zone, any type of land development is prohibited. In the second zone, development is allowed only when a development permit with more environmental protection requirement is granted by a responsible government agency. Three compensation programs are then proposed to reduce the resistance from landowners of the properties to be included into the source-water-protected areas. Thus, how to delineate a protected area subject to different levels of regulations is of concern while initiating the compensation programs.

In this paper, we demonstrate a case study on the Kao-Ping River Watershed, a subtropical watershed with highly heterogeneous land uses and widely ranged elevations. Through this case study, we empirically examined the proposed management and compensation systems. In which, the targeted abatement of water pollution was regarded as the basis for determining the needed areas of two types of protected zones. Following this principle, computer models were utilized as the planning tools. Arc/View GIS software and two water quality prediction models, QUAL2E and GWLF, were applied to repeatedly simulating the effects of pollution reduction after the development restrictions are imposed with different acreage of two development-restricted zones. The modeling results for the optimal areas of the protected zones were further used for estimating compensation amounts based on one of the three following schemes: land banking, conservation easement, and transferable development rights.

KEYWORDS. QUAL2E, GWLF, water quality modeling, riparian buffers.

INTRODUCTION

In Taiwan, water resources protection is currently executed through legal regulations. In order to protect the sources of drinking water from pollution, many regulations require the delineation of protected areas in the watersheds where drinking water is acquired. For drinking safety purposes, the protected areas are usually defined as the upstream watershed above the water intake points. However, this definition results in more than 25% of the land of Taiwan is classified as the area under protection. In fact, delineating protected areas and the subsequent enforcement are difficult to proceed due to the conflicts between responsible agents and the local residents for the land use restrictions imposed on their properties. Therefore, other strategies have been proposed, such as different levels of protecting controls considering the distance factor to the water intakes.

---

[1] Associate Research Fellow, Institute of Environment and Resources, F7, 45 Hanko Street, Taipei, Taiwan. chlin@ier.org.tw.

[2] Assistant Researcher, Institute of Environment and Resources. erica_huang@ier.org.tw.

The goal of this project is to provide an example of management framework, as Figure 1, to financially compensate the watershed residents for their opportunity income loss from watershed protection implementation.

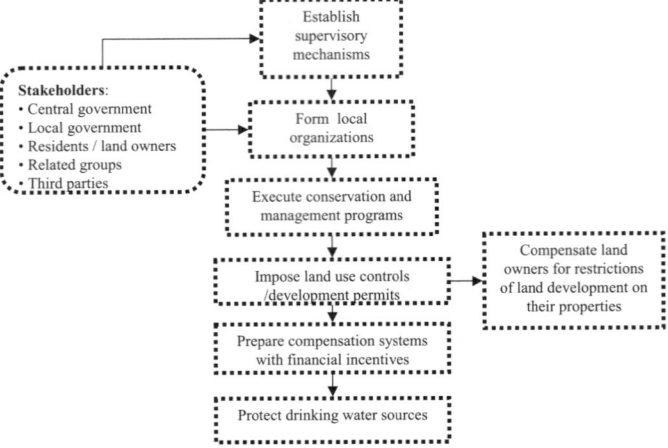

Figure 1. Management framework for protected watersheds

METHOD

Ciriacy-Wantryp (1968, 1964) proposed the "Safe Minimum Standard" (SMS) approach as the principle for renewable resource utilization. Beyond the threshold point of resource utilization (SMS), resources do not recover naturally at a reasonable social cost. This standard provides a conceptual limitation on land development but also allows the utilization of resources to a certain extent, a balancing point between conservation and development. The research steps of this project are shown as Figure 2.

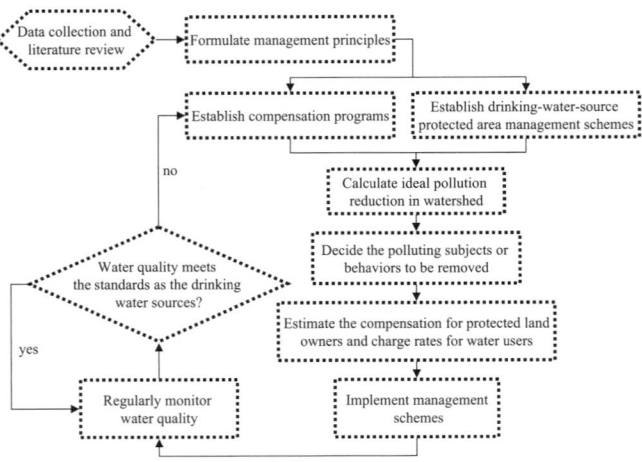

Figure 2. Research steps

Since this project is expected to establish a watershed management framework for social equity, several issues in drinking-water-source watersheds are involved as follows.

- Setting up the goals of water quality improvement and protection.
- Determining the range and location of the protected area.
- Sub-zoning in the protected area for different levels of development restrictions.
- Establishing the compensation programs for the residents of protected areas.
- Financial resources planning for the expenditure of compensation programs.

Although many issues are involved, this paper is more focused on the zoning in the protected watershed and the effects of protection efforts on water quality. We used water quality modeling as the tool to flexibly adjust the boundaries of different management zones.

Zoning in the protected area

We suggest defining the upstream watershed area above the water intake as the area of protection as most water protection laws require. Beyond this definition, the protected area should be further divided into two zones for different management scenarios. Zone 1 is the area of waterway or water body plus 30meter-wide buffer zones on both sides of the river (Figure 3). Any type of land development in Zone 1 is strictly prohibited. For the rest of the protected watershed, denoted as Zone 2, we do not suggest unconditionally prohibition of land development. Instead, the "no-development" area in Zone 2, or the extension of Zone 1, is only the upstream area within a certain distance from the water intake. The certain distance from the water intake is determined depending on the intended extent of water quality improvement after the protection enforcement is executed.

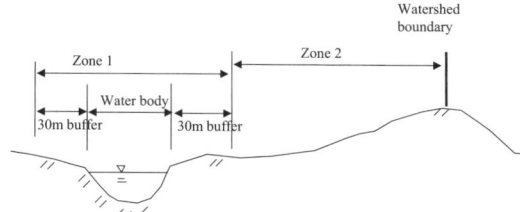

Figure 3. Cross sectional view of a protected watershed

Water quality improvement goals

Maintaining acceptable water quality of the water bodies as drinking water sources is the objective of setting up protection areas in watersheds. By water quality modeling, current water quality is compared with the predicted water quality after the watershed protection enforcement. If the predicted water quality after the watershed protection enforcement is not meeting the water quality standard for drinking water sources, the Zone 1 will be further expanded until the predicted water quality is acceptable.

WATERSHED BACKGROUND

Being the largest watershed in Taiwan occupying 3,257 Square Kilometers of land (Figure 4) with population of 600,000 people, the Kao-Ping River Watershed ever suffered with serious water pollution from swine farming business. After more than 500,000 hogs were removed from this watershed in 2001 and swine farming has been banned since then, water quality of this river has dramatically improved. However, based on the water quality records from the monitoring stations of Taiwan Environmental Protection Administration (Taiwan EPA), with industries and fruit farming remained in the watershed, water quality of the Kao-Ping River is still unsatisfactory especially in dry seasons. Currently this river provides one million cubic meters of water per day for domestic uses in Southern Taiwan. To improve the quality of drinking water from this watershed, a sanitary sewage plan for cities in the watershed has been proposed but the construction progress is slow. In addition to the engineering approach, various means of optimal land management were also proposed and discussed in the past decade.

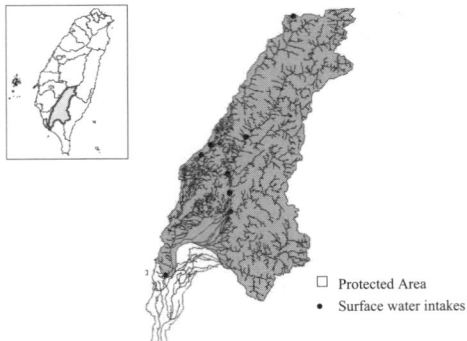

Figure 4. The Kao-Ping River Watershed, Taiwan

MANAGEMENT SCHEMES

The following data sets from Taiwan EPA in the digital format serve as the sources of basic information of the Kao-Ping River Watershed for this project:

- Watershed boundary map.
- Contour map with 10-meter-elevation interval.
- Stream system map.
- Industry distribution map.
- Water Intakes map.
- Wastewater collection system map.
- Land use map.

The first task in this project was determining the riparian buffer zones along the river. In the past decades, researchers proposed the required widths of buffer zones for different purposes and different site conditions. For simplicity in future implementation, we suggest a fixed width of 30 meters, also as the most common suggestions in literature, for our project. The total area of Zone 1 is therefore the 30-meter-wide buffer zones plus the area within a certain radical distance from the water intake. We started the "radical distance" with 1000 meters then gradually extended the radius of protection by increasing 1000 meters each time (Figure 5). After each expansion of Zone 1, the effects of pollution reduction through the restrictions on land uses are estimated by water quality modeling. For data availability and examining several key pollution indicators, we simulated point source pollution by the "Enhanced Stream Water Quality Model" (QUAL2E) of U.S. EPA (2004). For non-point source pollution, the Generalized Watershed Loading Function (GWLF) (Haith and Wu, 1989) was used for the modeling.

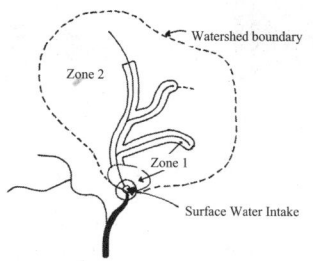

Figure 5. Illustration of sub-zoning in a protected watershed

Point Source Pollution Modeling

Using Taiwan EPA's GIS data of the industry distribution and wastewater discharge in the Kao-Ping River Watershed, we estimated the amounts of daily pollutant loading from Zone 1 of 30m-wide buffer zones plus the area within 6000m of radical distance from the water intake as Table 1. The pollutants interested include Biological Oxygen Demand (BOD), Chemical Oxygen Demand (COD) and Suspended Solid (SS).

Table 1. Estimated pollutant loading with different ranges of Zone 1

Zone 1	Total waste water discharge (m³/day)	BOD discharge (Kg/day)	COD discharge (Kg/day)	SS discharge (Kg/day)
Without protected area (currently)	4,439	33	169	61
With 30m buffer protected	4,079	32	136	57
With 30m buffer plus 6000m radius protected area	2,140	15	71	38

After the reduction of pollutant loading with different range of protected Zone 1 was estimated, the variations of water quality along the river system were then simulated by QUAL2E for both dry and wet seasons. The simulation results of Dissolved Oxygen (DO), BOD, Ammonia Nitrogen (N-NH₃) and SS concentrations were compared to the designated water quality standard for each section of the river system. Examples of simulation results are as Figure 6.

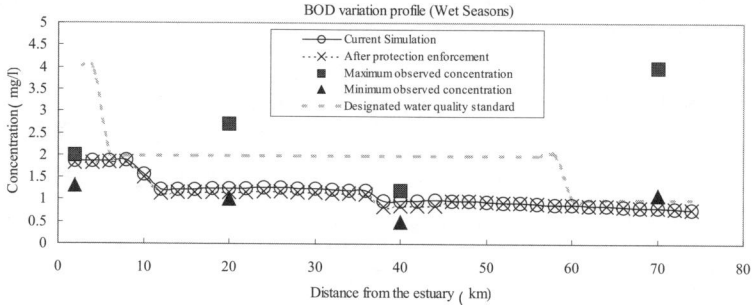

Figure 6. Estimated BOD concentration in wet seasons

After the protection enforcement, water quality improvement is expected to be as listed in Table 2.

Table 2. Major pollutant abatement effects from Kao-Ping river watershed protection

Pollutants	Seasons	Currently meeting water quality standard?	Simulated effects of water quality improvement
DO	Dry	Yes	Slightly improved

	Wet	Yes	Much improved
BOD	Dry	No	Much improved
	Wet	Yes	Slightly improved
NH3-N	Dry	Yes	Slightly improved
	Wet	Yes/No	Much improved

Non-Point Source Pollution Modeling

In addition to point source pollution, non-point source pollution in the Kao-Ping River watershed is also a serious problem caused by fruit and sugarcane farming in the upstream areas. Considering the size of the watershed and current land uses, we utilized the model of GWLF in our project. The GWLF model was developed in 1980's using the SCS curve number method to estimate runoff and using the Universal Soil Loss Equation to predict non-point source pollution. Following Chang's (et al., 1999) estimation, we applied the same input values of the parameters of nutrient concentration in GWLF modeling as the following table.

Table 3. The inputs of nutrient concentrations from different land uses for GWLF modeling

Parameter	Total Phosphorous	Total Nitrogen
Nutrient concentration in runoff (mg/L):		
Fruit farmland	1.709	0.46
Rice farmland	17.74	9.56
Sugarcane	10.69	4.66
Forest	1.925	0.18
Grassland	0.30	0.15
Bare land	0.1	0.05
Accumulated nutrient in urban areas（mg/ha-day）:	0.0436	0.0076
Nutrient concentration in groundwater (mg/L):	0.0	0.0
Nutrient concentration in sediment （mg/Kg）:	397.0	141.0

In the simulations, it was assumed that any land development in the protected area should be converted into woodlands or other Best Management Practices (BMPs) after the protection enforcement. The loading of non-point source pollutants should therefore be changed for the land use changes accordingly. Besides, based on previous studies, BMPs can reduce 70-90% of nutrients and 80-95% of soil particles entering into water bodies. With all of these assumptions, the modeling results of non-point pollution abatement effectiveness for different extents of protected areas are listed in the following table. For non-point source pollution reduction, the ratios of reduction are expected to be 42.6% of sediment, 45.7% of total nitrogen and 51.1% of total phosphorus.

Table 4. Simulated water quality improvement (non-point source pollution)

Pollution type	currently	With 30m buffer zones	With 30m-wide buffer zones and 6000m radius of protection area
Sediment (Kiloton/year)	1449	830	830
Total Nitrogen (ton/year)	1430	776	775
Total Phosphorus (ton/year)	383	187	187

Financial compensation calculation

After these two models estimated the reduction of pollutants, the financing for the compensation programs proposed was then proceeded. Based on the principles of equity (Rawls, 1971) and efficiency, three types of financial compensation programs are proposed in this study: land banking, conservation easement, and transferable development rights. Of which, land banking is a mechanism letting governments acquire preemption rights. Conservation easement lets landowners keep their properties as long as they do not develop their lands. Transferable development rights entitle landowners to trade their lands in the protected area for another unit of land outside of the protection. Since the drinking-water-source protected area in the Kao-Ping River Watershed covers the lands across six counties, three compensation schemes result in different amounts of cost based on the land prices of different land uses in each county.

Depending on the areas to be included into the protection and the official land price information, we estimated the cost is ranged from 82 to 545 billion NT dollars (approximately 35 NT dollars = 1 US dollar) for restricting land development in the protected area as Table 5.

Table 5. Financial costs for three financial compensation programs

Compensation program	Estimated Cost (Million NT dollars)
Land banking	275,690 ~ 545,520
Conservation easement	82,700 ~ 163,658
Development right transfer	0

Among the three financial compensation programs, conservation easement is the more feasible approach. The land banking system requires high amount of financial resources. The transferable development right system involves complicated administrative issues. For the "easement" approach, the financial resources can be collected from additional special fees charged through water bills or property tax of water users. As Table 6, if the special fees are collected through water bills, for every cubic meter of water used, the charge is estimated to range from 0.21 to 72.6 NT dollars depending on the scope of beneficial water customers and different assumptions on land prices. If the fees are charged through property taxing, the property tax for each landowner is estimated to increase 65%.

Table 6. Financial sources for conservation easement programs

Charging mechanism	Rate
Charged with water bills (for domestic water users only)	$0.6 \sim 72.6$ NT dollars/m^3
Charged with water bills (including agricultural water users)	$0.21 \sim 25$ NT dollars/m^3
Charged with Property Tax	1.33~1.65 times of original property tax

CONCLUSION

The purpose of this project is to build a management system for the areas under drinking-water-source-protection. We suggest dividing a protected watershed into two zones. In the first zone, any development is strongly restricted. The development control mechanism in the second zone is development permission.

We suggest land banking, easement and development transfer as the compensation options for watershed protection. After the case study on the Kao-Ping River watershed protection, the "easement" mechanism is considered the most workable approach. With this approach to compensate the landowners in the water-protected area, the total compensation amount is estimated between 82 billion and 163 billion NT dollars. The government can only pay the rent to landowners by year to lower the compensation budget. The fund for the compensation can be collected from tap water using by 1 to 72 NT dollars per cubic meter. If the special fee is charged through property taxes, the tax rate will be increased 33% to 65%.

All of the management mechanisms and the boundaries of protected areas are determined based on the QUAL2E and GWLF modeling results and verifications with water quality data of the Kao-Ping River Watershed. Marginal costs for water quality improvement can also be derived from modeling activities.

REFERENCES

1. Chang, N., D. Shaw, C. Wen, M. Hsu and L. Yang. 1999. *Management and Optimized Strategy for Industrial Wastewater Pollution Abatement in Watersheds.* Final Report to Taiwan Environmental Protection Administration, Taipei, Taiwan (in Chinese).

2. Ciriacy-Wantrup, S. 1964. The 'New' Competition for Land and Some Implications for Public Policy. *Natural Resources Journal* 4: 252-267.

3. Ciriacy-Wantrup, S. 1968. *Resource Conservation: Economics and Policies*. 3rd ed. Berkeley, CA: University of California Press.

4. Haith, D. and R. Wu. 1989. GWLF: Generalized Watershed Loading Functions-User's Manual. Cornell University, Dept. of Agricultural Engineering, Ithaca, NY.

5. Rawls, J. 1971. *A Theory of Justice*. Cambridge, MA: Harvard University Press.

6. U.S. EPA. 2004. Water Quality Models: Enhanced Stream Water Quality Model, Windows (QUAL2E) at http://www.epa.gov/waterscience/QUAL2E_WINDOWS/. Accessed on 25 June 2004.

USING A BANK EROSION HAZARD INDEX (BEHI) TO ESTIMATE ANNUAL SEDIMENT LOADS FROM STREAMBANK EROSION IN THE WEST FORK WHITE RIVER WATERSHED

M.A. Van Eps[1], S.J. Formica, T.L. Morris, J.M. Beck, A.S. Cotter

ABSTRACT

The Arkansas Department of Environmental Quality (ADEQ), through an EPA Section 319 NPS grant, utilized a bank erosion hazard index (BEHI) and data collected from surveys of streambank profile measurements to develop a graphical model to estimate streambank erosion rates and to estimate the annual sediment load due to accelerated streambank erosion in the West Fork White River (WFWR) watershed. The WFWR watershed, located in Northwest Arkansas, has a watershed area of 31,700 Ha (79,400 ac) and is a tributary to the White River which eventually drains to the primary drinking water source for the region, Beaver Lake. Sediment is a contaminant of concern because the WFWR has been designated as impaired due to "excessive turbidity and siltation" (ADPC&E, 1998). As part of a comprehensive project to assess the various sources of sediment in the WFWR watershed, ADEQ utilized methods, developed by Rosgen (2001), to estimate sediment (bedload and suspended) contributions from accelerated streambank erosion. A streambank erosion inventory was conducted in 2002 to determine the bank erosion potential of streambanks along 64 river kilometers (40 river miles) of the main stem and tributaries of the WFWR watershed. Using ranking criteria consisting of bank angle, root depth, bank material, and other variables, streambanks were evaluated and scores were assigned based on erosion potential. Toe pins were installed at permanent survey sites and lateral erosion was measured over a one-year period. A graphical model to predict streambank erosion rates based on relationships between BEHI, near-bank sheer stress, and observed annual erosion was developed. For the WFWR watershed, it was estimated that on an annual basis, a total of 21,455 metric tons of sediment enter the river network from streambanks where accelerated streambank erosion was observed. The mass of bedload and suspended load was 7,493 metric ton/yr and 13,962 metric ton/yr, respectively.

KEYWORDS. Sediment, watershed assessment, streambank erosion, sedimentation, erosion, river stability

INTRODUCTION

Sediment is the second leading cause of impairment of monitored rivers and streams in the United States according to the U.S. Environmental Protection Agency (U.S. EPA, 2000). Sources of sediment often cited include agriculture, urban runoff, construction, and silviculture. Streambank erosion contributions of sediment have been found to constitute a majority of total sediment supplies in some watersheds (Rosgen, 1976). Lateral streambank erosion may be accelerated in systems that have been hydraulically affected by changes in land-use, removal of riparian vegetation, and/or changes in channel dimension from activities, such as, in-stream gravel removal. Accelerated lateral erosion contributes additional sediment to the stream network that can impact water quality and increase the potential for river instability. The focus of this investigation was on the West Fork White River (WFWR) in Northwest Arkansas, having a total watershed area of 31,700 Ha (79,400 ac). The State Water Quality Inventory Report of

---

[1]Arkansas Department of Environmental Quality, Environmental Preservation Division, 8001 National Drive, Little Rock, AR 72219

1998, prepared by ADEQ pursuant to section 305(b) of the Federal Water Pollution Control Act, had assessed the aquatic life use as "not supported" in 53.8 km (33.4 miles) of the WFWR. The cause cited was 'high turbidity levels and excessive silt loads.' The probable sources listed were: (1) agricultural land clearing; (2) road construction and maintenance; and (3) gravel removal from stream beds. Based on the results of the inventory report, the WFWR was added to the State's list of impaired waters known as 303(d) list by the ADEQ in 1998 (ADPC&E, 1998). The work presented in this paper was a component of an overall effort to determine the relative annual loads from various sediment sources in the WFWR watershed (ADEQ, 2004).

During the overall WFWR watershed assessment, accelerated lateral streambank erosion was identified as a potential sediment source contributing to water quality problems in the WFWR watershed. For the purposes of this paper, sediment from streambank erosion is defined as consisting of bedload, particles with mean diameters greater than 2 mm, and suspended load, particles with mean diameters equal to or less than 2 mm. Using methods proposed by Rosgen (2001), both the annual bedload and suspended load of sediment resulting from accelerated streambank erosion in the WFWR watershed was estimated. The general method used to estimate sediment loads from excessive stream bank erosion in the WFWR involved: 1) Conducting an inventory of streambanks for erosion potential based on a bank erosion hazard index (BEHI) and the near-bank shear stress (NBSS), 2) Developing a graphical model to predict streambank erosion rates in the watershed by measuring erosion rates at permanent survey sites representing the various BEHI and NBSS values observed during the streambank erosion inventory, and 3) Applying the graphical model to the streambank erosion inventory.

STREAMBANK EROSION INVENTORY

An inventory of eroding streambanks in the WFWR watershed was developed by traveling the entire length of the main stem and several kilometers of tributary streams of major subwatersheds. The banks inventoried or evaluated were streambanks where there were indications of accelerated erosion including hanging roots, exposed bank material, or sod mats at the toe of the bank. The erosion potential was estimated for each inventoried bank by estimating ratings for erosion risk (BEHI) and NBSS. BEHI variables included bank angle, bank height ratio, root density, rooting depth, percent of bank protected by boulders or logs, and bank materials. The height of the streambank was measured with a survey rod and the length of the streambank was determined using a range finder. A rating for NBSS was estimated for each inventoried streambank based on the general cross-section shape of the channel and local stream slope conditions. All of the BEHI variables and NBSS information were electronically cataloged using ArcPad GIS software on a water-resistant, Cassiopeia EG-800 handheld PC. Forms were developed for the ArcPad software which allowed for the input of the streambank BEHI and other data. The general locations of streambanks were created in the GIS environment by adding a

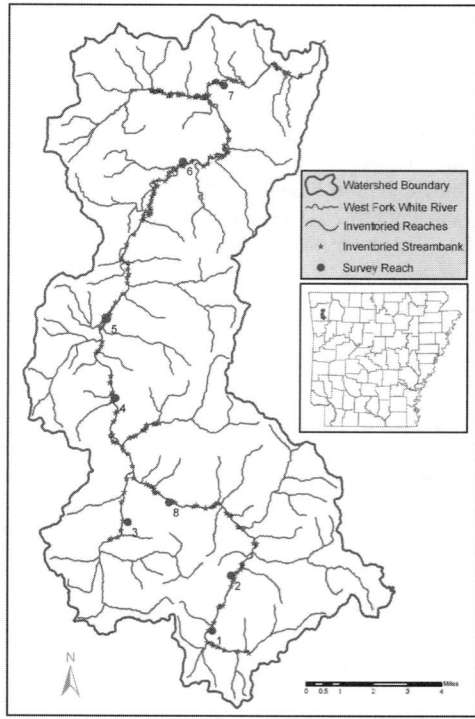

Figure 1. Locations of inventoried streambanks and reaches where erosion rates were measured in the WFWR watershed.

126

feature to a streambank line shapefile previously loaded into ArcPad. In the office, the data was downloaded from the handheld PC and then managed in ArcGIS. This approach reduced the amount of time required to transfer raw field data into a digital format. It also allowed for rapid manipulation and presentation of the results of the field work. Photographs of each of the eroding banks that were inventoried were taken using a Kodak DC5000 water resistant 2.1 MP digital camera.

During the spring of 2002, the main stem of the WFWR, 48.8 km (30.3 miles), and the lower 3.9 km (2.4 miles) of Winn Creek, a major tributary, were inventoried for eroding streambanks. The lower 2.3 km (1.4 miles) of Mill Creek was inventoried in February of 2003 and the lower 3.7 km (2.3 miles) of Town Branch were inventoried in January of 2004. A map highlighting the areas of the WFWR stream network where the streambank inventory was performed is shown in Figure 1. Once the field data had been collected, a spreadsheet was used to convert the recorded BEHI variable values of each streambank into points using the scoring system proposed by Rosgen (2001). Based on the total number of points a streambank received, a general rating of the erosion risk was assigned. As the number of points increased, the erosion risk increased. BEHI risk rating categories included low, moderate, high, very high, and extreme. Some streambanks that did not display obvious signs of active erosion were included in the inventory to allow comparison of erosion rates between streambanks of lower and higher erosion risk ratings. Evaluation of NBSS was based on rating categories that ranged from low to extreme.

Inventory Results: During the inventory process, 192 individual streambanks were evaluated. Based on the field evaluated BEHI variables, the erosion risk of the streambanks along the main stem of the WFWR and selected tributaries was estimated. Table 1 indicates the number of streambanks within each erosion risk rating category that were cataloged.

The estimated ratings for NBSS of inventoried streambanks are also shown in Table 1. The combination of a streambank erosion risk rating (BEHI) and local NBSS affects the degree of lateral migration observed for an eroding streambank. For streambanks with similar erosion risk ratings, higher NBSS will result in greater amounts of lateral erosion.

Table 1. Counts of inventoried streambanks, erosion risk rating, NBSS rating

Number of Inventoried Streambanks	Erosion Risk Rating	Number of Inventoried Streambanks	NBSS Rating
4	Low	34	Low
44	Moderate	64	Moderate
113	High	56	High
28	Very High	33	Very High
3	Extreme	5	Extreme

GRAPHICAL MODEL TO PREDICT STREAMBANK EROSION RATES

To estimate the lateral erosion rates of inventoried streambanks using the BEHI and NBSS ratings, a graphical prediction model based on physical measurements of streambank erosion was developed for the WFWR watershed. The graphical model was developed based on the methods described by Rosgen (2001).

Methods: Eight reaches were selected on the main stem of the WFWR where annual lateral erosion rates could be measured. Within these reaches, 24 permanent survey sites were established. The permanent survey sites were selected based on various combinations of BEHI and NBSS ratings representing different streambank conditions along the main stem of the WFWR observed during the streambank inventory process. The general locations of the eight reaches with permanent survey sites are shown in Figure 1.

127

Annual lateral erosion rates at the permanent survey sites were determined by installing vertical pins at the toe of the streambanks. The toe pins were installed by driving sections of 1.2 m (4 ft.) long 1.9 cm (¾ inch) thick rebar vertically into the channel bed immediately adjacent to the streambank of interest. The BEHI variables and NBSS condition for each bank where toe pins had been installed were evaluated and recorded. Using a pair of flat-edged survey rods and a framing level, the profile of the eroding streambank was surveyed by measuring the horizontal distance from the landward side of the toe pin to the streambank for various heights above the toe pin cap, depending on the shape of the bank profile (Figure 2). The toe pins were resurveyed after one year to determine annual erosion rates. The range of stream discharge during the one-year period was monitored by using data from a USGS gage station (07048550) at the downstream end of the watershed. This allowed for a determination of the discharge conditions represented by the graphical model.

Figure 2. Measurement of lateral streambank erosion in the WFWR watershed.

Bank profile survey data from 2002 and 2003 were placed into a spreadsheet and graphed. Using the graphed data, the average lateral erosion for the entire height of the streambank was calculated by taking the average of the lateral erosion data measured every two tenths of a foot of vertical elevation above each surveyed toe pin. By relating the BEHI rating, the local NBSS, and the measured erosion rate at each permanent survey site, a graphical model to predict streambank erosion rates was developed. Using the graphical model, erosion rates were predicted for all the streambanks included in the streambank erosion inventory. The volume of sediment generated due to erosion of individual streambanks was calculated by multiplying the predicted annual lateral erosion rate by the length and height of the bank from the original streambank erosion inventory.

Results: A total of 24 streambanks were surveyed in 2002 and 2003 within the eight reaches evaluated for this study. Graphical representations of the results of 2002 and 2003 streambank profiles for selected toe pins at permanent survey sites are shown in Figure 3. The graphical model for predicting streambank erosion rates in the WFWR watershed is shown in Figure 4.

Using the recorded BEHI and NBSS values determined during the streambank inventory, and the graphical model developed based on the toe pin surveys at the permanent survey station sites, the lateral erosion rates of inventoried streambanks were estimated. The maximum erosion rate predicted was 3.9 m/yr (12.9 feet/yr). The average erosion rate for inventoried banks where the rated erosion risk was moderate or greater was 0.2 m/yr (0.6 feet/yr).

The maximum measured flow during the period between surveys was 117 m³/s (4120 cfs) as measured on the WFWR at the USGS gage station. This discharge exceeds the bankfull discharge based on regional curves (ADEQ, 2002) by approximately 27%. In addition, project team members observed bankfull discharge and slightly greater than bankfull discharge at several of the reaches during the time that the 117 m³/s discharge was recorded. Since the discharge during the period between the surveys was at or slightly above bankfull, the survey data should represent erosion rates for years where bankfull flow is approached, equaled, or slightly exceeded. In years where the discharge is either well below or greatly exceeds the bankfull discharge, the graphical model will lose accuracy.

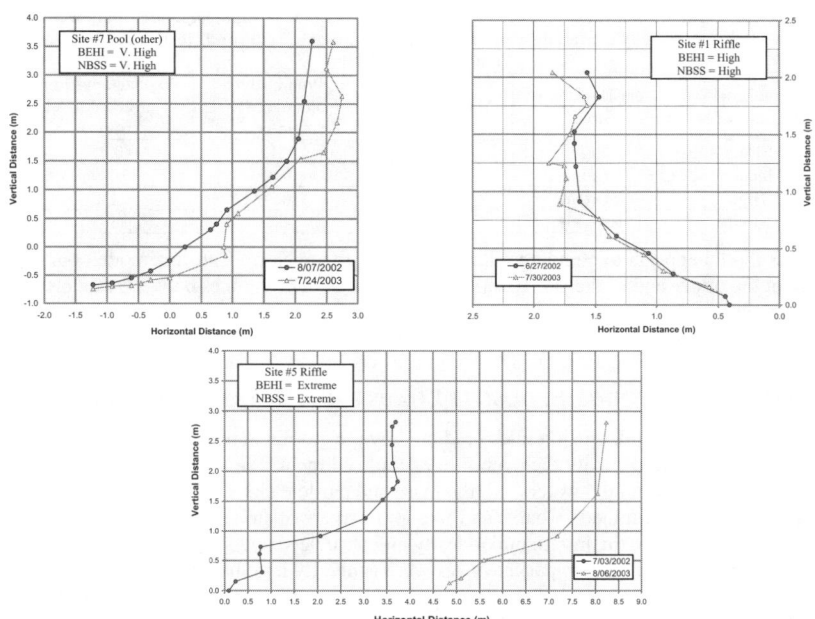

Figure 3. Examples of lateral erosion measurements taken at selected locations in the WFWR watershed.

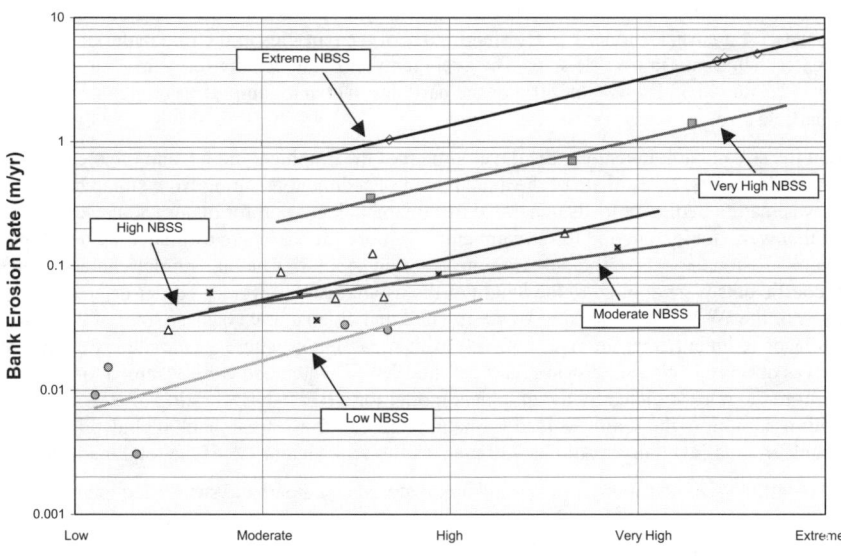

Bank Erosion Hazard Index Rating

Figure 4. Graphical model for predicting streambank erosion rates in the WFWR watershed

A comparison of the predicted erosion rates developed through this work to erosion rates based on models developed by Rosgen (2001) as well as Jessup and Harmon (2004) is shown in Table 2. Lateral erosion rates predicted by the model were less than half the rates predicted by the Colorado model for a BEHI and NBSS combination rating of moderate and high. However, for other combinations of BEHI and NBSS, erosion rates predicted by the WFWR model were higher than those predicted by the other models by a factor ranging from 1.3 to 2.8 times.

129

Table 2. Lateral erosion rates predicted by the WFWR watershed graphical model compared to erosion rates predicted by models developed in Colorado (Rosgen, 2001) and North Carolina (Jessup and Harmon, 2004)

Bank Erosion Hazard Index	Near-Bank Shear Stress	Predicted by WFWR Model (m/yr)	Predicted by Colorado Model (m/yr)	Predicted by North Carolina Model (m/yr)
Moderate	High	0.05	0.12	0.05
Moderate	Extreme	0.60	0.45	0.8
High	Extreme	1.4	0.76	1.1
Extreme	Extreme	7.0	4.27	3.0

The WFWR model appears to compare relatively well with the North Carolina model, but begins to diverge at the upper end of the BEHI and NBS categories. Differences between models may be due to variation in watershed size or characteristics of the various physiographic regions, such as, soils and precipitation.

ESTIMATE OF ANNUAL SEDIMENT LOADS FROM STREAMBANK EROSION

In order to relate streambank erosion rates predicted by the graphical model to water quality impacts and to be able to compare the overall impact of sediment from streambank erosion to other sources of sediment in the watershed, the mass and particle size distribution of sediment in the streambanks was determined. Coarse and fine grain layers of streambanks in the watershed were sampled to determine the in-situ bulk density and particle size distribution. Fine grain sediment streambank layers where particle sizes were generally less than 2 mm in size, were sampled using a hammer-driven Shelby tube. To characterize streambank layers generally composed of coarse materials, methods developed by Brye et. al.(2004) were utilized to collect samples. The in-situ bulk density and particle size distributions were then related to inventoried streambanks by comparing sampled bank strata to photos of banks inventoried during the streambank erosion inventory process. The average in-situ bulk density for fine grain material samples was 1.4 g/cm³ (1.2 ton/yd³). By weight, 8% of the particles in the fine material samples were greater than 2 mm in particle size. The average in-situ bulk density for coarse samples was 2.6 g/cm³ (2.2 ton/yd³). By weight, 80% of the particles in coarse samples were greater than 2 mm in particle size.

Based on the streambank inventory, the development of the graphical model, and the measurement of in-situ bulk density, an estimate of the annual load of sediment resulting from streambank erosion was made. Sediment loads generated by streambank erosion for major tributaries of the WFWR that were not included in the streambank inventory process were estimated by developing streambank erosion export coefficients from inventoried tributaries having similar characteristics. Export coefficients were applied to the length of the tributaries that were 3rd order or greater streams. For the WFWR watershed, it was estimated that on an annual basis, a total of 21,455 metric tons of sediment enter the river network from streambanks where accelerated streambank erosion was observed. Natural erosion rates for the WFWR watershed were assumed to be equivalent to the rate predicted by the graphical model for a BEHI-NBSS rating of low-low. Using this assumption, the sediment load for natural erosion from streambank included in the streambank erosion inventory would be 739 metric tons/yr, which is 3% of the total load estimate.

Using the particle size distribution of streambank materials, bedload and suspended loads were determined and are shown in Table 3. The mass of bedload and suspended load from streambanks included in the watershed inventory was 7,493 metric ton/yr and 13,962 metric ton/yr, respectively. Suspended sediment represented 65 percent of the estimated total sediment load. The sediment load that consisted of particles less than 0.02 mm in size was 6,563 metric ton/yr.

The estimated sediment load resulting from erosion of streambanks along the main stem of the WFWR that were included in the inventory was 16,812 metric ton/yr. Of that amount, 11,227 metric tons or 67% of the load consisted of sediment 2 mm or less in size. 80 percent of the estimated suspended sediment load for the watershed resulted from erosion of streambanks along the main stem of the WFWR that were included in the inventory. One reach along the main stem approximately 1.6 km long contributed 25% of the total load of particles less than 2 mm in size.

130

Table 3. Estimated sediment loads from eroding streambanks in the WFWR watershed

Main Stem and Tributaries	Length		Watershed Area		Sediment Load (metric ton/yr)		
	km	mi	km²	mi²	Particles > 2 mm	Particles ≤ 2 mm	Total
Main Stem	48.7	30.3	321.16	124	5,585	11,227	16,812
Wilson Branch	2.1	1.3	8.5	3.3	18	241	259
Dye Creek	3.5	2.2	9.4	3.6	166	190	356
Riley Creek	4.0	2.5	10.2	4.0	196	94	290
Cato Springs	3.1	1.9	11.9	4.6	26	347	374
West Mtn Creek	4.1	2.6	12.1	4.7	196	224	420
Sinclair Creek	3.6	2.3	13.1	5.1	227	127	355
London Creek	3.7	2.3	13.3	5.1	174	199	372
Rock Creek	4.2	2.6	15.3	5.9	199	228	426
Hutchins Creek	5.1	3.2	15.3	5.9	244	279	523
Mill Creek	4.1	2.5	19.0	7.3	195	223	418
Town Branch	4.2	2.6	30.2	11.7	35	474	509
Winn Creek	4.8	3.0	37.4	14.4	231	110	342
					7,493	13,962	21,455

CONCLUSION

A bank erosion hazard index (BEHI) and estimated near-bank shear stress (NBSS) was used to inventory eroding streambanks in the WFWR watershed. Twenty-four permanent survey sites were established on 8 reaches to develop a relationship between BEHI and NBSS ratings and lateral streambank erosion rates. From these relationships, a graphical model to predict streambank erosion rates was developed for the WFWR watershed. The graphical model should be appropriate for use during years where the bankfull discharge is approached, equaled, or slightly exceeded. The graphical model was used to estimate erosion rates for inventoried streambanks in the WFWR watershed. Streambank materials were characterized to determine the in-situ bulk densities of various layers of the streambanks. Based on the data collected, a total of 21,455 metric tons/yr of sediment enter the WFWR from streambanks where accelerated streambank erosion was observed. The sediment load was estimated to be 739 metric tons/yr if the same banks were in a stable condition. The model developed through this work yielded lateral streambank erosion rates as much as 2.8 times greater than those predicted by existing models that were developed for other physiographic regions. The differences between the models indicate the importance of collecting data specific to the physiographic region of interest. The work performed as part of this study was conducted in one 124 mi² watershed that was in portions of the Boston Mountain and Springfield-Salem Plateau physiographic regions. The model may be effective for predicting streambank erosion rates outside of the WFWR watershed if applied to the same physiographic regions. However, more streambank erosion data should be collected to validate and improve the model for other watersheds within those physiographic regions.

REFERENCES

1. Arkansas Department of Environmental Quality. 2004. West Fork White River Watershed - Data Inventory and Nonpoint Source Pollution Assessment. Little Rock, Arkansas

2. Arkansas Department of Environmental Quality, Included as course materials for Applied Fluvial Geomorphology training course, Fayetteville, AR, 2002.

3. Arkansas Department of Pollution Control and Ecology. 1998. Water Quality inventory Report. Agency Report Prepared pursuant to Section 305(b) of the Federal Water Pollution Control Act.

4. Brye, K.R., T.L. Morris, D.M. Miller, S.J. Formica, M.A. Van Eps. 2004. Estimating bulk density in vertically exposed stoney alluvium using a modified excavation method. *Journal of Environmental Quality*. In Press.

5. Jessup, A. and W. Harmon. 2004. USDA-NRCS. Personal Communication regarding research findings from work in progress monitoring streambank erosion in North Carolina

6. Rosgen, D. L., The Use of Color Infrared Photography for the Determination of Suspended Sediment Concentrations and Source Areas. In: Proceedings of the Third Inter-Agency Sediment Conference, Water Resources Council. Chap. 7, 30-42. 1976.

7. Rosgen, D L. A Practical Method of Computing Streambank Erosion Rate. Proceedings of the Seventh Federal Interagency Sedimentation Conference. Vol. 2, pp. II - 9-15, March 25-29, 2001, Reno, NV

8. Schueler, T. R. and H. K. Holland. 2000. The Practice of Watershed Protection. Center for Watershed Protection. Article 14.

9. U.S. Environmental Protection Agency (USEPA). 2000. National Water Quality Inventory: 2000 Report to Congress. http://www.epa.gov/305b/2000report/. Last Accessed July 9, 2004.

10. Wildland Hydrology, Inc: Research and Educational Center for River Studies. River Assessment: Field Book. Pagosa Springs. 2001.

WEST FORK WHITE RIVER WATERSHED - SEDIMENT SOURCE INVENTORY AND EVALUATION

S.J. Formica[1], M.A. Van Eps, M.A. Nelson[2], A.S. Cotter, T.L. Morris, J.M. Beck

ABSTRACT

Understanding the causes and sources of water quality problems is critical to developing practical solutions and long-term strategies that can result in watershed restoration. The West Fork of the White River (WFWR) located in Northwest Arkansas has been identified by the Arkansas Department of Environmental Quality (ADEQ) as an impaired stream and has been placed on the Arkansas 303(d) list, because its "aquatic life" use designation was not being supported due to "high turbidity levels and excessive silt loads." Through the U.S. Environmental Protection Agency's 319 grant program, administered by the Arkansas Soil and Water Conservation Commission, a comprehensive watershed assessment was performed to identify probable sources of contamination and to estimate pollution potential of identified sources. Field and GIS data along with modeling methods were used to estimate sediment source loads from stream bank erosion, roads, pastures, and other land uses in the watershed. Also, a long-term, strategic water quality monitoring program was initiated that included the installation of a continuous water quality monitoring station and the collection of baseline data near the mouth of the river. Data collected at the monitoring station included flow; turbidity; and total suspended solids. Management practices that will effectively control sediment loading to the WFWR were identified. The results of this study are being used by a local stakeholder group to prioritize source reduction efforts and to develop restoration strategies as part of a WFWR watershed management plan.

KEYWORDS. Sediment, watershed assessment, sediment sources, erosion, sediment loads, land use

---

[1] Arkansas Department of Environmental Quality, Environmental Preservation Division, 8001National Drive, Little Rock, AR 72219

[2] Arkansas Water Resource Center, 112 Ozark Hall, University of Arkansas, Fayetteville, AR 72701

Channel Alternatives to Enhance Ecosystem Function of Drainage Canals in Eastern North Carolina

R.O. Evans   R.D. Hinson   R. Johnson   M. Doxey   K.L. Bass   J.T. Smith

Abstract

Drainage is an important and necessary component of land management in eastern North Carolina where more than 50 percent of soils require improved drainage for efficient production and other uses. For more than 250 years, drainage practices have focused on straightening and deepening natural channels to increase their hydraulic capacity and minimize upstream flooding. Today, there are very few un-channelized streams remaining. In most cases, traditional channel improvements have disassociated the channel from the natural floodplain degrading riparian floodplain ecological functions. Wetness continues to be a major concern to many landowners, but intensive drainage systems sometimes remove more water than necessary especially during drier periods. Pilot studies were begun in 1994 to investigate, evaluate, and demonstrate alternative channel design geometries and management to enhance ecological and water quality functions while maintaining the necessary drainage function. Channel alternatives included: establishment of in stream wetlands, lowering of the floodplain to reconnect the channel with the floodplain, redesign of channels using natural channel design principles, and establishment of conservation easements to encourage establishment of perennial riparian vegetation. This paper discusses several recent projects that have been implemented to provide better management of drainage water and to restore or enhance ecological functions of large drainage canals.

Early Drainage

Settlers migrated to North Carolina from Jamestown, Va. and established one of the first permanent settlements along the Chowan River in 1626. The explorations from Jamestown to eastern North Carolina required travel around or through the Dismal Swamp. As early as 1728, William Byrd surveyed the Dismal Swamp and saw the opportunity to drain it noting that the swamp was higher in elevation than the surrounding land (Boyd, 1929). In 1763, George Washington along with five associates formed the "Adventurers for Draining the Great Dismal Swamp" (also know as the Dismal Swamp Canal Company) and obtained 16,188 ha with the intention of draining them. They dug one 7.5 km canal to Lake Drummond known as the Washington Ditch (Brown, 1970). The Dismal Swamp canal was proposed in 1784 to provide navigation from the Chesapeake Bay to the Albemarle Sound. Construction began in 1793 and was completed in 1805 (Brown, 1970).

While the Dismal Swamp Canal was dug primarily to provide navigation, it had a profound effect on water movement through and subsequent drainage of the Dismal Swamp. The natural overland flow through the swamp was generally from northwest to southeast. The spoil material from the canal placed along one side of the channel formed a dyke that effectively blocked the natural overland flow. As a result, lands to the west of the canal became wetter while lands to the east became drier (Ruffin, 1861). With the construction of additional ditches, the lands to the east were soon extensively cultivated. Edmond Ruffin traveling the area in 1836 described an extensive parallel ditch drainage system that is still widely utilized in eastern North Carolina today (Ruffin, 1861).

The authors are Robert O. Evans, Professor, Kris L. Bass, Extension Associate, and Jonathan T. Smith, Extension Assistant, Department of Biological and Agricultural Engineering, North Carolina State University; R. Dwane Hinson, District Conservation, USDA Natural Resources Conservation Service; Rodney Johnson, Albemarle RC & D Coordinator, USDA-NRCS; and Mike Doxey, District Technician, Currituck Soil and Water Conservation District.

Today, there are very few un-channelized streams remaining. In most cases, traditional channel improvements have disassociated the channel from the natural floodplain degrading riparian floodplain ecological functions. Woody riparian vegetation along the sides of the streams have been removed and ditch banks are routinely mowed to provide access for periodic clean-out and removal of silt. In many channelized streams, most storm flows are now confined predominately to the main channel. The riparian floodplain that once routinely remained soaked or inundated during the winter and spring for months at a time now only flood during very large storms. In bypassing the floodplain, there is less opportunity for potential pollutants in the drainage water to be filtered and assimilated. While wetness is still a major concern to landowners, intensive drainage systems sometimes remove more water than necessary especially during drier periods, leading to over drainage.

Drainage Canal Alternatives

Pilot studies were begun in 1994 to investigate, evaluate, and demonstrate alternative channel design geometries and management to enhance ecological and water quality functions while maintaining the necessary drainage function. Channel alternatives have included establishment of in stream wetlands, lowering of the floodplain to reconnect the channel with the floodplain (priority II restoration), redesign of channels using natural channel design principles to reconnect the channel with the natural floodplain (priority I restoration), and establishment of conservation easements to eliminate traditional ditch bank mowing to encourage establishment of perennial riparian vegetation.

Tulls Creek Project – Currituck County

The first drainage alternative project was initiated in the Tulls creek watershed in Currituck County in 1995. The project involved a one mile reach of the main canal system draining approximately 80 ha of cropland. One reach of the canal (450 m) was managed in the traditional "free" drainage mode, a second reach was managed in the "controlled" drainage mode, and the third reach was planted with a variety of wetland plants (Fig 1) to provide a combination in-stream wetland/controlled drainage system. Approximately 4500 plants were planted in the wetland section. Based on visual observations made during September, 1996, over 90 percent of all species except cattail survived. Annual re-growth of wetland vegetation was excellent throughout the evaluation period (1997 - 1999. The outlet of each reach was instrumented to continuously measure outflow (Fig 2). Grab samples were collected monthly to evaluate treatment effects on water quality. In general, nitrogen concentrations were higher while phosphorus concentrations were lower with drainage control both with and without the addition of wetland plants. Total flow, phosphorus and sediment transport were significantly lower with both controlled drainage treatments. The addition of plants did not appear to provide an additional water quality benefit other than an improvement in water clarity over drainage control alone. Habitat benefits may have been enhanced by the plants; but, habitat benefits were not evaluated. This initial project demonstrated that it was possible to maintain wetland plants in drainage ditches without adversely impacting the drainage performance of the ditch.

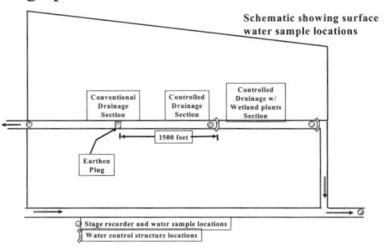

| Figure 1. In-stream wetland reach of Tulls Creek project. | Figure 2. Schematic of water management treatments and sample locations. |

Liza's Bottom – Chowan County, Town of Edenton

The Edenton urban project, initiated in 1997, involved construction to lower the floodplain along approximately 400 m of stream known as Liza's Bottom and create approximately 1 ha of in stream wetlands. The stream carries drainage waters from agricultural lands and runoff from commercial and urban sources such as a solid waste processing site and a former farm supply facility. The channel was altered prior to 1900, so that the riparian floodplain was rarely functional. A 1 ha wetland was constructed in April-May, 1997. The wetland was built by excavating the hydraulically dysfunctional floodplain area down to the stream base-flow level and raising the stream bottom (Fig 3). The soils found in the floodplain and used for the wetland substrate were variable, with some reduced and high in organic matter and others clayey in content. Islands were built in the interior to minimize transportation of cut/fill material and create a more sinuous flow path. A low head, wooden bulkhead was installed at the outlet to maintain water depths of 0.1m to 0.5m (6-18"). The wetland bottom was graded for a mixture of shallow and deeper pool areas. Native plants were used in the wetland and transplanted on a 1m x 1m spacing. The wetland intercepts drainage waters from approximately 240 hectares of surrounding watershed. One hundred and sixty hectares are attributed to agricultural and natural forested area, sixty ha to urban area, and twenty ha to intensive commercial areas. The resulting watershed/wetland area ratio is 100:1, which is less than half the minimum size typically recommended in the literature (Scheuler, 1992).

Figure 3. Constructed in-stream wetland, Edenton, N.C. construction was completed May,

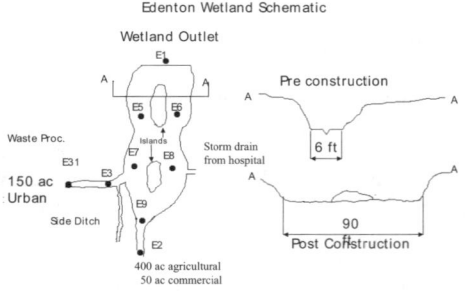

Figure 4. Schematic of monitoring locations at Edenton in-stream wetland.

The wetland was instrumented to continuously measure inflow and outflow (Fig 4). Flow measurements were made at the two main inlet streams (E2 and E3) and at the wetland outlet (E1) using continuous water level recorders. Stage measurements were combined with weirs and calibrated discharge curves to generate a continuous mass flow record. A comparison between predicted drainage and runoff from the watershed and measured wetland outflow volumes lead to the conclusion that the overall flows observed were reasonable for use with concentration data to predict nutrient mass transport (Bass, 2000). Water quality samples were taken over time and at various flow stages. Background water samples were acquired from January 1996 until construction began in May, 1997. After planting, grab samples were taken on bi-weekly intervals. Automatic water samplers were utilized at the two main inlets and at the outlet. Over 1000 samples were acquired during the monitoring period.

During base flow conditions, attenuation within the Liza Bottom constructed wetland is approximately 7 days with no significant attenuation during large storm flows. Prior to wetland construction, there were no changes in ammonium (NH4) concentrations between up and downstream monitoring locations while nitrate showed a 40% decrease indicating some nitrate was likely being removed along the stream by denitrification. Over the four-year monitoring period which ended December, 2000, NO$_3$-N concentrations were reduced by 60%, NH$_4$-N by 30%, and TKN by 9.5% resulting in a flow-weighted total nitrogen reduction of 20% . Yearly and seasonal means indicated that significant improvements in NO$_3$-N and NH$_4$-N concentrations can be achieved with relatively small wetlands. Initially, total- and ortho- phosphorus concentrations were unchanged by the wetland. However, by the end of year 2 phosphorus concentrations began to increase. During the last two years of monitoring, there was a net increase in phosphorus discharge from the wetland of 55%. This project, located near the county high school, has provided living labs for both biology and physical science classes.

Table 1. Overall nutrient reductions on concentration and mass basis, Edenton wetland

	NO$_3$-N	NH$_4$-N	TKN	TN	TP	OP
Concentration basis	60%	30%	9.5%	20%	-55%	-55%
Mass basis	55%	16%	6%	18%	-50%	-50%

Guinea Mill Watershed – Currituck County

Currituck County is one of the fastest growing counties in North Carolina resulting largely from urban sprawl originating in Tidewater Virginia. The Guinea Mill watershed project (approximately 2,000 ha) was initiated in 1999 to address drainage and water quality issues arising from rapid urbanization of a predominately rural county (Fig 5). Over half the watershed is projected to be converted from farmland to residential by 2010. Drainage systems that were generally adequate for agricultural land uses are not adequate for residential development (Fig 6).

Figure 5. Residential development is rapidly displacing agricultural crop fields in Currituck County.

Figure 6. Agricultural drainage is often inadequate after urbanization.

As part of the Guinea Mill project, a tax-supported Water Management Service District, one of the first in North Carolina, was formed by the Currituck County commissioners. The purpose of the Service District is to generate revenue to assure the future maintenance and persistence of the project components. There are 289 parcel owners within the watershed with the majority in eight (8) major subdivisions. Permanent conservation easements involving 49 acres along both sides of the canal were purchased and are managed by the county utilizing an advisory board comprised of five landowners in the watershed. The advisory Board is charged with the duty of investigating, studying and making recommendations to the Board of Commissioners pertaining to the construction, enlargement, improvement, maintenance, operation and regulation of the Service District. A county ordinance was established requiring all new subdivisions and any landowners encroaching on the easement and canal to submit a plan for that encroachment (i.e., culvert, drainage swale, etc) to the Service District prior to installation.

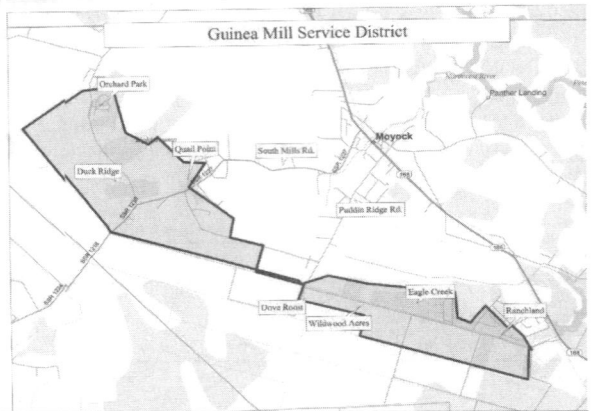

Figure 7. Shaded area represents guinea Mill Water Management Service district.

Guinea Mill Canal dissects the watershed and runs the entire length of approximately 13 km (Fig 7). The fourteen kilometers of riparian buffers were established along the Guinea Mill Canal. Where farming use to occur next to the stream bank (Fig 8), buffers are established with vegetation maintained 0.3 to 1 m high (Fig 9). In-stream constructed wetlands (Fig 9) were installed on 3.4 ha within the Guinea Mill Canal. In the wetland section, the canal was widened from 10 to 20 m. Two in-stream wetland cells were constructed one approximately 2 km and the second about 1 km in length. The wetland cells were planted with a variety of wetland plants on a 1 m by 1 m spacing.

Figure 8. Upstream section of Guinea Mill with neither buffers or in-stream wetlands.

Figure 9. Lower section of Guinea Mill after installation of riparian buffer and instream wetland. Note path was moved away from canal bank.

A rock weir water control structure was installed to enhance hydrologic function at low flows. The rock weir (1 m high by 1.2 m top width by 21 m length) was located just downstream of a 15 ha hardwood swamp (Fig 10). The hydrology of the swamp had been significantly altered by channelization of Guinea Mill Canal. The rock weir raised the base flow elevation by approximately 1 m, restoring some hydrologic function to the riparian swamp (Fig 11). Annual inspections of all conservation easements are made by the Advisory Board and the Currituck SWCD Board with their respective reports submitted to the County commissioners.

Figure 10. Rock weir soon after installation.

Figure 11. Rock weir raises upstream water level at base flow enhancing hydrologic function of upstream swamp.

Newland Watershed Project – Pasquotank County

A similar tax supported Service District project was initiated in the Newland Watershed in Pasquotank County in 1998. The watershed Service District encompasses approximately 7,000 ha. The US Hwy158 Canal and Shepard Ditch are the primary outlets for the southeastern section of the Dismal Swamp Wildlife Refuge. In recent years, landowners downstream of the refuge have been subjected to flooding resulting from failure and overtopping of the refuge dyke.

This project involved development of a conceptual water management plan for the refuge that balanced the water management needs of the refuge with the drainage and water quality needs of the downstream landowners and citizens. Debris was cleared and snagged in the upper Pasquotank River which is the outlet for both the 158 and Shepard Ditch canals. Both canals were excavated to stabilize stream banks. The ditch bottoms were excavated to create a ledge for establishment of 2.6 acres of in stream constructed wetlands (Figs 12 and 13).

Figure 12. Section of Shepard Ditch showing bank shaping, wetland ledge and rock weir soon after installation.

Figure 13. Section of Shepard Ditch during dry period after establishment of wetland plants on constructed ledge (right bank).

Five associated rock weir water control structures (Figs 14 and 15) were installed to enhance base flow hydrology and ecological function. At the rock weirs, the channel is widened to maintain the same cross sectional area as existed prior to the project so that the channel capacity will not be reduced at high flows. Vegetative buffers were established along 4.2 km of the 158 Canal and 3.6 m on Shepard Ditch. Annual inspections are made by the Advisory Board and the Pasquotank SWCD Board with their respective reports submitted to the County commissioners.

Figure 14. One of five rock weirs installed to elevate the water level at base flow and enhance hydrologic function.

Figure 15. Vegetative buffers were established along ditch banks and the channel was widened to maintain capacity at the rock structures.

Edenton Airport and Industrial Park – Chowan County

The Edenton Airport and Industrial Park restoration and enhancement project in Chowan County was initiated in 2000. As part of the construction and development of the Edenton Army Base in the 1930's, the lower stream reach of the watershed was channelized with the spoil deposited in the adjacent floodplain and a short circuit cutoff constructed that shortened the flow path of drainage water to the Albemarle Sound by more than a kilometer. This project involved three hydrologic enhancements. The first involved restoration of stream and riparian floodplain functions in the lower stream segment. Spoil piles were removed to restore some hydrologic functions to the floodplain (Fig 16). Re-growth maples were replaced with cypress and mixed bottomland hardwood species. At several locations, the straightened, channelized stream was re-routed back through it's original floodplain (Fig 17).

Figure 16. Restored swamp and floodplain reach of Edenton airport project.

Figure 17. View of lower floodplain after third growing season.

The second hydrologic enhancement involved construction of a 300 m reach of stream and riparian floodplain. Final design consisted of an 250 m stream/wetland valley with floodplain width varying from 7 to 10 m. The stream was designed to meander within the wetland valley (Fig 18). The stream/wetland system was designed to be from 0.5 to 1.5 below original grade. Lastly a series of 3 storm water wetlands were constructed between the constructed and restored stream reaches (Fig 19). Hydrologic and water quality functions of the three wetland system continues to be evaluated.

140

| Figure 18. Constructed stream and floodplain of small headwater stream. | Figure 19. Constructed stormwater wetland in series in drainage system. |

Golf Course Stream and Wetlands – Chowan County

An innovative approach was initiated in 2003 to manage poor drainage and stormwater (Fig 20) at the Chowan Golf Course. Existing drainage ditches (Fig 21) were redesigned utilizing natural channel design concepts and interconnected to a network of constructed stormwater wetlands (Fig 22). All stormwater conveyance channels and wetlands are protected by a permanent conservation easement under the control of the County Commissioners.

| Figure 20. Poor drainage and ponding at Chowan Golf Club prior to project. | Figure 21. Typical drainage ditch on Chowan Golf course prior to stream project. |

Nutrient and water management plans are being developed for the golf course. Once the management plans are adopted, modifications will require the approval of the County Commissioners. The wetland system is designed to treat and retain the first two inches of runoff which is recycled back onto the course through irrigation resulting in a nearly "closed" system except for very large events. The hydrologic and water quality performance of this system is being evaluated over the next three years. Construction was completed in May and the course is scheduled to reopen in September.

| Figure 22. Constructed stream channel with riparian floodplain after construction. | Figure 23. Constructed stream and wetland one month after planting. |

Summary

Drainage is an important and necessary component of land management in eastern North Carolina. For more than 250 years, drainage practices have focused on straightening and deepening natural channels to increase their hydraulic capacity and minimize upstream flooding. Today, there are very few un-channelized streams remaining. Pilot studies were begun in 1994 to investigate, evaluate, and demonstrate alternative channel design geometries and management to enhance ecological and water quality functions while maintaining the necessary drainage function. Channel alternatives have included establishment of in stream wetlands, lowering of the floodplain to reconnect the channel with the floodplain, redesign of channels using natural channel design principles to reconnect the channel with the natural floodplain, and establishment of conservation easements to eliminate traditional ditch bank mowing to encourage establishment of perennial riparian vegetation. This paper discussed several recent projects that have been implemented to provide better management of drainage water and to restore or enhance ecological functions of large drainage canals.

The projects discussed herein were supported by a combination of grants from the North Carolina Clean Water Management Trust Fund, N.C. Department of Environment and Natural Resources 319(h) Program; N.C. Agricultural Cost Share Program, N.C. Conservation Reserve and Enhancement Program, along with local government and landowner in kind contributions.

References

Bass, K.L. 2000. Evaluation of a small in-stream constructed wetland in North Carolina's Coastal Plain. MS Thesis. N.C. State University, Raleigh. 113.p.

Boyd, W.K., ed., 1929. William Byrds' histories of the dividing line betwixt Virginia and North Carolina. The North Carolina Historical Commission, Raleigh, NC. 341 p.

Brown, A.C. 1970. The Dismal Swamp Canal. Norfolk County Historical society, Chesapeake, Va. 234 p.

Ruffin, E. 1861. Sketches of lower North Carolina and the similar adjacent lands. Institute for the Deaf and Dumb, Raleigh, NC. N.C. State Univ. Microfiche No. 17,243-17,246a. 296 p.

Scheuler, Thomas R. 1992. Design of Stormwater Wetland Systems: Guidelines for Creating Diverse and Effective Stormwater Wetlands in the Mid-Atlantic Region. Metropolitan Washington Council of Governments, Washington,DC 20002-4201.

THE EFFECTIVENESS OF A COMBINATION WEEP BERM-GRASS FILTER RIPARIAN CONTROL SYSTEM FOR REDUCING FECAL BACTERIA AND NUTRIENTS FROM GRAZED PASTURES

J.R. Barnett[1], R. C. Warner[2], and C. T. Agouridis[3]

ABSTRACT

Much of the pollution in our lakes and streams has been attributed to agricultural practices, with bacteria, nutrients, and sediment being the primary pollutants. Runoff from grazed pastures and manure-applied lands can contain high concentrations of fecal coliforms and nutrients. Riparian grass filters have proven successful in reducing pollutants reaching streams and wetlands but effectiveness is dependent upon achieving shallow, uniform flow. Most landforms have undulations that will concentrate flow, reducing grass filter efficacy. In an attempt to enhance the effectiveness of the riparian zone as a pollution control area, a low-cost control system, consisting of a combination weep berm-grass filter, was developed and tested under simulated continuous grazing and rotational grazing practices. Three replicate tests were conducted on three field plots subjected to simulator-generated rainfall. Plots were instrumented to enable monitoring of surface runoff up-gradient of the weep berm and down-gradient of the grass filter. The system achieved average reductions in fecal coliform concentrations (99%), total nitrogen (87%), total phosphorus (44%), and total suspended solids (90%). The control system also reduced peak runoff rate from high intensity, short duration rainfall events by 92%. Based on these results, the weep berm-grass filter system affords the following advantages over simple grass filters: 1) peak flows are highly dampened, 2) due to short-term storage, some settling and infiltration occurs above the berm, and 3) flow is passively and uniformly released through the weep berm to the grass filter at a slower rate, thereby allowing the grass filter to perform more effectively.

KEYWORDS. stream and wetland protection, riparian BMP, management of riparian zones.

INTRODUCTION

The National Water Quality Inventory: 2000 Report stated that 39% of rivers and 45% of lakes surveyed across the United States had pollution problems, primarily due to high levels of bacteria, nutrients, and sediments. The US EPA has associated these pollutants with agricultural activities and hydrologic modifications (USEPA, 2000).

To effectively meet the goals of the Clean Water Act, non-point source pollutants must be properly managed. Best Management Practices (BMPs), such as grass filter strips, have been shown to improve water quality; however, a 4.5 m grass filter strip alone cannot sufficiently improve runoff to meet water quality standards (Coyne et al., 1998). Additional BMP practices are necessary to sufficiently improve water quality. The addition of a simple structure, such as a weep berm, may prove to be effective in reducing runoff contaminants.

Grass Filters

Grass filter strips have been studied extensively for their effectiveness in improving the water quality in runoff. The parameters identified to impact grass filter performance are vegetation type

---

[1] Engineer Associate, [2]Extension Professor, and [3]Engineer Associate, Biosystems and Agricultural Engineering Department, C. E. Barnhart Building, University of Kentucky, Lexington, KY 40546-0276.

and height, terrain area and slope, soil type and infiltration rate, rainfall intensity and duration, and antecedent moisture conditions (Deletic, 2001). A study by Gharabaghi, et al. (2001) suggests that the first five meters of a grass filter are the most efficient, providing 95% of the removal of aggregates larger than forty microns. A 4.5 m filter strip can reduce sediment concentrations by 96%, fecal coliform concentrations by 75% and fecal streptococci by 68% (Coyne, 1998). Dillaha et al. (1989) found that a 4.6 m strip reduced total suspended solids by 70%, phosphorus by 61%, and nitrogen by 54%. Although the pollutant reduction within a grass filter is significant, water quality standards may still be exceeded in the effluent, especially in areas where landform undulations concentrate flow, reducing the effective area of the grass filter.

Weep Berm

A weep berm is a small berm constructed perpendicular to the direction of runoff that performs as a temporary detention structure.

A weep berm – forested riparian area was used to retain sediment, reduce peak flow and infiltrate runoff at an active construction site at Alpharetta, Georgia (Warner and Collins-Camargo, 2001). Monitoring was conducted only immediately up and down-gradient of the weep berm since, by design, all discharge emanating from the weep berm infiltrated within the forested area. For the storm event of 3.7 inches occurring August 31 through September 1, 2000 the weep berm's effluent suspended sediment concentration (SSC) ranged between 78 and 30 mg/L. It was noted that runoff originating from small, highly intense rainfall events were contained by the berm and a large reduction in sediment loads was observed. The infiltrated water was incorporated as groundwater, which increased stream baseflow and improved conditions for aquatic life. Since the seep berm is a passive dewatering system, settled sediment was cost effectively removed.

A weep berm – grass filter control system was installed down-gradient of two elongated gradient sediment ponds placed in series at a site receiving sediment-laden flow in the mountains of Peru. The ponds were equipped with a flocculation system that was used during the wet season to enhance sediment-trapping efficiency. The reported effluent sediment concentration entering the adjacent stream was always below 50 NTU and averaged approximately 15 NTU during storm events. The passive weep berm – grass filter system was viewed as a control system that successfully achieved effluent requirements of the project (Warner and Torrealba, 2003).

Some parameters of concern when considering a grass filter as a single control are vegetation type, terrain characteristics, soil type, rainfall characteristic, and antecedent conditions (Deletic, 2001). The overall efficiency of the grass filter will be dependant on the area and slope of the land, the amount and rate of infiltration, the type and height of grass, the intensity and duration of the storm event, and the antecedent moisture condition of the soil.

Design Considerations for Combination Weep Berm –Grass Filter Control System

The design of a combination weep berm – grass filter control system to protect the water quality of a stream requires consideration of numerous parameters. The spatial location of the control system is a major consideration and should be integrated with stream geomorphic characteristics and cost components specifically associated with lost, or reduced use, of agriculturally productive lands. Design parameters include the design rainfall event, flow characteristics of the stream, contributing watershed area, pollutant load and concentration, type of pollutant to be treated, desired treatment efficiency, soil type, height and length of the weep berm, location, type and configuration of the outlets, and internal check dam spacing and height. It is important to understand the hydraulic properties of soils that exist both up and down-gradient of the weep berm. The infiltration characteristics of the soils help to define the period of time required to dewater the detained portion of the runoff. The infiltration rate will affect treatment efficiency, sizing of the weep berm and length of the down-gradient grass filter.

Preliminary design guidance was obtained from two combination weep berm – riparian filter systems that were implemented at an active construction site in Georgia and along the access road of a copper and zinc mine in Peru (Warner and Collins-Camargo, 2001 and Warner and Torrealba, 2003). At the Georgia site an experimental weep berm was designed, constructed, and tested to determine its performance with respect to sediment removal. The weep berm was designed to provide a controlled release of discharge to a forested riparian zone. Four discharge configurations were investigated: (1) a perforated riser wrapped in stone, (2) a perforated riser wrapped with a large-opening geotextile and stone, (3) a fixed siphon, and (4) an internal sand filter located within the side wall of the berm itself. A simple straight pipe configuration was employed as the Peruvian site. Approximately 60, 25-cm PVC pipes were located at various elevations along the length of the weep berm in Peru. At both projects the design storm to be retained with controlled release through the various outlet configurations was the 2-year 24-hour event. Larger storm events would be partially retained with the excess peak flow overflowing the weep berm. The weep berm was stabilized by vegetation on the contour; therefore, the overflow would simply enter the down-gradient riparian zone without eroding the weep berm.

The design discharge rate from the weep berm to the down-gradient natural filter is a function of the infiltration rate and length (in the flow direction) of the filter and the desired systems effectiveness. The highest efficiency is often a function of the amount of discharged water that is infiltrated within the filter. Generally, the longer the filter and the higher the infiltration rate in the filter, the better the performance. For the Georgia construction site the riparian zone consisted of a second growth mixed forest that was well established and had an infiltration rate that was expected to exceed 7.5 cm/hr. The Peruvian site had a natural grass filter that ranged in length from approximately 30m to 60m and had an infiltration rate estimated to be 0.5 cm/hr. If the discharge rate exceeds the infiltration rate then removal efficiency becomes a function of the filtering action of the riparian material.

The length and slope of the weep berm are two additional design parameters of concern. Longer weep berms enable discharging detained runoff to larger riparian filter area, thereby potentially enhancing system effectiveness. The slope of the weep berm is related to the weep berm height and the spacing and height of porous check dams located along the length of the weep berm. For the construction and mining projects, the weep berms were segmented into elongated chambers through use of porous check dams that were located along the length of the weep berm. In the case of the Georgia construction site, chambers were 40m in length and for the Peruvian site each chamber was approximately 80m in length.

OBJECTIVES

The objective of this project was to investigate the overall effectiveness of the combination weep berm - grass filter control system in reducing effluent concentrations of fecal coliforms and nutrients. The project also evaluated the hydraulic performance of the control system.

METHODS

The study was conducted at the University of Kentucky's Main Chance Farm located in Lexington, Kentucky. The predominate soil type was Maury Silt Loam. Three plots measuring 12.2 m in length and 2.4 m in width (Figure 1) were used. The average slope of the plots was 4% with a cross-slope of 1.3%. Each plot was bound by rust proof metal borders to insure that all runoff was contained in the desired area. Down-gradient of the grass filter, a wooden collection gutter intercepted surface runoff.

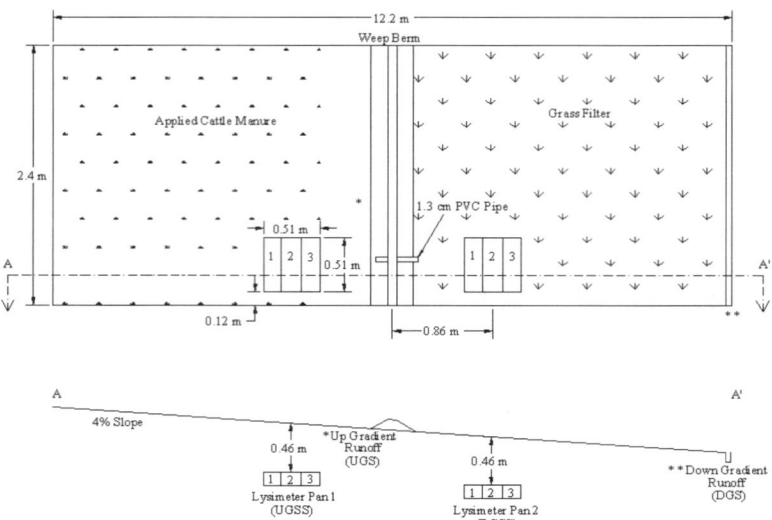

Figure 1. Plot Plan and Section Views

A weep berm 14 cm in height, with 2:1 side slopes, was constructed in the middle of each plot. The berm was constructed with moderate compaction using loamy topsoil. Fescue was sowed on the berm and covered with a coconut mat to promote growth and reduce the potential of erosion. A 12.7 mm diameter straight PVC pipe was installed through the berm, at an invert height of 8.9 cm, to facilitate dewatering at high flows. The weep berm and the outlet pipe were surveyed and an elevation-discharge relationship determined.

The desired rainfall rates and intensities were produced with two 2.7 m by 6.4 m rainfall simulators. Lexington municipal water served as the source for the rainfall. Since fecal coliforms were being monitored, it was necessary to eliminate the chlorine from the municipal water. This was accomplished by using 10 mg of anhydrous sodium thiosulfate per liter of source water.

Prior to each experiment, soil samples were collected from each plot above and below the berm to determine moisture content. Cattle manure was applied once at a rate of 0.94 kg/m2 to the upper 5.9 m of the plots following the background assessment and just prior to the first simulation. The manure was distributed evenly throughout the plots.

Rainfall was initially applied to the plots at an intensity of 5.1 cm per hour for a 20 minute duration. The intensity was then increased to 7.6 cm per hour for approximately 40 minutes. If no substantial runoff was detained by the berm, the intensity was increased to 10.2 cm per hour and maintained until discharge through the weep berm outlet occurred at full pipe flow for approximately 10 minutes.

Three runoff samples were taken uniformly throughout each simulation. The first sample was collected when runoff initially began, the second approximately twenty minutes later, and the third at the end of the experiment. Samples were collected at two locations: one from the backwater of the weep berm and the other down-gradient of the grass filter. A volume weighted composite sample was created from the samples collected at each sampling location.

A total of four simulations were conducted consisting of an initial background assessment and three replicated experiments. The initial background assessment was conducted to determine the existing level of contamination. Prior to the first experiment cattle manure was applied to each plot; therefore, the first simulation mimicked continuously grazed cattle pastures. The second and the third simulations were conducted in the same manner as the first; however, no additional

146

manure was applied. These experiments thus simulated rotational grazing practices in which the cattle had been removed from the pasture.

Parameters monitored during the study included: fecal coliforms, total nitrogen, and total phosphorous.

EXPERIMENTAL RESULTS

Surface flow data was measured to assess the control system's hydraulic performance. Rainfall up-gradient of the berm was detained by the berm, infiltrated, evaporated, evapotranspirated, or discharged through the straight pipe at high stages. Stage data was recorded to determine both the volume retained up-gradient of the berm and the flow dewatered through the straight pipe located through the weep berm. The runoff rate was also measured down-gradient of the grass filter.

Fecal bacteria are naturally occurring bacteria found in animal excrement; therefore, a large increase in fecal coliforms was observed following the application of cattle manure. After the initial application of the cattle manure, the fecal coliform count increased to an overall average of 2.3×10^5 counts per 100mL.

Comparing the runoff up-gradient of the berm and the runoff down-gradient of the grass filter, high reduction rates were observed. The control system effectively reduced fecal coliform counts 99%, 64%, and 61% during each sequential rainfall simulation. After the third rainfall experiment, the average fecal coliforms exiting the weep berm – grass filter control system was 340 counts per 100mL illustrating nearly a 1,000 times decrease in concentration compared to the average of the first rainfall simulation, which was 230,000 counts per 100mL. Figure 2 illustrates the average fecal coliform concentrations up-gradient and down-gradient of the control system and the overall reduction achieved by the control system. Table 1 lists the averaged fecal coliform concentrations reported from each experiment.

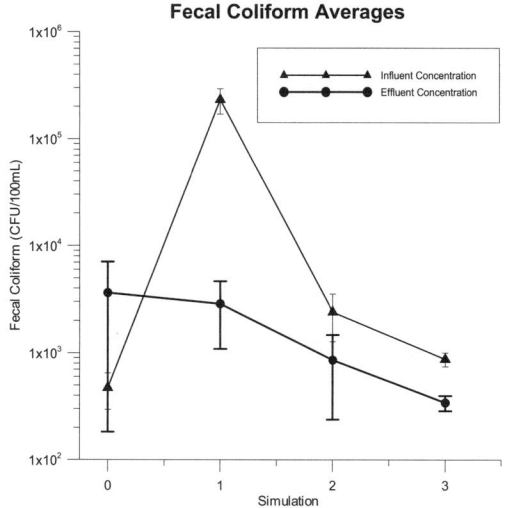

Figure 2. Fecal Coliform Concentrations

Table 1. Fecal Coliform Data

Run	Up-Gradient Concentration (CFU/100mL)	Down-Gradient Concentration (CFU/100mL)	Overall Reduction
1	2.9×10^5	2.9×10^3	98.8%
2	2.4×10^3	8.6×10^2	64.3%
3	8.8×10^2	3.4×10^2	60.8%

Nutrient reductions are listed in Table 2 and graphically displayed in Figure 3. The weep berm - grass filter control system was highly effective in reducing total phosphorous and total nitrogen by 44% and 87%, respectively during the first rainfall simulation. As anticipated, the nutrient data illustrated a diminishing incremental reduction in efficiency for the second and third experiments due to a large reduction observed during the first experiment. It is important to note that the weep berm – grass filter control system achieved a high reduction for the rainfall simulation immediately following application of waste material; therefore, the potential for off-site pollution is largely eliminated.

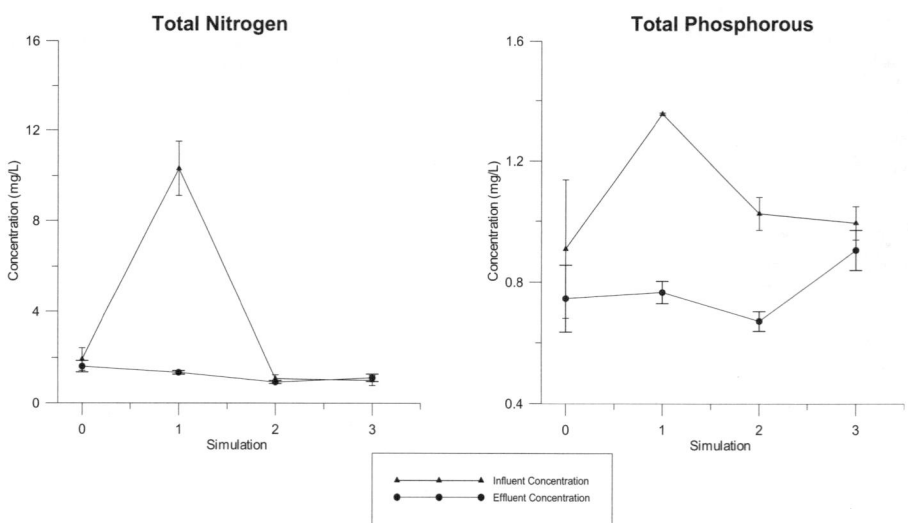

Figure 3. Runoff Nutrient Concentrations

Table 2. Nutrient Reductions

	Total Nitrogen		
Run	**Up-Gradient Concentration (mg/L)**	**Down-Gradient Concentration (mg/L)**	**Overall Reduction**
1	10.32	1.13	86.8%
2	1.09	0.94	13.9%
3	1.02	1.36	0.0%
	Total Phosphorous		
Run	**Up-Gradient Concentration (mg/L)**	**Down-Gradient Concentration (mg/L)**	**Overall Reduction**
1	1.35	0.77	43.5%
2	1.03	0.67	34.7%
3	0.99	0.91	9.0%

CONCLUSION

The effectiveness of a combination weep berm – grass filter control system was field-tested using three replicate plots subjected to three rainfall events. The three plots were instrumented to enable monitoring of surface runoff up-gradient of the weep berm and down-gradient of the grass filter. In this study, the control system was highly effective in reducing critical non-point source pollutants. Overall, the control system performed well hydraulically and improved effluent water quality, even under worst-case conditions. The control system achieved average reductions of fecal coliforms, total nitrogen, and total phosphorous 99%, 87%, and 44%, respectively. Based on these results, the weep berm-grass filter system affords the following advantages over simple grass filter when attempting to protect streams and wetlands: 1) peak flows are highly dampened, 2) due to short-term storage, some settling and infiltration occurs above the berm, and 3) flow is passively and uniformly released through the weep berm to the grass filter at a slower rate, thereby allowing the grass filter to perform more effectively.

REFERENCES

1. Coyne, M.S., R.A. Gilfillen, A. Villalba, Z. Zhang, R. Rhodes, L. Dunn, R.L. Blevins. 1998. Fecal Bacteria Trapping by Grass Filter Strips During Simulated Rain. *Journal of Soil and Water Conservation* 53(2): pp140
2. Deletic, A. 2001. Modeling of water and Sediment Transport over Grassed Areas. *Journal of Hydrology* 248(1-4): pp168-182
3. Dillaha, T.A., R.B. Reneau, S. Mostaghimi, D. Lee. 1989. Vegetative Filter Strips for Agricultural Non-Point Source Pollution Control. *Trans. ASAE* 32(2) pp.513-519
4. Gharabaghi, B., R.P. Rudra, H.R. Whiteley, W.T. Dickinson. 2001. Sediment-Removal Efficiency of Vegetative Filter Strips. ASAE Meeting Paper No. 01-2071 Sacramento, CA.:ASAE
5. U.S. Environmental Protection Agency. 2000. National Quality Inventory: 2000 Report United States Environmental Protection Agency Office of Water. EPA-841-R-02-001
6. Warner, R.C., and F.X. Collins-Carmargo. 2001. Erosion Prevention and Sediment Control Computer Modeling Project. Project Final Report, The Chattahoochee-Flint Regional Development Center Dirt II Committee. June. pp51-54, 133
7. Warner, R. C., and S. Torrealba. 2003. Design and Evaluation of Stormwater, Erosion and Sediment Control Systems for the Antamina Mining Complex. Project Report, Compania Minera Antamina, Lima, Peru.

BOKES CREEK WATER QUALITY ENHANCEMENT PROJECT
POWDERLICK RUN CHANNEL RESTORATION

Steve Phillips, Dan Binder, Ed Rankin [*]

ABSTRACT

Hydromodification (channelization) of stream channels is the leading cause of water quality impairment in the State of Ohio. Hydromodification typically results in impaired water quality and biology and a reduction in the ability of the channelized stream to assimilate pollutants when compared with natural stream channels. The Bokes Creek Water Quality Enhancement Project demonstrates how restoring channelized streams and agricultural ditches using Natural Channel Design (NCD) methods improves water quality and the ability of a stream to assimilate pollutants and nutrients. This project is part of the Bokes Creek TMDL implementation plan and an excellent example of a significant source water protection initiative. A Section 319 Water Quality Implementation Grant from Ohio EPA funded the restoration of Powderlick Run.

This project demonstrates scientifically sound methods of restoring ecosystem benefits in hydro-modified stream channels while maintaining drainage benefits. An Adaptive Management process was utilized in data collection and synthesis, design, implementation and the development of monitoring performance standards. A design-build approach took the project from data collection and design through construction of the channel and re-vegetation. The channel morphology and hydraulics of the hydromodified stream channel were studied to determine baseline conditions. A comparison of baseline and reference reach data helped define restoration goals and the morphological characteristics for the new restored channel.

The final design was a multidisciplinary product that was reviewed by the project partners and various agencies. The project was constructed in early 2003 and the initial biological monitoring that was performed immediately after construction indicated significant biological recovery.

KEY TERMS: Natural channel design, ecosystem restoration, multi-stage channels, assimilative capacity, adaptive management, and design-build.

INTRODUCTION

Bokes Creek is an 84.2 square mile sub-basin located in the Eastern Corn Belt Plains eco-region of Ohio within the Upper Scioto River watershed, which is a major drinking water supply for the City of Columbus, Ohio. Powderlick Run has a designated life-use of warm water habitat (WWH) and was in chronic non-attainment due to water quality impairments such as nutrient enrichment, low dissolved oxygen and extensive hydromodification and habitat deficiencies (OEPA, 1999).

---

[*] Steve Phillips, CPESC, Oxbow River & Stream Restoration, Inc., 2905 Klondike Road, Delaware, Ohio 43015, Phone: (740) 362-4134. Dan Binder, City of Columbus Water Quality Assurance Lab, Phone: (614-645-7691) and Ed Rankin, Center for Applied Bioassessment and Biocriteria, Columbus Ohio 43221, Phone: 740-517-2275.

The ultimate test of success for river restoration activities in Ohio is attaining the aquatic life goals set forth in the Ohio Water Quality Standards and the measurable sub-components of that process (Yoder 2004). Ohio adopted numerical biological criteria into the Water Quality Standards (WQS) as part of an existing water quality management process in 1990. Biological criteria are based upon measurable characteristics of aquatic communities that are structured in Ohio's WQS regulations in a system of tiered aquatic life uses.

Aquatic life-use designation describes a stream's water quality and physical attributes based upon key indices that would be expected in a given eco-region. The Qualitative Habitat Evaluation Index (QHEI) measures the physical integrity and habitat of the stream channel including the floodplain. This index ranks features necessary to support aquatic life such as riffles, substrate embeddedness, pool depths and riparian vegetation. The Invertebrate Community Index (ICI) is a combination of 12 metrics collectively used to describe the health (production, function, tolerance, and reproduction) of the macroinvertebrate community in the stream. Finally, the Index of Biological Integrity (IBI) is a combination of 10 metrics collectively used to describe the health (production, function, tolerance, and reproduction) of the fish community in a stream. The ICI and IBI indices quantitatively measure the diversity and density of the aquatic biology and are direct indicators of water quality.

Biological integrity is the integrated result of chemical, physical and biological processes in the aquatic environment. There are very strong correlations between the physical integrity of a stream system and both the biology present in the stream system and the way the stream system cycles pollutants and nutrients (Rankin et al., 1999). A comparison of total phosphorus (TP) with a measure of habitat quality using QHEI shows that TP concentrations are typically lowest at locations in a stream with high quality habitat (Rankin et al., 1999).

In small streams, the relationship between the channel and floodplain plays a key role in the processes of nutrient cycling (uptake and assimilation) laterally within the riparian area (Malanson 1993), especially with regards to how these areas act as sources and sinks of nutrients. Wooded riparian floodplains are key attributes in effectively reducing the total phosphorus (TP), sediment loadings, and nitrates (Fennessy and Cronk 1997).

Maximum phosphorus retention is attained by floodplains containing a mix of herbaceous and woodland areas. High quality stream habitats with intact riparian zones and natural channel morphology assimilate excess nutrients directly into both plant and animal biomass and reduce sunlight through shading (a principal limiting factor in algal production). (Malanson 1993; Barling and Moore 1994). Sunlight becomes the factor limiting algal growth (Cummins 1974) and small streams with a dense closed canopy have cooler water temperatures and higher measures of dissolved oxygen (DO). These connections are reduced or altogether eliminated in streams that have been physically altered for flood control or agricultural drainage purposes, such as Powderlick Run.

Both Ohio EPA and the City of Columbus had collected extensive chemical and physical water quality data for the Bokes Creek watershed. Ohio EPA's data is summarized in several Technical Support Documents; Biological and Physical Habitat Study 1994, Biological and Water Quality Study of the Upper Scioto River Basin 1997 and Biological and Water Quality Study of Bokes Creek and Selected Tributaries 1999. The City of Columbus had also studied the nitrate levels in the Upper Scioto River watershed in an attempt to better understand causal relationships and available alternatives to mitigate nitrate spikes that historically had exceed Safe Drinking Water standards. Columbus has been active in monitoring source water and in developing partnerships with watershed stakeholders and began discussions about watershed management with local watershed groups and Universities as early as 1996. Source water protection is an emerging area of management for many drinking water utilities utilizing surface water sources. Ohio EPA has implemented a Source Water Assessment Program (SWAP) as a requirement of the 1996 Amendments to the Safe Drinking Water Act. The SWAP not only maps drinking water sources but also recommends watershed management activities. The existing data by Ohio EPA and the City of Columbus was critical in determining the feasibility of intervention and developing alternatives for restoration that would improve water quality in the Upper Scioto River watershed.

PROJECT GOALS AND APPROACH

This project involved the restoration of 3,600 linear feet of Powderlick Run, which is a 5.1 mile long tributary to Bokes Creek with a drainage area of 3.84 square miles. Powderlick Run, like many agricultural drainage ditches, has been straightened and modified to promote artificial sub-surface drainage from agricultural fields. This modification has removed the riparian corridor and natural channel attributes essential to the physical integrity, aquatic habitat and water quality associated with healthy stream ecosystems.

The goal of the project is to improve the ability of Powderlick Run to assimilate pollutants, particularly nitrogen and phosphorus, and to restore the Warm Water Habitat (WWH) life use-designation as set forth in the Ohio Water Quality Standards. The project approach involved restoring WWH attributes and reducing or eliminating negative modified attributes as identified and measured using Ohio EPA's Qualitative Habitat Evaluation Index (QHEI). QHEI measures the physical integrity and habitat of the stream channel including the floodplain. Powderlick Run was in non-attainment for all three bio-criteria indices, QHEI, ICI, and IBI with significant departure in ICI and IBI. A comparison of pre-existing, 3 months post construction and target indices for this project can be found in **Table 2**.

The restoration project on Powderlick Run is at river mile 2.7 with a drainage area of 1.05 square miles. Upstream of the restoration site, Powderlick Run is essentially a straightened drainage ditch with very little riparian cover for the entire 2.4 miles. Downstream of the restoration site, Powderlick Run has more natural characteristics meandering through several woodland tracts and open pasture areas. At the restoration site, the channel was last modified in 1985-86 when the ditch was dipped, the side slopes dressed and the tile outlets stabilized. In 1995, filter strips were installed extending 25 feet on both sides of the ditch. These buffers were enlarged to 75 feet on each side of the ditch and planted to warm season native switch grass in 1997. Most recently, the channel had evolved into a "two-stage" ditch with some vertical sinuosity, significant lateral erosion and small benches forming in the bottom of the ditch. The lateral erosion was the cause of significant mass bank failures and slumping of soils into the channel and the failure of sub-surface tile outlets. The Lateral Recession Rate (LRR) prior to restoration through the project area was measured as severe and contributed more than 150 tons per year of sediment to the water column creating an increase in total suspended solids (TSS), substrate embeddedness and nutrient export.

This project is a Priority 2 restoration (Rosgen, 1997) designed to restore a functional floodplain at the lower elevation of the existing channel with a minimum meander width not exceeding the current buffer area. The design emphasized the restoration of natural features that will improve the stream's ability to assimilate nutrients and pollutants and therefore, improve water quality. In an effort to restore the stream channel, great effort was made to address not only the morphology of the channel but the biology that supports a healthy, functioning stream. In addition to properly designed morphology and in-stream features, a well vegetated, connected floodplain was deemed paramount. There are also several significant secondary (incidental) benefits derived from this project that have economic benefits to agricultural or commercial land use with regard to improved conveyance and subsurface tile function, reduced channel maintenance, and regional detention of storm flows.

DESIGN SOLUTION

This project was a demonstration of how an Adaptive Management process can allow disparate disciplines to develop a successful design solution. This project incorporated the disciplines of engineering, morphology and biology and was an example of how these three disciplines must integrate their data, language and understanding of stream resources to better make decisions in projects incorporating Natural Channel Design.

This multi-disciplinary approach was used in data collection to assess current conditions and develop a channel design that involves the following three principal categories, all characterized by the assumption of equilibrium:

- Analog approach: adopts templates from historic or adjacent channel characteristics
- Empirical approach: uses equations that relate various channel characteristics derived from regionalized or universal data sets
- Analytical approach: makes use of hydraulic and sediment transport models to derive equilibrium conditions

The assessment of physical and morphological conditions was developed based on methods outlined by Harrelson et al. (1994), Rosgen (1996) and Ohio EPA. Field data, such as field survey and morphological data, was collected at 10 selected locations over 2.5 miles of Powderlick Run and at four locations on a nearby tributary, Brush Run. Morphological characteristics included channel gradient, bankfull width, bankfull depth, width of flood-prone area, and pool depth. In addition, substrate was measured using both the Wolman Pebble Count and a volumetric sieve methodology.

Horizontal and vertical controlled field survey was conducted by the Union County Soil and Water Conservation District on 3,600 linear feet of Powderlick Run. Cross sections were developed every 100 feet for the length of the proposed project starting at the eastern property line and working west. Cross sections included top of bank, toe of slope, edge of bank and thalweg of the existing channel. This information was imported to CAD to generate the cross-sectional and gradient profile drawings. The survey data was then used as an overlay for a digital ortho-photo site map obtained from the County for presentation purposes and public discussions.

The cross-sectional data was also used to build a flow model using the U.S. Army Corps of Engineer's HEC-RAS computer program to calculate discharge volumes, channel velocities, shear stress and water surface elevations during various storm events. The survey data used in the model, demonstrated that the 10-year storm event was predominantly confined within the existing trapezoidal channel and that all larger storm events begin to flood out into the adjacent fields. Several of the tile outlets were submerged during the 1-2 year storm events and shear stress within the channel were high due to the confined width and relative depth during storm events. In a more natural channel type, these flows would be displaced out into the floodplain area, reducing the available in-stream energy and lowering the calculated flood elevations. The model results for the restored channel and floodplain indicated that the 100-year storm event would be contained within the newly constructed floodplain area and that the tile outlets would now have a free outlet even during the 10 year flow events.

The existing channel was a G type channel with a cross sectional area that was 60% undersized, had an entrenchment ratio of 1.5 to 3, and provided approximately 20% of critical natural features (riffle/pools) with no riparian trees or shrubs. The existing riffle areas averaged 6 feet in width and less than 14 feet in length, or less than 20% of the total channel length. These features had concentrated channel flows and velocities that ranged from 3 to greater than 5 feet per second which resulted in some sorting of bed materials and movement of fine sands. The pools were glide dominated with a mean depth of .5 feet with 75% of the total surface area being half as deep as the plunge pool area immediately downstream of the riffle. The D-84 as defined by a weighted pebble count (both pool and riffle) averaged 11 mm. The riffle D-84 using a pebble count measured 16 mm. and the riffle D-84 from a volumetric sieve analysis measured 32 mm.

In keeping with the knowledge incorporation aspect of the Adaptive Management process, the design-build approach utilized for this project played a key role in the design implementation phase of the project. The same professionals that were involved in the assessment and design of this project also physically constructed this project. This approach provided for greater flexibility and adaptation as knowledge from observations were implemented directly during the construction process. The following design elements were identified as having the highest potential for improving the natural ability of Powderlick Run to assimilate pollutants:

- Stable channel/floodplain morphology
- Substrate that provides subsurface interaction
- Riparian vegetation that provides nutrient conversion
- Filter strips that trap soil movement from fields

The project restored 3,600 linear feet of ditch to a meandering channel and floodplain. Restored channel widths varied from 14 feet at the start of the project to 16 feet at the downstream end of the project with an average maximum depth of 1.3 feet. The floodplain was excavated to a belt-width of 40 to 50 feet with 3:1 side slopes transitioning back up to the existing farm fields, resulting in an overall top width of between 80 and 90 feet. The constructed belt-width represents a 4 to 1 relationship to channel width. This is on the low end of stable C type stream channels observed in Central Ohio. This relationship will favorably increase over time as vegetation establishes and increases the shear resistance within the channel allowing for significant channel narrowing and more "E" channel characteristics. The two limiting factors to the belt-width in the design process were the funds available for mass excavation and the desire to keep the overall width of the project within the existing filter strip area, therefore requiring no more land acquisition. The design required approximately 16,000 cubic yards of soil to be excavated to accommodate the floodplain width at an average depth of 6 feet. Table 1 summarizes the morphologic information for the project.

The design incorporated riffle features that required approximately 750 tons of substrate materials that consisted of approximately 20% sands (.5-2 mm) and 20% gravel (2-8 mm), 20% gravels (8-64 mm), 30% gravel (32-64 mm), 10% cobbles (64-128 mm). This design feature increased the available riffle habitat to average 40% of the overall channel length with riffle and pool spacing occurring every 70 to 100 linear feet. Riffle hydraulics were calculated based upon crest height, gradient and width to appropriately size the substrate material, develop a stable down slope gradient and provide hydraulic heterogeneity for benthos production (Newbury 1994). Sub-surface interstitial flow characteristics were enhanced by over-excavation of the channel at the riffle area to 2.5 times the surface area of the channel riffle. The construction of these riffles provides oxygen saturation of the water column to aid in the volatilization of nitrogen and ammonia while the subsurface substrate materials provides anaerobic sub-surface interstitial flows designed to convert or provide uptake of nitrogen and phosphorus pollutants.

Re-vegetation of the floodplain/riparian area consisted of planting 10,200 bare root seedling shrubs in the flood plain area along the entire restored channel length. Topsoil was re-graded onto the floodplain bench after the mass excavation, subsoil ripping and placement of the substrate. In addition, 2,700 bare root hardwood trees were planted in the upper bank or terraced areas along with warm season grass communities that blend back to the existing buffer strip area. Subsurface tile outlets were repaired with solid extensions to prevent tree root growth and stabilized with rock outlets in the floodplain bench.

CONCLUSION

As with any activity based planning approach, there is a natural tendency to measure success in terms of the activity and structural outputs of a process. But this measure stops short of determining the ultimate success (i.e., the biology) of the process. The management of non-point sources of pollution and determining the assimilative capacity needs to include more than dilution dynamics alone. As a pollution control strategy, the physical integrity, or habitat, of the stream and its floodplain are the lowest common denominator and the controlling variable with respect to aquatic life use-attainment. Because habitat is a critical component to stream function and can strongly influence water quality, habitat data must be considered an integral part of any attempt to restore aquatic life in streams if such restoration efforts are to succeed.

Headwater streams (less than 5 square miles) such as Powderlick Run are very important to the assimilation of nutrients and sediment from runoff and can be instrumental in meeting total maximum daily loads goals and the overall quality of downstream resources. Headwater streams also pose a prime opportunity to effectively address aquatic life use-attainment goals due to the scale of the projects undertaken and the relative ease of creating habitat that is critical to stream function.

It is well documented that wooded riparian floodplains are a vital functional component of the stream ecosystem. Wooded riparian floodplains are instrumental in the detention, removal and assimilation of nutrients from or by the water column and the floodplain soils are essential in the cycling process that sequesters and converts excess nutrients directly into plant biomass. In headwater stream systems, floodplains can produce twice the biomass compared with the in-stream aquatic biomass production (Newbury et al., 2000)

For any given stream restoration project, the channel morphology must allow for frequent inundation and saturation of the floodplain soils. These highly organic soils, if removed, must be replaced during the construction process and/or must be introduced to the floodplain prior to the planting of riparian vegetation. While it is imperative to construct a proper floodplain during restoration efforts, the temporal aspect of restoration must be kept in mind. It can take 10 to 15 years for newly constructed floodplains to develop into functional wooded riparian areas that effectively cycle nutrients.

Bio-criteria, and especially the QHEI metric provides a quick and reliable method to assess project alternatives, designs and the final success of restoration projects. As a rule of thumb, goals for stream restoration projects should strive to reduce the number of modified attributes to four or fewer and eliminate all high influence modified attributes (Rankin et al., 1999).

The Powderlick Run Restoration Project demonstrates a reasonable approach to meeting both the water quality goals set forth in the Ohio Water Quality Standards and conventional "drainage" or conveyance goals that support sub-surface drainage needs. The Powderlick Run restoration project was successful in reducing the total number of negative attributes to four, as measured by QHEI. The one high influence negative attribute remaining is that of "sparse cover" or shading of the channel. That attribute will be eliminated in several years as the vegetation in the restored riparian floodplain develops and becomes more dense and diverse.

ACKNOWLEDGMENTS

We would like to acknowledge the following individuals who contributed to the project concept, technical support and monitoring efforts: Daniel Binder, City of Columbus Water Quality Assurance Lab, Andrew Rogowski, Ohio Department of Agriculture and Ed Rankin and Chris Yoder, Mid-West Biodiversity Institute.

Table 1. Pre and post construction channel morphology

Characteristic	Pre-construction	Post-construction
Bankfull Width (ft)	6.1	16.0
Bankfull Mean Depth (ft)	0.9	0.9
Width/Depth Ratio	7	17.3
Cross-sectional Area (sq ft)	5.3	14.8
Bankfull Maximum Depth (ft)	1.7	1.30
Flood Prone Width (ft)	18	60
Entrenchment Ratio	2.9	3.8
Channel Slope (%)	.27	.25
Valley Slope (%)	.28	.28
Sinuosity	1	1.2-1.3
Rosgen Stream Type	G	Ce

Table 2. Comparison of conditions 3 months post-restoration with target goals

Metric	Pre-condition	Post-condition	Target (ECPB ecoregion, WWH Use Designation)
QHEI	39	58	60
ICI	< 6	NA	36
IBI	18	30	40

REFERENCES

Barling, R.D. and Moore, I.D., 1994. Role of buffer strips in management of waterway pollution; A review. Env. Mgmt. 18(4): 543-558

Bunte, Kristine, Abt, Steven, 2001. Sampling Surface and Subsurface Particle-Size Distributions in Wadable Gravel and Cobble-Bed Streams for Analyses in Sediment Transport, Hydraulics and Streambed Monitoring, USDA Forest Service General Technical Report RMRS-GTR-74

Cummins, K.W. 1974. Structure and function of stream ecosytems. BioScience 24: 631-641

Fennesy, M.S. and Cronk, J.K., 1997. The effectiveness and restoration potential of riparian ecotones for the management of nonpoint source pollution, particularly nitrate. Critical Reviews in Environmental Science and Technology 27(4): 285-317

Harrelson, Cheryl C., Rawlins, C.L., Potyondy, John P. Stream Channel Reference Sites: An Illustrated Guide to Field Technique. USDA, USFS Technical Report RM-245.

Malanson, G.P. 1993. Riparian Landscapes. Cambridge University Press, Cambridge, Great Britain.

Michigan Department of Environmental Quality, 1999. Pollutants Controlled Calculation and Documentation for Section 319 Watersheds Training Manual.

Newbury, Robert W., Gadbury, Marc N., 1994. Stream Analysis and Fish Habitat Design

Newbury, Robert W., Gadbury, Marc N., Bates, David, 1996. Constructing Riffles and Pools in Channelized Streams, Proceedings of the River Restoration '96 Conference, Silkeborg, Denmark.

Yoder, Chris O., Rankin, Edward T., 1995. The role of Biological Criteria in Water Quality Monitoring, Assessment, and Regulation, Ohio EPA

Ohio EPA, 1999. Biological and Water Quality Study of Bokes Creek and Selected Tributaries.

Rankin, E.T. 1989. The Qualitative Habitat Evaluation Index (QHEI): Rationale, Methods, and Application. Ohio EPA, Columbus, Ohio.

Rankin, E.T., Miltner, Bob, Yoder , Chris, Mishne, Dennis 1999. Association Between Nutrients, Habitat, and the Aquatic Biota in Ohio Rivers and Streams, Ohio EPA Technical Bulletin MAS/1999-1-1

Rosgen, D.L. 1997. A Geomorphological Approach to Restoration of Incised Rivers, Proceedings of the Conference on Management of Landscapes Disturbed by Channel Incision.

Yoder, Chris O., Rankin, Edward T. 1997. The Effects of Hydromodification for Agricultural Drainage on Streams and Rivers in Western and Northwestern Ohio.

Sediment and Erosion Control Techniques on Stream Restoration Projects

D. R. Clinton, G. D. Jennings, R. A. McLaughlin, D. A. Bidelspach[1]

ABSTRACT

Erosion control, sediment loss and turbidity control are important considerations during construction of stream restoration projects. Standard practices such as basins, traps, check dams, and silt fences may reduce sediment losses but may not be effective at reducing turbidity impacts to downstream water. Innovative techniques for improved sediment control and turbidity reduction include flocculants, sediment basin enhancements, baffles, outlet modifications, micro-grading, and construction phasing. The purpose of this paper is to describe the application of innovative erosion and sediment control techniques for stream restoration construction projects.

KEYWORDS Fluvial geomorphology, turbidity, sedimentation, stability, watershed

INTRODUCTION

Stream restoration projects typically involve earth moving activities such as channel relocation, shaping of streambanks and floodplains, and installation of in-stream structures. These projects often result in large disturbed land areas with the potential to cause downstream water quality impacts due to erosion and sedimentation. Traditional engineering approaches to controlling sediment impacts may not meet water quality goals, especially with regard to turbidity impacts. Innovative techniques for reducing sediment loss and turbidity control during construction activities may be necessary to effectively control sediment and turbidity. This paper describes the application of innovative erosion and sediment control techniques for stream restoration in three categories: (1) site preparation; (2) construction; and (3) post-construction stabilization.

Site Preparation

Typical site preparation practices include stone construction entrance, demarcated staging/stockpiling areas, and haul road establishment (Smolen et al, 1988). Maintenance is critical with all erosion control but the following can aid in the effectiveness of each. A wash station added to construction entrance can add to its effectiveness and can aid in a positive public perception particularly in urban settings. Wash stations typically include hoses, a pump, a splash station, a collection facility, a settling area and a return channel. In addition to public perception, maintaining a wash station reduces soil from leaving the construction site therefore reduces potential turbidity and sediment damage off site. Sediment on the surrounding roads can also be a safety (and liability) hazard if it accumulates enough to become slick. Connecting the wash station return channel to a sediment basin can increase its effectiveness by allowing the sediment laden water time to settle out larger particles prior to it returning to the stream. Polyacrylamide (PAM) can be added to the pump water to further reduce turbidity by flocculating out finer particles.

Demarcating staging and stockpile areas can benefit a construction site by minimizing disturbed areas and focusing stockpile areas to controllable limits. Water flow coming from stockpile areas should be directed into a sediment basin to reduce the risk of sediment reaching the stream.

---

[1] Biological & Agricultural Engineering Department, North Carolina State University, Raleigh, NC 27695. dan_clinton@ncsu.edu

Construction haul roads are areas of high sediment movement because they tend to intercept runoff and bring it downslope at high velocities. Intermediate road berms which direct sediment laden water off the road prior to it accumulating into significant volumes can reduce potential stream sedimentation. Extra care should be taken in areas where haul roads cross stream channels. In these areas, diversion berms should be established to direct sediment laden water toward sediment basins where it can be clarified prior to entering the stream channel. A stable, clean pad of stone should be maintained at crossings to reduce sedimentation into stream channels.

Project clearing can enhance or counter sediment and erosion control efforts on a restoration project. Recommendations for typical clearing techniques include phased clearing in conjunction with construction activities, salvaging of transplant materials and spot clearing to maintain a viable seed bank post construction. Phasing of clearing operations can reduce open area thus reducing sedimentation risk. Phased clearing can also improve the viability of transplant materials by keeping the transplants in the ground for a longer period of time.

Prior to construction on most stream restoration projects, a diversion is typically required if in-channel work is necessary. Although no national standards have been established, channels with drainage areas less than 4 square miles are typically required to be diverted in North Carolina. The diversion can be accomplished through an open channel, pipe, or by pumping.

Within the diversion area, water that has infiltrated into the construction area from ground water or seepage through diversion dikes needs to be pumped out in order to work. The pumped water often has high sediment concentrations which should be removed prior to release downstream. Sediment filter bags are often used for this purpose because they are simple and do not require a lot of space. However, they are only partially effective and the water can still be quite turbid after passing through the bag. A stilling basin may be more effective, particularly if it is designed with porous baffles (Thaxton et al, 2004). The most successful material for these has been a combination of jute matting and coir erosion control blanket, strung across the basin on hog wire or similar. The effect is to force the water to spread across the entire cross-section of the basin and to reduce turbulence that interferes with settling. Providing a floating outlet, or skimmer, attached to a solid riser or similar, can also greatly improve sediment capture (McLaughlin, 2003; Millen et al., 1997; Figure 1)

The addition of polyacrylamide (PAM) has substantial potential for reducing turbidity (McLaughlin, 2003). We have had some success in metering in PAM solutions at the pump as well as running the water across PAM logs or PAM powder sprinkled in a pipe or on jute matting. However, each sediment material is different and the correct PAM needs to be matched to the sediment on the project. There is no one way to introduce the PAM to the water, but the system will require high turbulence to form the flocs followed by a settling area with minimum turbulence. One method we've used is to add PAM solution at the pump intake and release the pumped water into a corrugated pipe followed by a stilling basin (Figure 2). PAM injected into the pumped water can improve the performance of sediment bags, but there may be problems with the flocs clogging the bag long before it is filled. This may be acceptable, however, since the bags are relatively inexpensive.

The outfall area of the diversion should be adequately protected to avoid unnecessary scour from the diversion water. Typical outfalls consist of rip-rap pads or stilling basins.

Construction Activities

Erosion control during stream restoration construction involves stabilizing stream banks immediately after grading and prior planning of construction sequence. Stabilizing stream banks involves installing erosion control blankets (ECBs) and developing a temporary herbaceous cover as soon as possible while the woody materials are becoming established. ECBs should cover all open stream banks to provide adequate protection until vegetation becomes established. Various grades of ECBs can be utilized in higher and lower stress areas of the stream banks. For example, a tougher grade of ECB is needed on the outside meander bend compared to the lower stress point bar areas. The point bar area ECB should decay at a much higher rate since that area of the stream scours and redeposits during large storm events.

ECBs should overlap the top of the stream bank by a minimum of one foot to reduce the potential of rills and gullies forming under the matting. Upstream and top of bank edges of the ECB should be well keyed-in to reduce lifting of the matting during storm events. ECBs should be trimmed such that they are tightly secured around rootwads. This minimizes potential scour.

In our experience, straw mulch placed under an ECB with large openings will enhance the protection of the bank and increase the herbaceous cover success rate. Herbaceous plugs inserted through the ECB can also be used to provide faster ground cover on stream banks.

ECBs, particularly on riffle and outside meander banks, should be rated to withstand average predicted velocities. Securing the ECB with either metal sod staples or biodegradable wood stakes should follow manufacturer's guidelines to ensure stability.

The sequence of events is critical to provide effective sediment and erosion control on stream restoration projects. Stream restoration projects often involve moving significant volumes of soil. Soil should be placed in its final location or secured in stockpile areas behind silt fence at the completion of each day's construction or before the threat of rain.

The clearing sequence is also a critical part of the erosion control plan. Minimizing the duration of cleared land reduces the risk of sedimentation downstream. Typically 500 feet of cleared stream length is adequate to maintain construction timeline. Reduced clearing also increases the viability of potential transplants because the reduced amount of time they are removed from the ground.

Sediment control basins may be necessary depending on the size of the disturbed area. The sediment capture rate can be greatly increased through the enhancement mentioned previously (floating outlet, porous baffles). A standing pool has also been demonstrated to increase sediment capture (Fennessey and Jarrett, 1997). To further reduce turbidity, PAM can be used to flocculate the suspended fines. It is important to allow a zone of mixing and contact time followed by a settling zone prior to entering the stream channel.

Alternative bank stabilization methods include root wrap and brush mattresses used for higher stress areas along outside meander bends. Root wrap, where root wads and tightly packed along an outside meander, can provide excellent bank protection and enhance the habitat values of the pool area. Brush mattresses, although labor intensive to install, provide an almost instantaneous living stream bank capable of withstanding high stream velocities, if constructed properly.

Post-Construction Stabilization

Beyond establishing a stand of herbaceous cover upon completion of construction activities, ensuring that the floodplain is graded smoothly and free of high pockets can strongly aid in the bank stability. High areas within the floodplain can cause water to concentrate and enter the stream in large volumes in local areas. These overland flow interception points often form gullies where vegetation cannot become established. Often ECBs are not sufficient to protect the bank in these areas. Attention should be directed to areas where water is likely to intercept the stream channel. Hardened inlet features such as rock-lined channels can be used to ensure erosion will not occur.

CONCLUSION

Effective control of soil erosion, sediment loss, and downstream turbidity impacts is critical for the success of stream restoration construction. Designers must look beyond standard practices and consider innovative techniques such as flocculants, sediment basin enhancements, baffles, outlet modifications, micro-grading, and construction phasing. These approaches are being evaluated in several projects in North Carolina and should be considered for further research and development as components of stream restoration projects.

REFERENCES

1. Smolen, M.D., R.G. Jessup, L.C. Wyatt, D.W. Miller, J. Lichthardt, A.L. Lanier, W.W. Woodhouse, and S.W. Broome. 1988. North Carolina Erosion and Sediment Control Planning and Design Manual. NC Department of Environment, Health, and Natural Resources, Division of Land Resources, Land Quality Section, Raleigh, North Carolina. 572 pp.

2. Thaxton, C. S., J. Calantoni, and R. A. McLaughlin. 2004. Hydrodynamic assessment of various types of baffles in a sediment retention pond. Transactions of the ASAE Vol. 47(3): 741-749.

3. McLaughlin, R. A. 2003. The potential for substantial improvements in sediment and turbidity control. Pp. 262-272 in Total Maximum Daily Load (TMDL) Environmental Regulations II, Conference Proceedings, 8-12 November 2003 (Albuquerque, New Mexico, USA), ed. Ali Saleh. ,8 November 2003 . ASAE Pub #701P1503

4. Millen, J. A., A. R. Jarrett, and J. W. Faircloth. 1997. Experimental evaluation of sedimentation basin performance for alternative dewatering systems. Trans. Am. Soc. Agr. Eng. Vol. 40: 1087-1095.

5. Fennessey, L. A. J., and A. R. Jarrett. 1997. Influence of principal spillway geometry and permanent pool depth on sediment retention of sedimentation basins. Trans. ASAE 40(1):53-59.

6. Mostaghimi, S., T.M. Gidley, T.A. Dillaha, and R.A. Cooke. 1994. Effectiveness of different approaches for controlling sediment and nutrient losses from eroded land. J. Soil and Water Cons. 49 (6) 615-620.

7. NCDENR (North Carolina Department of Environment and Natural Resources), NC Sediment Control Commission, and NC Agricultural Cooperative Extension. 1993. Practice Standards and Specifications – Surface Stabilization. In North Carolina Sediment and Erosion Control Planning and Design Manual.

8. NCDENR (North Carolina Department of Environment and Natural Resources), Division of Water Quality. 2002. Surface Water and Wetland Standards (NC Administrative Code 15A NCAC 02B.0100 & .0200). p.22. Accessed at http://h20.enr.state.nc.us/admin/rules/rb010102.pdf

Figure 1. Example of a floating outlet design for a sediment basin.

Figure 2. Example of a combination of PAM treatment and a large stilling basin at a borrow pit pumping operation. This can be scaled down considerably for lower pumping volumes.

DESIGN CONSIDERATIONS FOR THE RESTORATION OF SAND BED STREAM CHANNELS AND ASSOCIATED RIPARIAN WETLANDS

Kevin L. Tweedy[1], PE

ABSTRACT

Restoration projects that address stream and wetland functions concurrently require special design considerations that are not typically evaluated for projects that restore stream or wetland functions alone. The design should take into account the interrelated functions of the system as a whole. Stream and floodplain functions include the transport and storage of water and sediment. These functions, in turn, provide diverse aquatic and riparian habitats. This systems-based approach allows for the restoration of self-sustaining riparian ecosystems.

This paper discusses design considerations for the restoration of sand bed stream channels and associated riparian wetlands. Data are presented from two completed mitigation projects and four reference sites. Sites were located on first or second order headwater stream systems whose associated floodplain areas contain hydric soils. Mitigation sites were used for agricultural production prior to restoration of stream and wetland functions, and reference sites were forested reference wetland and stream systems that exhibited little evidence of recent disturbance.

KEYWORDS. Headwater systems, natural channel design, mitigation, reference sites.

INTRODUCTION

The federal Clean Water Act (CWA) requires compensatory mitigation for certain impacts to waters of the US, where an impact is considered to be any disturbance in which the nature and/or functions of the waterbody are negatively affected. In the state of North Carolina, mitigation is regulated by the US Army Corps of Engineers (USACE) and the North Carolina Division of Water Quality (NCDWQ), under Sections 404 and 401 of the CWA, respectively. When impacts to streams and wetlands under authority of these agencies occur, several options are available to mitigate for the impacts. Options include a state managed in-lieu fee program, on-site mitigation, and off-site mitigation. Under each of these three options, mitigation of stream impacts in the state has focused on restoring stream functions to degraded systems in such a way that streams are stable and self-sustaining over time. Natural channel design techniques can be used to achieve this restoration goal.

Natural channel design concepts generally relate to the physical structure of the stream channel and associated floodplain areas. The underlying premise of natural channel design is creation of a stream channel that transports water and sediment delivered from the watershed such that the channel does not aggrade or degrade. This is achieved by designing stream channels that are appropriate for the current valley type and watershed conditions. The design of the restored stream channel is often based on ratios developed from nearby reference reaches. Rosgen (1998) defines a reference reach as a portion of a river that represents a stable channel within a particular valley morphology. Basing the design on stable characteristics exhibited by the reference reach helps to restore a self-sustaining, functional stream channel.

In riparian wetland systems, the stream channel has a significant effect on the hydrology of adjacent wetland areas. Studies conducted to determine the effects of riparian wetland systems on non-point source pollution have reported that much of the pollution entering streams is transported via groundwater discharge and overland runoff (Jacobs and Gilliam, 1983; Cooper and Gilliam, 1987). Overbank flooding can also contribute significant amounts of water in certain systems (Cole et al, 1997). When restoring riparian wetland and stream systems

---

[1] Buck Engineering, 8000 Regency Parkway, Suite 200, Cary, NC 27511. ktweedy@buckengineering.com.

concurrently, additional considerations are needed during the natural channel design process to restore a self-sustaining wetland system.

METHODS

The information presented in this paper was collected during the design and construction of three stream and wetland mitigation projects across the Coastal Plain of North Carolina. One site was completed in the spring of 2003, another was completed in the spring of 2004, and the last site is in the final design stage at the publication of this paper. The projects were undertaken to fulfill mitigation requirements of the North Carolina Department of Transportation. Sites are located within the Neuse River Basin and involve restoration of wetland and stream functions to prior-converted farm fields. Prior to restoration, the sites were primarily used for row crop agriculture. All sites contain hydric soils typical of low gradient floodplain areas of the North Carolina Coastal Plain. Streams flowing through the sites had been channelized in the past to promote drainage and reduce flooding of adjacent farm fields.

Pre-restoration assessments were conducted on the three mitigation sites during 2001 and 2002. Assessments included procedures to document the geomorphic and functional condition of streams on the sites, including channel cross-sections, longitudinal profiles, bed material analyses, existing vegetation identification, and potential causes of impairment. Cross-section and longitudinal surveys were conducted with total station surveying units and all surveyed points were tied to a relative benchmark on-site. Cross-sections were measured along the degraded stream reach to document the dimension of the existing channel. Any apparent bankfull features (Rosgen, 1996) were surveyed as part of the cross-sections and longitudinal profiles. Bankfull features typically included an alluvial bench or prominent scour line. Longitudinal profiles were conducted to determine overall stream and valley slope.

Bed material analyses were collected from riffles and pools along the stream reaches. Samples included 5 cm diameter cores to a depth of 5 cm from the streambed along the surveyed stream reach. Samples were combined for each stream reach and dry sieved in a lab using ASTM standard testing sieves. Vegetation assessments were performed to identify the primary canopy, sub-canopy, shrub, and herbaceous species present on each site.

Soil profiles across the site were examined and saturated hydraulic conductivity measurements were conducted in-situ using the auger-hole method, as described by vanBeers (1970). Monitoring wells were installed and monitored for at least one growing season to document existing site hydrology. Wells were installed to a depth of one meter, constructed of 5 cm diameter PVC pipe, and screened over the entire one meter depth. Some wells were instrumented with automated loggers (Remote Data System WL-40), recording water levels twice per day, while other wells were recorded once per month using manual readings.

Four reference sites were surveyed as part of the described work. The sites all contained first or second order sand bed channels with associated riparian wetland systems along the adjacent floodplain. Drainages areas for the reference reaches ranged from 0.67 to 7.8 km^2. Sites were selected based on the presence of a stable, unincised, meandering stream channel and a mature, hardwood canopy. All reference sites also exhibited adjacent wetland floodplains with hydric soils. Measurements were similar to those described for the existing condition of the project sites, and included cross-sections, longitudinal profiles, bed material analyses, and riparian vegetation assessments. Detailed cross-sections were conducted for at least two riffles and one pool at each reference site. Detailed longitudinal profiles were conducted for a length equal to at least 20 bankfull channel widths. The longitudinal profile recorded thalweg, water surface, top of both stream banks, and any bankfull feature elevations at each of the following facet locations: head of riffle, head of pool, and maximum pool depth. Bed material analyses and vegetation assessments followed the same procedures described for pre-restoration assessments on the project sites. Monitoring wells were installed across two of the reference sites to document reference wetland hydrology. Wells included both automated recording wells and manually read wells.

After restoration of a site is completed, mitigation monitoring guidelines (USACE et al., 2003) require that the following parameters be collected on stream and wetland restoration sites for a five year period: yearly cross-sections, bi-yearly longitudinal profiles, periodic photographs throughout the year, monitoring wells (both automated and manual), and survival and identification of vegetation in monitoring plots. Portions of these data have been collected on the two completed mitigation sites and are presented in the following discussion.

RESULTS AND DISCUSSION

The following discussion is based on experiences gained from restoring small, headwater sand bed stream systems with associated riparian wetland floodplains in the Coastal Plain of North Carolina, and data collected from reference sites. The discussions assume that the primary technique for restoring wetland hydrology is to raise the bed elevation of an incised stream to reconnect the stream to its floodplain.

Floodplain Considerations

Most impaired stream channels have been channelized, either directly through the fall of the valley, or moved to the toe of an adjacent hillslope to allow for more efficient agricultural production. Restoration of these systems should focus on restoring the channel back to the topographic low point of the valley. In so doing, high energies that can occur during flooding events are more evenly spread across the floodplain and there is less potential for channel avulsion and cut-offs.

Flooding frequency, duration, and depth in key areas can be controlled to some degree through topographic manipulations. Data collected from reference sites indicate that hydrology is variable across a site, as has been reported by others (Bledsoe and Shear, 2000). Topographic depressions in the floodplain, both minor and major, store rainfall and overbank flows for longer periods of time. The resulting increased surface storage has a significant effect on local groundwater hydrology. Localized high spots are formed by tree falls and fluvial deposits and contribute to the properties of forest soils and the diversity and patterns of plant communities (Lutz, 1940; Stephens, 1956; Bratton, 1976; Ehrnfeld, 1995). Restoration attempts should incorporate microtopography treatments to provide increased surface storage, topographic variability, and biodiversity. Data collected from reference sites in this study indicate that minor topographic undulations from 10 to 20 cm across distances of 2 to 3 meters are typical. Topographic differences as low as 10 cm have been linked to significant differences in plant community composition and structure (Bledsoe and Shear, 2000). Such small scale topographic manipulations can be achieved through the use of common tillage equipment used in agricultural operations. For the projects described in this study, a 2.4 meter wide disk was configured such that the disk blades cast spoil to one side when pulled behind a tractor. The sites were disked in an irregular, random fashion, such that subsequent passes with the disk crossed previous passes. The resulting topographic pattern was similar to that observed in reference sites.

Remnant and overflow channels are common in reference sites. These channels help distribute flows across the floodplain, provide increased surface storage, and increase habitat diversity. Overflow channels can be incorporated into a restoration design, with several considerations. Remnant and overflow channels that direct water away from the restored channel and then carry the flow back to the restored channel can create the potential for channel avulsion. These should be used with caution and should also incorporate grade control to prevent cutoffs. Channels can also be built that distribute water across the floodplain, but do not tie back into the restored channel. This reduces the risk of channel cutoff; however, the design must assess the potential for excess water flowing back to the restored channel during a high flow event and causing bank erosion. Consideration should also be given to the long-term effectiveness of using overflow channels to distribute water. In high sediment yield systems, deposition of sediment at the mouth of the overflow channel may eventually seal the channel, effectively creating a floodplain depression that only receives stream flow during floodplain flow events.

Design of Stream Dimension (Cross-section)

It is critical to understand the effect that baseflow stream conditions have on local riparian hydrology. Data collected during this study indicate that sand bed reference streams on low gradient floodplains typically exhibit high width-to-depth ratios (10 to 14). Width-to-depth ratio is defined as the bankfull width of the channel divided by the mean bankfull depth (Rosgen, 1996). Similar ratio values were documented by Sweet and Geratz (2003) as part of a study to develop bankfull hydraulic geometry relationships for the North Carolina Coastal Plain. Higher width-to-depth ratios are the result of sediment transport and fluvial processes that control the morphology of channels. These low slope systems are likely at the transition point between transport systems and aggradational or swamp systems. The effect on local water levels is two-fold: 1) a higher baseflow water elevation in relation to the ground surface, as compared to lower width-to-depth channels, and 2) more frequent overbank flooding.

The baseflow water elevation in a stream channel is typically the lowest discharge point for groundwater moving laterally through its floodplain. Therefore, the stream has a direct effect on groundwater elevations, especially in locations immediately adjacent to the channel. This effect has been documented on reference sites and restoration sites, as illustrated in Figure 1. Figure 1 demonstrates that during the dormant season (approximately mid November to mid March) and during wet periods, stabilized groundwater elevations compare closely with the approximate baseflow water level in the adjacent stream. Therefore, the restored stream dimension, which defines the depth of the restored channel and the associated baseflow water level, can have a pronounced effect on groundwater levels immediately adjacent to the restored stream channel. The magnitude and extent of the effect are functions of the hydrologic gradient, lateral conductivity of the soil, evapotranspiration losses, and seepage losses of groundwater to deeper aquifers.

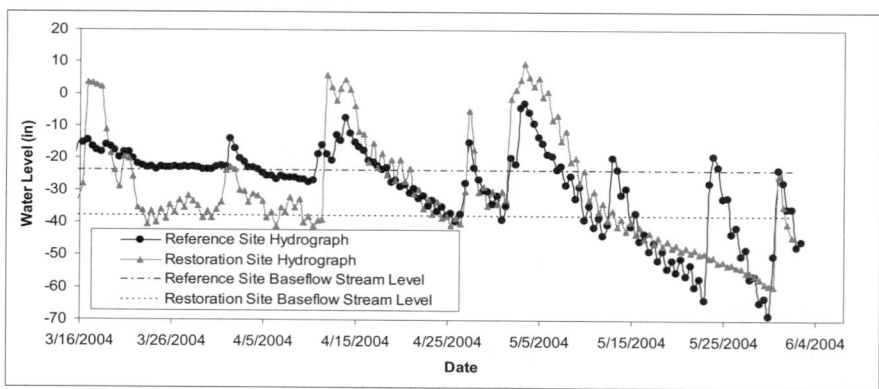

Figure 1. Groundwater hydrographs from a reference and restoration site, comparing hydrograph fluctuations to the approximate baseflow water level in the nearby stream channel.

In light of the above discussion, streams should be designed with high width-to-depth ratios when restoration of wetland hydrology is a project goal. Stream design becomes complex when considering the direct effects of width-to-depth ratio on sediment transport processes, stream stability, and wetland hydrology. Data collected from reference sites indicates a correlation (n = 4, R^2 = 0.94) between width-to-depth ratio and channel slope, that can be used as empirical data on which to base width-to-depth design values (Figure 2). In general, stream channel width-to-depth ratios increase as valley and channel slopes decrease.

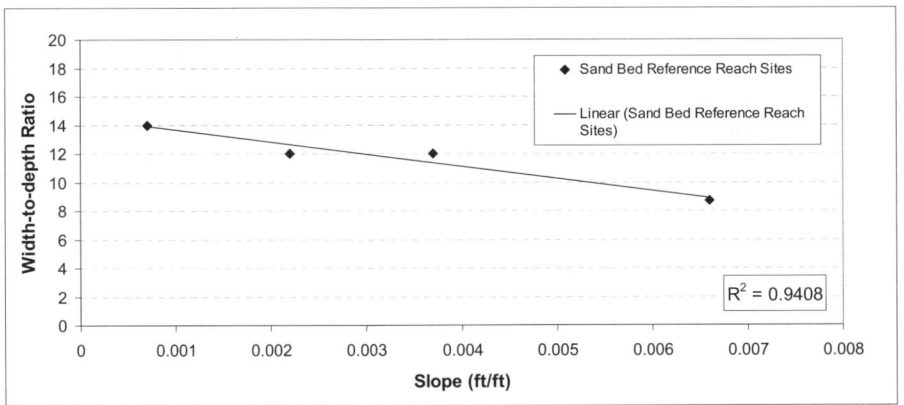

Figure 2. Width-to-depth versus slope relationship for sand bed reference reaches.

In forested wetland systems, such as the four reference sites in this study, a dense mature canopy suppresses the growth of herbaceous vegetation and stream channels typically exhibit relatively high width-to-depth ratios. On restoration sites involving agricultural fields, restored channels initially experience little shading and show the tendency to narrow due to herbaceous vegetation growth along the channel banks and sediment deposition. The increased stream power and shear stresses that result from channel narrowing can lead to subsequent channel downcutting and incision, adversely affecting wetland hydrology. Grade control structures should be incorporated into the restored stream design to maintain design stream bed elevations and protect against potential channel incision. A variety of grade control structures have traditionally been used, consisting primarily of structures constructed from rock and logs. Structures should be well sealed using filter fabric and a gradation of stone on the upstream side to reduce the potential for piping around or beneath the structure. Structures should be designed and constructed such that there is no more than 15 cm of drop in water surface over the invert of the structure. Larger drops create excessive shear stress that can lead to bed and bank scour downstream of the structure, possibly undermining the structural integrity.

Flooding processes can also be manipulated to some extent through the design of stream dimension. Natural channel designs typically have the objective of restoring a channel in which flows larger than bankfull begin to breakout onto the adjacent floodplain. This equates to a bank height ratio (BHR, ratio of stream bank height to bankfull height) of 1.0 for the design stream channel. However, in reference sites, the BHR typically ranges from 1.0 to 1.3 at various locations along the channel. This means that certain areas flood more frequently and to a greater extent than other areas. When the BHR increases in the downstream direction, the effect is that flooding increases in the lower BHR areas during out of bank events. This is due to the constriction of flows in the higher BHR areas where flood flows are routed back toward the channel. Varying BHR in restoration designs can be used to design different flooding regimes for different areas of the restoration site, depending on the habitat and water quality goals of the project.

Design of Stream Slope

The water surface slope of a stream during channel forming flows is one of several variables used to determine the stream's ability to move its sediment load. A stream channel that is designed with a slope that is too low may begin to aggrade and become unstable as the channel loses sediment transport capacity. On the other hand, a stream that is designed with a slope that is too high may experience excess shear stresses on the bed, which can lead to channel instability through incision and downcutting. A stable stream transports its sediment load without aggrading or degrading over time. Therefore, determining the proper slope for the design channel is crucial to restoring a self-sustaining system.

Water surface slopes of reference reaches surveyed as part of this work varied from 0.001 m/m to 0.0066 m/m, as averaged along the entire surveyed reach. Of the four reference stream systems surveyed, channel slope was highest for the smaller streams and decreased with increased drainage area. In the smaller headwater stream systems, abrupt drops in water surface elevation, as a result of debris jams, logs, and root mass, resulted in stable stream systems with an overall slope higher than would be expected for a stream with sand substrate.

To further investigate the range of stable sand bed stream features, longitudinal profile data were analyzed in detail to determine the range of slopes at which stable sand bed riffles formed without any visible grade control structures. The surveyed reference reaches exhibited maximum stable riffle slopes of 0.001 to 0.0055 m/m without any visible grade control in the surveyed channel section. In reference reaches that exceeded a slope of approximately 0.005 m/m, natural grade control features, such as debris jams, logs, and dense root masses, formed knick-points in the channel that allowed abrupt drops in water surface elevation over short distances. Between these natural grade control points, reach slopes fell below the 0.005 m/m maximum slope documented on other reaches that did not exhibit grade control structures.

These results indicate that for stream systems similar to those surveyed in this study, grade control is necessary when the water surface slope of the stream exceeds approximately 0.005 m/m. The number of grade control structures required for a given project can be determined by the overall stream slope of the restored channel and factoring in a maximum stream slope of 0.005 m/m between grade control structures. As discussed previously, drops should be less than 15 cm per grade control structure, to reduce the stress placed on the structure and excess scour that could undermine the integrity of the structure.

Sediment Transport Considerations

To restore a stable stream channel, the stream must be designed to carry its sediment load without aggrading or degrading overtime. While numerous research studies have been conducted to describe sediment transport in sand bed stream channels, much of the work has been conducted on larger sand bed rivers or in laboratory flume experiments. There is little information available that describes sediment transport relationships in natural headwater stream systems such as those described in this study.

Sediment transport is typically assessed by computing channel competency, capacity, or both. Sediment transport competency is a measure of force per unit area (N/m^2) that refers to the stream's ability to move a given grain size. In sand bed systems, all particle sizes are mobile during bankfull flows; therefore, there is little need to determine the maximum particle size that a stream can transport. However, comparing the design shear stress values of a restored channel to those calculated from sand bed reference reaches provides a useful comparison to determine if the stresses predicted for the design channels are within the range of those found in stable systems.

Shear stress placed on sediment particles within a stream channel may be estimated by the following equation:

$$\tau = \gamma RS, \text{ where} \tag{1}$$

τ = shear stress (N/m^2)
γ = specific weight of water $(9,810 \text{ N/m}^3)$
R = hydraulic radius (m)
S = average channel slope (m/m)

Shear stress values were calculated for the four reference reaches and are plotted versus stream slope in Figure 3.

168

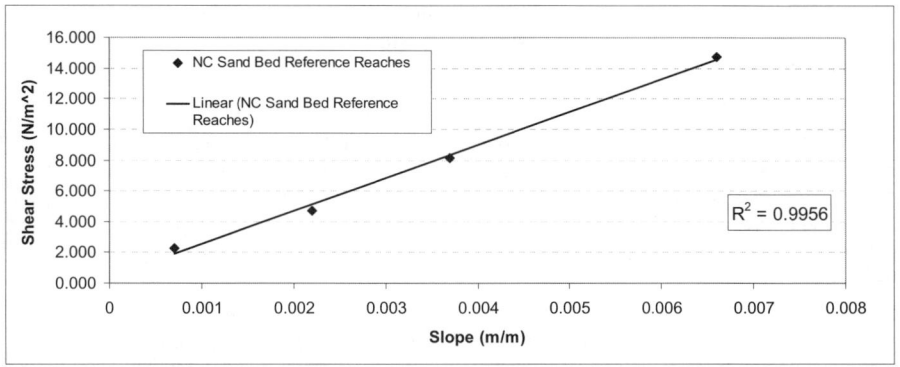

Figure 3. Shear stress versus slope relationship for sand bed reference reaches.

Sediment transport capacity refers to the stream's ability to move a mass of sediment past a cross section per unit time. Stream power is commonly used to estimate transport capacity. Stream power can be calculated a number of ways, but the following equation was used for this study:

$$\omega = \gamma QS/W, \text{ where} \qquad\qquad (2)$$

ω = mean stream power (W/m^2)
γ = specific weight of water (9,810 N/m^3). $\gamma = \rho g$ where ρ is the density of the water-sediment mixture (1,000 kg/m^3) and g is the acceleration due to gravity (9.81 m/s^2)
Q = bankfull discharge (m^3/s)
S = design channel slope (m/m)
W = bankfull channel width (m)

Equation 2 does not provide a sediment transport rate. However, it does describe the stream's ability to accomplish work, i.e., move sediment. Equation 2 was used to calculate stream power for the four reference reaches, and the results are plotted against stream slope in Figure 4.

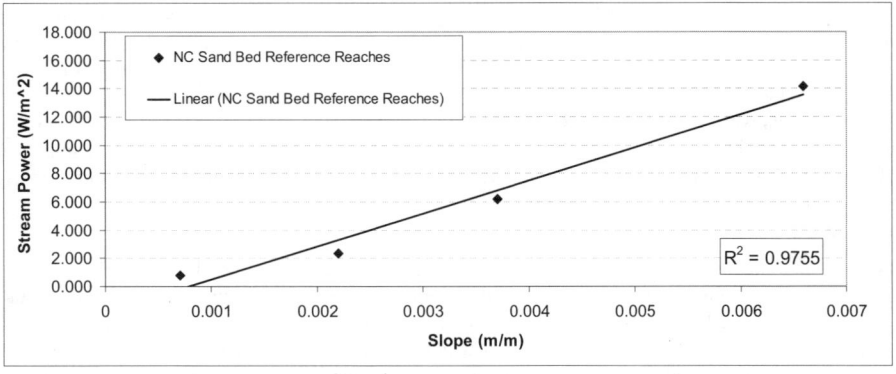

Figure 4. Stream power versus slope relationship for sand bed reference reaches.

The empirical information presented in Figures 3 and 4 can be used as a tool in the design of sand bed channels similar to those described in this study. Estimates of shear stress and stream power values for designed channels can be compared with reference reach values to assess the potential for restoring a channel that will transport its sediment load. If the design channel exhibits shear stresses or stream power values higher than those observed on the reference reaches, there is the potential that the design channel will downcut and incise. If the design channel exhibits shear stresses or stream power values less than those observed on the reference reaches, there is the potential that the design stream will not be able to transport its sediment load and may aggrade over time.

<u>Aquatic Habitat Features</u>

In channelized stream systems, aquatic habitat is often severely degraded due, in part, to large sediment loads and a lack of bed form diversity. Habitat features of restored streams can be improved by incorporating structures that provide stream bank stability, shade, holding areas and cover, and improved pool formation. The two completed mitigation sites described in this paper incorporated a variety of such structures, including root wads, log vanes, transplants, and log weirs.

Root wads are installed into the stream bank around an outer meander bend to provide stability, cover, and improved pool formation. A root wad consists of the root ball and approximately two to three meters of trunk from a tree approximately 25 cm or greater in diameter. The trunk of the root wad is buried into the stream bank, such that the exposed root mass along the stream bank deflects stream flows. The roots which extend into the channel and under the water surface increase scour in the pool, promote development of the thalweg near the outer stream bank, and provide cover for fish and other aquatic life.

A log vane is a 6 to 9 meter length of tree trunk, approximately 25 cm or greater in diameter, that is installed in the pool formed around a meander bend. The log is buried into the stream bank and angled upstream and down into the stream bed. The primary purpose of the log vane is to provide a scouring obstruction within the meander pool to maintain deep pool depths and provide cover for aquatic life.

Transplants are masses of existing vegetation that are harvested from an area of the restoration site and transplanted along the stream bank of an outer meander bend. Species that have been used for transplants include native cane (*Arundinaria gigantea*), hornbeam (*Carpinus caroliniana*), alders (*Alnus spp.*), and other species with dense, relatively shallow roots that can be moved to a new location and thrive in the wet conditions of a restored stream bank. Transplants provide immediate root mass and stability, shading, and aquatic cover and habitat. Seed bearing species develop seeds that stream flows disperse along the restored stream channel and floodplain areas.

Measurements were recorded relating maximum pool depths to the structures placed in meander bends of a restored stream channel. Along the restored stream channel completed in the spring of 2003, maximum pool depth was measured relative to baseflow water level in the summer of 2004, after the stream had experienced numerous bankfull flow events during the previous year. Measurements were recorded on 106 meander bends with structure treatments falling in one of four categories: 1) only erosion control matting placed around the meander bend, 2) transplants placed around the meander bend, 3) root wads and transplants placed around the meander bend, and 4) a log vane placed in the pool and transplants placed around the meander bend. Results are summarized in Table 1.

The results indicate that root wads and log vanes maintain deeper pool depths relative to meander bends that are simply stabilized with transplants or erosion matting alone. These results were expected based on observations from reference sites. In the surveyed reference sites, the deepest pools were associated with woody obstructions that increase local scour and maintain pool depths. Without a physical obstruction to increase scour, sand bed systems do not maintain deep pools around meander bends, as are often noted in gravel bed systems.

Table 1. Comparison of maximum pool depths related to structure placements in a restored stream channel.

Treatment	Minimum Recorded Pool Depth (m)	Maximum Recorded Pool Depth (m)	Average Recorded Pool Depth (m)
1. Erosion matting only around the meander bend (n = 19).	0.60	1.73	0.97
2. Transplants placed around the	0.5	1.45	1.02

meander bend (n = 40).			
3. Root wads and transplants placed around the meander bend (n = 11).	0.85	1.55	1.13
4. Log vane in the pool and transplants placed around the meander bend (n =38).	0.80	2.23	1.38

CONCLUSION

The information presented in this paper is provided as an aid to those undertaking restoration efforts in riparian wetland areas. Data collected from completed mitigation projects demonstrate the effect of the stream channel on associated wetland areas. Between rainfall events, baseflow water surface elevations have a strong influence on the groundwater level in adjacent wetland areas. The extent of this effect is a function of hydrologic gradient, lateral conductivity of the soil, and the amount of water lost from the system by other pathways (e.g., evapotranspiration and deep seepage). For this reason, high width-to-depth ratio channels are preferable when restoration of wetland hydrology is a project goal. Considering the effects of width-to-depth ratio on sediment transport and, ultimately, stream stability, the design process for restoring these systems becomes an iterative process in which stream stability and function must be balanced with the hydrologic goals of wetland restoration.

More information is needed regarding the physical processes which govern the trends observed in collected data. The design considerations and recommendations presented in this paper are based on empirical data collected from reference sites, and lessons learned by restoring mitigation sites. The information has been collected for relatively small sand bed stream systems (first and second order) in the Coastal Plain of North Carolina. Data from other physiographic regions and larger riparian wetland systems are needed to determine if the recommendations are appropriate for other similar wetland and stream systems.

REFERENCES

Bledsoe, B. P. and T. H. Shear. March 2000. Vegetation along hydrologic and edaphic gradients in a North Carolina Coastal Plain Creek Bottom and Implications for Restoration. WETLANDS, Vol. 20, No. 1, pp. 126-147.

Bratton, S. P. 1976. Resource division in an understory herb community: responses to temporal and microtopographic gradients. The American Naturalist 110 (974):679-693.

Cole, A. C., R. P. Brooks, and D. H. Wardrop. December 1997. Wetland hydrology as a function of hydrogeomorphic (HGM) subclass. WETLANDS, Vol. 17, No. 4, pp. 456-467.

Cooper, J. R. and J. W. Gilliam. 1987. Phosphorus redistribution from cultivated fields into riparian areas. Soil Science Society of America Journal 51:1600-1604.

Ehrnfield, J. G. 1995. Microsite differences in surface substrate characteristics in Chamaecyparis swamps of the New Jersey pinelands. Wetlands 15(2):183-189.

Jacobs, T. C. and J. W. Gilliam. 1983. Nitrate loss from agricultural drainage waters: implications for nonpoint source control. Report 209. Water Resources Research Institute of the University of North Carolina, Raleigh.

Lutz, H. J. 1940. Disturbance of forest soil resulting from the uprooting of trees. Yale University School of Forestry Bulletin No. 45.

Rosgen, D. L. 1996. Applied River Morphology. Wildland Hydrology, Pagosa Springs, CO.

Rosgen, D. L. 1998. The reference reach – a blueprint for natural channel design. In Proceedings of the ASCE Wetlands and Restoration Conference, March 1998, Denver, CO.

Stephens, E. P. 1956. The uprooting of trees: a forest process. Soil Science Society of America Proceedings 20:113-116.

Sweet, W. V. and J. W. Geratz. 2003. Bankfull hydraulic geometry relationships and recurrence intervals for North Carolina's Coastal Plain. Journal of American Water Resources Association, August 2003. pp. 861-871.

US Army Corps of Engineers, US Environmental Protection Agency, NC Wildlife Resources Commission, and the North Carolina Division of Water Quality. April 2003. Stream Mitigation Guidelines.

van Beers, W. F. J. 1970. The auger-hole method: a field measurement of hydraulic conductivity of soil below the water table. Rev. ed. ILRI Bulletin 1, Wageningen, 32 pp.

East Fork of the Blue Earth River Realignment: Design and Construction of a River Realignment Utilizing Bioengineered Channel Stabilization Techniques

Walter Eshenaur, SRF Consulting Group, Inc., (member ASAE)
David Filipiak, SRF Consulting Group, Inc.,
John McDonald, Faribault County, MN

Reconstruction of the East Street bridge over the East Fork of the Blue Earth River involved removing the original bridge and returning the river to its natural meander. This paper presents the research, design, modeling, permitting and construction details of the river realignment. Bank stabilization techniques of fillstone toe protection to counter effects of high velocity and potential scouring, bendway weirs that force the thalweg into the center of the river together with vegetative techniques were combined in an integrated, bioengineered solution. Three years of monitoring results will also be presented.

Presenter's Contact Information: Walter Eshenaur, P.E., M. Ag. Eng.
                                     David Filipiak, P.E.
                                     SRF Consulting Group, Inc.
                                     Plymouth, MN
                                     Phone: 763-475-0010
                                     E-mail: weshenaur@srfconsulting.com

Conservation Development and Ecological Stormwater Management:

An Ecological Systems Approach™

D. M. Mensing[1], MS, PWS and K. A. Chapman, PhD[1]

ABSTRACT

Conventional development practices often eliminate or significantly compromise natural resources and their associated ecological systems. Native habitats are typically lost, fragmented, or degraded, and hydrologic systems modified significantly, reducing water quality and causing more volatile flows and water level fluctuations. Increasing and inevitable development across the country requires that approaches to development address and reduce these adverse ecological impacts if conservation objectives and sustainable communities are to be achieved.

Many of these challenges can be overcome when we take an Ecological Systems Approach™ to site design. Applied Ecological Services, Inc. (AES) developed this approach, which integrates the science of ecology harmoniously with project design to emulate natural systems in all phases of development and restoration projects. Employing conservation development principles can reduce or avoid many adverse ecological effects and better integrate healthy ecosystems with the built environment, to the mutual benefit of people and nature. At AES our approach to conservation development focuses on preserving, restoring, connecting, and managing healthy native ecosystems in perpetuity, while still achieving development goals. Ecological stormwater management plays an important role in most of our projects. We employ a Stormwater Treatment Train™ approach, which first addresses source management and then utilizes a combination of different natural systems (e.g., vegetated swales, prairies, wetlands) to provide rate and volume control and to increase water quality while enhancing infiltration, recharging groundwater systems and baseflow, and creating and connecting aesthetically pleasing wildlife habitat. This approach is in sharp contrast to conventional stormwater management techniques, which typically employ curb, pipe, and Nationwide Urban Runoff Program (NURP) ponds. Case studies are presented to illustrate the application of conservation development and ecological stormwater management principles to site design, implementation, and management to achieve conservation and stormwater objectives.

KEYWORDS. Conservation development, low impact development, storm water management, ecology.

INTRODUCTION

Humans are inescapably dependent on our ecological resources, and the sustainability of these resources is one of the greatest challenges of the 21st century (Costanza et al. 1997; Dailey 1997). Ecosystem services (e.g., providing stable soils, filtering and absorbing surface water, removing nutrients, cleaning air) have long been ignored or undervalued in many developing countries, but they are now receiving greater recognition as their importance becomes more understood. People have begun to acknowledge the link between the sustainability of human society and economic systems, and the sustainability of natural systems.

The adverse impacts of conventional development practices on ecological systems can be, and often are, severe. Consequences of conventional development include significant loss and

---

[1] Applied Ecological Services, Inc. 21938 Mushtown Road, Prior Lake, MN 55372

fragmentation of native plant communities (e.g., prairie, native forest) with associated declines in wildlife populations (Askins 1995; Herkert 1994), species composition changes and invasion by non-native species in remnant natural areas (Noss & Peters 1995), and negative effects on the integrity of lakes, rivers, streams, and wetlands (Naiman et al. 1995). These direct and indirect impacts typically lead to a loss of native biodiversity and species endangerment (Naidoo & Adamowicz 2001), a decrease in soil stability and consequent increases in erosion and sedimentation rates, an increase in nutrient loadings (Miller & Nudds 1996), and the diminishment of ecosystem services.

One of development's greatest effects on ecological resources is the alteration of the local hydrologic cycle, primarily from land clearing and the replacement of natural vegetation with impervious surfaces (Groffman et al. 2003). Yet accepted mitigation practices do not effectively treat the most damaging outcomes of development. Roofs, driveways, roads and other impervious surfaces typically are designed to concentrate runoff in conventional stormwater infrastructure, which consists of conveyance, detention, and retention structures. Pond design criteria of the Nationwide Urban Runoff Program (NURP) relies on sedimentation for stormwater treatment, typically resulting in relatively deep pools that provide limited aesthetic and wildlife values and moderate nutrient removal efficiencies. Studies have shown (EPA 1999) that following NURP guidelines does not ensure nutrient capture and retention, especially when removal of accumulated sediment is not conducted regularly. NURP ponds also are often anaerobic by design, which favors the development of a bacterial community that releases soluble phosphorus into the water column. In the Midwest, phosphorus is often the nutrient of greatest concern in surface waters and aquatic systems (Kroening and J. Stark 1997, Sturgul et al.).

CONSERVATION DEVELOPMENT PRINCIPLES

At Applied Ecological Services, Inc., we have developed the Ecological Systems Approach™ to site design, where we integrate the science of ecology harmoniously with project design to emulate natural systems in all phases of development and restoration projects. Our approach to conservation development focuses on preserving, restoring, connecting, and managing healthy native ecosystems in perpetuity, while still achieving development goals. We have developed this approach over the last fifteen years in hundreds of projects ranging from those with a pure conservation goal, to those where regulations constrained a project and required an ecological approach.

From these experiences we have derived several conservation design principles:

1. Preserve the integrity, vitality and sustainability of natural systems;

2. Integrate natural resource protection with development;

3. Employ environmental engineering principles to manage stormwater runoff (i.e., mimic natural systems);

4. Restore damaged ecological systems;

5. Buffer natural resources;

6. Ensure protection and management over the long term;

7. Encourage native landscaping;

8. Provide opportunities for ecological education and volunteer stewardship.

We have seen that proper application of these principles results in a unique, culturally rich human environment and a financially successful project for the client.

The application of these principles creates noticeable differences in residential design between convention and conservation developments. In conventional residential design, standard-sized lots are distributed evenly throughout a site's uplands and are large enough to accommodate individual septic systems (Figure 1). These design standards are required by most municipalities undergoing rapid suburbanization on the edges of metropolitan areas and in rural regions. The only non-private, and therefore common areas that can be used for conservation, are the outlots, which contain wetlands already protected under state and federal regulations. Stormwater runoff typically is concentrated and then managed with NURP ponds before being discharged to existing wetlands. The conservation outcomes are minimal, and this approach offers little opportunity for stormwater volume control and less than optimal treatment potential for stormwater constituents. Almost all the uplands are privately owned, and conservation is limited to wetlands which typically are in poor condition because of previous agricultural activities. Habitats are fragmented by private ownership, where a lack of interest, expertise, and resources fail to provide for restoration and long term management.

Figure 1. Conventional development design.

By contrast, conservation development (Figure 2) provides for the protection, restoration, and management of significant open space by concentrating homes in the most degraded land (left portion of site) and creating a large core habitat preserve (right portion of site). Conservation development often increases the number of homes above the conventional approach, but this increased density is offset by the inclusion of meaningful open space in the developed area. This open space offers greater opportunity for dispersed stormwater management with greater infiltration and volume control, and provides opportunities for amenities such as trails, native gardens, and increased aesthetic enjoyment of the neighborhood. Most lots face open space, giving each homeowner the sense of living on a larger lot. In conservation developments, streets are narrowed and shortened to increase open space and create space for the stormwater management system, but this also reduces the length of utility runs. This can result in savings that can be put towards natural resource restoration and management. This design does require connection to a municipal wastewater treatment system or construction of a local community wastewater treatment plant because the lots are too small for private septic systems. Conservation development thus integrates conservation and development goals to the mutual benefit of people and nature, and establishes larger protected areas that are restored and managed for conservation purposes, while creating a unique and more ecologically sensitive living space for humans.

Figure 2. Conservation development design (same site as Figure 1).

ECOLOGICAL STORMWATER MANAGEMENT

An important part of conservation development is to achieve ecologically meaningful mitigation of the damage to downstream aquatic resources resulting from impervious surfaces. Conventional stormwater management systems are designed to collect stormwater and convey it rapidly to holding and treatment ponds, while ecological stormwater management systems provide for infiltration and evapotranspiration along the length of the system. Consequently, a conventional stormwater approach produces a different hydrograph than an ecological approach (Figure 3). In contrast to conventional stormwater management, ecological stormwater management seeks to reduce the total volume of water leaving a site following a precipitation event. This affects the rate at which water leaves a site, creating more even flows and less dynamic water levels in downstream receiving waters. In addition, the length of the ecological stormwater management system removes phosphorus at high rates by capturing sediment before it reaches holding and treatment areas. Consequently, with ecological stormwater management, aquatic vegetation is less subject to rapidly fluctuating water levels and nutrient inputs, and this stability supports the persistence of native plant species, reduces erosion rates, sustains larger and more varied wildlife populations, and creates a more beautiful aquatic environment.

Annual Hydrographs and Normal Average Water Levels for Restored Wetlands

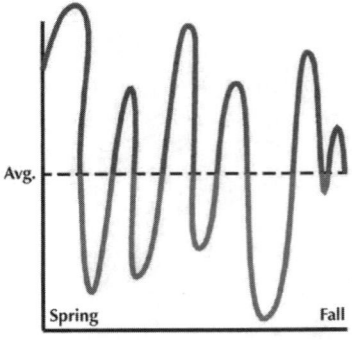

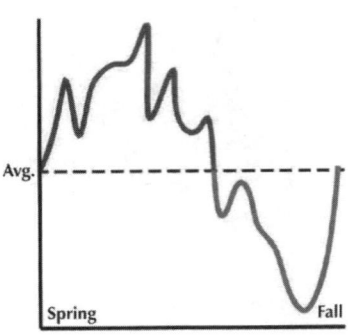

Conventional Approach to Hydrology

- Unpredictable Swings in Water Levels
- Creates Biological Instability
- Promotes Habitats for Weeds and Poor Aesthetics
- Promotes Poor Water Quality

Ecological Approach to Hydrology

- Annual Seasonal High and Low
- Predictable Hydraulics and Seasonal Trajectory
- Promotes Habitat for Stable yet Dynamic Plant Communities
  (Diversity of Plants and Animals)

Figure 3. Generalized hydrographs comparing a conventional stormwater pond (left) with an ecologically designed treatment wetland (right). The ecological approach provides source management prior to runoff reaching the treatment wetland, while the conventional design does not.

The environmental benefits of ecological stormwater management are achieved using the principles incorporated in the Stormwater Treatment Train™ (STT) (Figure 4). First and foremost, the STT seeks to reduce stormwater run-off volume by combining the infiltration and evapotranspiration effects of natural systems. With a reduction of overall volume comes better control of the rates at which stormwater moves through the system and exits the site. In addition, water quality is improved, groundwater is recharged, and aesthetically pleasing wildlife habitat is created.

The components of the STT are: 1) native, vegetated swales which provide conveyance and opportunities for infiltration and filtration; 2) upland and wet prairies which accomplish additional infiltration, remove suspended solids and capture nutrients, especially phosphorus; and 3) biofilter wetlands which are particularly effective at capturing nitrogen. The STT, like conventional systems, discharges to a lake, stream, or natural wetland. The effectiveness of the STT results from combining different treatment elements. This approach employs biological and mechanical processes, rather than relying on mechanical processes alone. The STT mimics the way natural watersheds work and provides a more robust stormwater treatment system than a conventional approach.

Besides water volume and rate control, and water quality benefits, we incorporate the STT into the project design as open spaces and interconnected greenways. Trails often run along the treatment train swales, prairies and wetlands. Groundwater recharge increases on the site and baseflows to nearby streams, lakes and wetlands are improved during drought. Lastly, by eliminating curbs, gutters, and storm drains, sewers, and outlets, the resulting savings can be applied to the costs of planting and establishing native vegetation throughout the stormwater system. We have found that the savings are greater the larger the site.

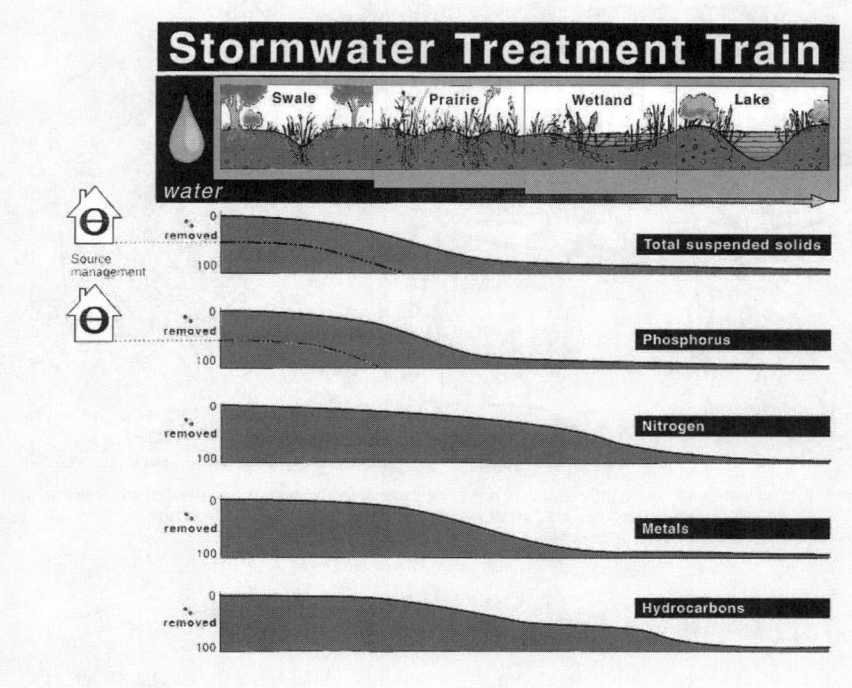

Figure 4. The Stormwater Treatment Train™ approach uses a combination of different natural systems to manage stormwater and restore more natural local hydrology.

Sometimes site constraints do not allow stormwater to be managed exclusively with ecological components. We have tailored the STT approach for dozens of sites, using the most appropriate elements in the space available. Occasionally hybrid systems are designed and constructed, which contain some STT elements along with conventional techniques.

Creating an ecological stormwater management system has implications for other parts of the development process. Because we work as much as possible with the natural contours, drainages, and depressions of a site, we find that less land needs grading, and the grading that is necessary is on average less extreme, making cut and fill depths shallower. Surface water is delivered through shallow swales to the open spaces (Figure 5). Rural road cross-sections are used and shallow swales are placed in the roads rights-of-way. Stormwater flows down these swales and along created lot line swales. Water is directed to the rear of the lots and into the adjacent open space consisting of upland prairie. During heavier rainfall or snowmelt events, water flows across the upland prairie areas into wet prairies, which eventually flow into shallow biofilter wetlands. The greatest challenge in employing this approach is the education of landowners who are used to mowed lawns up to their property lines. The swales are wet for much of the time in spring and early summer, making mowing difficult. We are learning that it is best to secure these lot lines with a stormwater utility easement and require, through a homeowners association by-law document, that the swales be planted with permanent vegetation or a combination of rock and vegetation with no underlying impervious landscaping material.

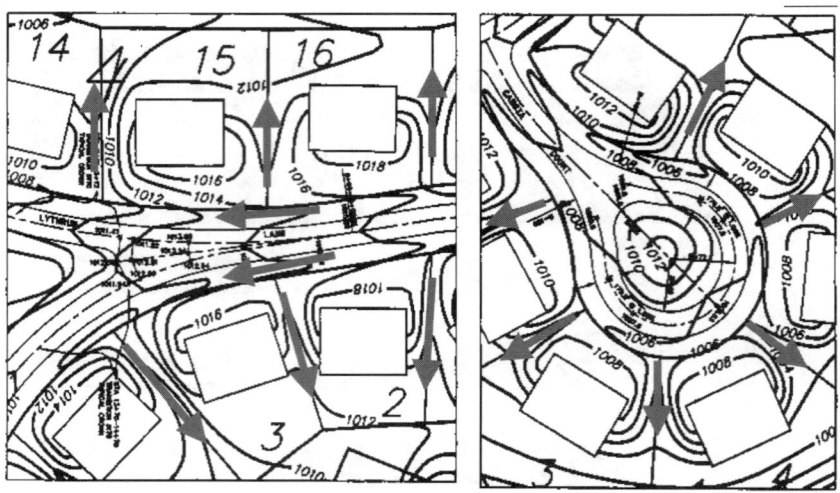

Figure 5. Conservation design includes appropriate grading to provide stormwater routing and management through native vegetated swales leading to other management elements.

CASE STUDIES

Wild Meadows

Wild Meadows is a conservation development located in Medina, Minnesota. The design of Wild Meadows was inspired by other successful AES conservation developments including Prairie Crossing, an award-winning development outside of Chicago. AES is the ecological consultant on the project team, which includes developers, planners, engineers, builders, and the Minnesota Land Trust, which will hold a conservation easement on the preserve portion of the development.

AES worked with the development team to design around and with the existing natural resources on the site. 200 acres of the 345-acre site (58%) are being restored to native open space designed by AES. This includes maple-basswood forest, oak savanna, upland prairie, wet prairie, and wetlands. AES is currently restoring and managing these areas, which will be maintained in perpetuity through homeowner association funds.

AES used the STT approach to alternative stormwater management at the site; no storm sewer, no curb and gutter, and no NURP ponds are used. Instead, stormwater is guided through gentle vegetated swales into a series of upland and wet prairie restorations and then into restored wetlands. AES is conducting a six-year hydrological monitoring program to provide field data regarding the effectiveness of the ecological stormwater management approach.

Other conservation development techniques employed at Wild Meadows include minimal road widths, planted cul-de-sacs, reduced land grading, reduced hard infrastructure, forest conservation through preservation and expansion of the maple-basswood forest remnant, establishment of conservation easements and other covenants, and preserving over half the site as open space that can be enjoyed by residents through the a network of walking trails.

Prairie Crossing

Prairie Crossing in Lake County, Illinois is a nationally-studied model of conservation development built in the early 1990's. Home sites were clustered to preserve the surrounding open space. An astounding 80% of the site consists of natural landscapes and working agricultural lands. To emulate the functions and aesthetics of the historical landscape, AES restored over 170 acres of the prairie and wetland communities that once occupied the site. Residents enjoy an impressive diversity of native birds, butterflies, and other wildlife, they partake in guided nature walks, and they assist with some ecological management activities. The majority of the management of restored areas is conducted by AES and funded through the homeowner association.

One of the main innovations at Prairie Crossing is the way stormwater runoff is treated. AES coined the term, Stormwater Treatment Train™ while designing this project, and the STT has produced outstanding water quality. Stormwater from the development is managed in the multi-component system and discharged into Lake Mascouten, a man-made lake. Residents enjoy the lake's swimming beach, and because of the exceptional water quality the lake was used by the Illinois Department of Natural Resources to rear fish species which are rare in Illinois because of water quality degradation in natural habitats.

CONCLUSION

The obvious and cumulative impacts of conventional development urge us to employ alternative approaches. Conservation development and ecological stormwater management have proven to deliver significant environmental benefits while simultaneously creating unique and livable communities for people. Both development and ecological goals can be achieved at reasonable cost by employing conservation design principles and ecological approaches to stormwater management early in the design process. The outcomes for society, economy and the environment will be measurably superior.

ACKNOWLEDGEMENTS

Applied Ecological Services, Inc. would like to acknowledge Dahlgren Shardlow and Uban, Inc. for providing Figures 1 and 2. We would also like to thank the numerous partners in AES and in the development and regulatory community who have allowed us to bring these ideas to fruition.

REFERENCES

1. Apfelbaum, S. I., J. D. Eppich, T. H. Price, and M. Sands. 1994. The Prairie Crossing project: attaining water quality and stormwater management goals in a conservation development. In *Using Ecological Restoration to Meet Clean Water Act Goals*, 33-38.

2. Askins, R. A. 1995. Hostile landscapes and the decline of migratory songbirds. *Science* 267:1956-1957.

3. Costanza, R., R. d'Arge, R. de Groot, M. Grasso, B. Hannon, K. Limburg, S. Naeem, R. V. O'Neil, J. Paruelo, R. G. Raskin, P. Sutton, and M. van den Belt. 1997. The value of the world's ecosystem services and natural capital. *Nature* 387:253-260.

4. Dailey, G. 1997. *Nature's Services: Societal Dependence on Natural Ecosystems.* Washington, D.C.: Island Press.

5. EPA. 1999. Storm Water Technology Fact Sheet, Wet Detention Ponds. EPA 832-F-99-048. Washington, D.C.: U.S. Environmental Protection Agency.

6. Groffman, P. M., D. J. Bain, L. E. Band, K. T. Belt, G. S. Brush, J. M. Grove, R. V. Pouyat, I. C. Yesilonis, and W. C. Zipperer. 2003. Down by the riverside: urban riparian ecology. *Frontiers in Ecology and the Environment* 1:315-321.

7. Herkert, J. R. 1994. The effects of habitat fragmentation on Midwestern grassland bird communities. *Ecological Applications* 4:461-471.

8. Kroening, S. and J. Stark. 1997. Variability of nutrients in streams in part of the Upper Mississippi River Basin, Minnesota and Wisconsin. Originally published as U.S. Geological

Survey Fact Sheet FS-164-97. Available at: http://wwwmn.cr.usgs.gov/164-97/text.htm. Accessed 18 June 2004.

9. Naidoo, R. and W. L. Adamowicz. 2001. Effects of economic prosperity on numbers of threatened species. *Conservation Biology* 15:1021-1029.

10. Naiman, R. J., J. J. Magnuson, D. M. McKnight, and J. A. Stanford. 1995. *The freshwater imperative*. Island Press, Washington, DC, USA.

11. Noss, R. F. and R. L. Peters. 1995. *Endangered Ecosystems: A Status Report on America's Vanished Habitat and Wildlife*. Washington, D. C.: Defenders of Wildlife.

12. Miller, M. W. and T. D. Nudds. 1996. Prairie landscape change and flooding in the Mississippi River Valley. *Conservation Biology* 10:847-853.

13. Robinson, S. K. and D. S. Wilcove. 1994. Forest fragmentation in the temperate zone and its effects on migratory songbirds. *Bird Conservation International* 4:233-249.

14. Sturgul, S., L. Bundy, and F. Madison. Phosphorus management practices. University of Wisconsin-Extension. Available at: http://ipcm.wisc.edu/pubs/pdf/phos1.pdf. Accessed 18 June 2004.

THE APPLICABILITY OF OPEN-SLOT GROUNDSILLS IN TRAINED RIVER TO ENCOURAGE THE FORMATION OF MEANDERING

C.-C. Wu[1] , J.-J. Lin[2], Y.-C. Lin[3], C.-M. Yeh[3]

ABSTRACT

The objective of this research is to experimentally assess the applicability of open-slot groundsills in trained streams to encourage the formation of meandering. Mechanics of meander formation as well as the sinuosity index in the settings of serial open-slot groundsills at different channel gradients, sediment diameters, offset-interval ratios, and flow discharges is discussed. Selective sediment transport as well as deposition occurred on the outskirts of thalweg and floodplain was observed after the installation of open-slot groundsills. Both processes encourage the formation of alternate bars and meanders, which result in greater diversity in trained streams.

KEYWORDS. Groundsills, Meandering, Sinuosity, River training

INTRODUCTION

In order to mitigate the disaster caused by flood and sediment, streams are often trained using dikes, revetments, etc. Grade-control structures such as groundsills and check dams are often implemented to adjust the channel gradient and prevent the channel bed from scouring. In the check-dam-controlled environment, safety of the surrounding is frequently the primary concern in stream training. However, greater effective fall heights of the structures result in discontinuity in the stream system that regularly segregates the passage for aquatic habits. On the other hand, low drop structures like groundsills; typically constructed in a submergence fashion, still likely produce the same environment (Figure 1) especially for streams with high high-low stage ratio.

Natural streams often follow the principles of self-adjustment. They tend to meander even after artificial training. Many factors control the formation of meanders in both natural and artificial channel reaches. Factors include sediment supply, properties of the sediment especially the size distribution, flow discharge, and the characteristics of the channel. As the results of meanders, riffle-pool sequences eventually occur.

Pool is a region of slow moving water with deep flow depth, it is often found in the meander bend. Riffle, on the other hand, is a region of fast moving water with shallow flow depth. It occurs crossover two sequential pools. Riffle-pool structure not only provides great diversity on bedforms, substrate materials, and local velocities, but also provides shelter and feeding ground for aquatic habits (Gordon et al., 1992).

Although low drop structures serve the purpose of channel control, the installation of groundsills alters both the flow fields and sediment transport in a channel reach, which results in a rather uniform appearance as compared to the riffle-pool setting in a natural environment. It may

[1] Professor, Dept. of Soil and Water Conservation, National Pingtung University of Science and Technology, TAIWAN. Contact e-mail address: ccwu@mail.npust.edu.tw

[2] MSc, Research assistant, Soil Erosion Research Unit, Dept of Soil and Water Conservation, National Pingtung University of Science and Technology, TAIWAN. E-mail: m9037010@mail.npust.edu.tw

[3] Master students, Dept. of Soil and Water Conservation, National Pingtung University of Science and Technology, TAIWAN. E-mail: m9237004@mail.npust.edu.tw, m9237002@mail.npust.edu.tw

therefore extend the self-adjustment process to a much longer period. Therefore, the objective of this research is to experimentally assess; from the viewpoint of sinuosity, the applicability of open-slot groundsills at different settings in a trained straight channel reach to encourage the formation of meandering.

Figure 1. Segregation of stream system caused by groundsills

LITERATURE REVIEW

L.B. Leopold and M.G. Wolman are probably the most important researchers in the history of river morphology especially in the quantification of river morphology using sinuosity index. Sinuosity index is defined as the ratio between flow length and meandering wavelength. Leopold and Wolman (1957) categorized the channels by sinuosity indexes into three types. They were straight, sinuous, and meandering channels respectively.

Under the classification of Leopold and Wolman, the sinuosity index for a straight channel was found to be less than 1.1; whereas, that for a sinuous channel was between 1.1 and 1.5. As sinuosity indexes reach beyond 1.5, channels tend to become meandering. The main reason that causing a straight channel to become meandering was the microstructures on the channel bed that generated fluctuation in channel elevations. These microstructures are known as riffles and pools that play an important role in the diversification of flow field.

Leopold et al. (1964) further analyzed the river morphology by assessing the relationship between channel gradients (S) and bank-full discharges (Q_b). The resultant empirical equation is written in Eq.[1]. By plotting Eq.[1] as that shown in Figure 2, one can quantitatively distinguish braided channels from meandering channels. For those having bank-full discharges and streambed gradients located underneath the curve in Figure 2, channels possess the nature of meandering; otherwise, they appear to be braided.

$$S = 0.0125 Q_b^{-0.44}$$ [1]

Rust (1978) classified the river channels into four groups based on river morphology. They were straight, meandering, braided, and anatomizing streams. Gordon et al. (1992) followed the same classification and summarized their findings with that from Leopold et al. (1964), Morisawa (1985), and Selby (185) in the aspect of channel pattern description, width-depth ratio, sinuosity, bank-full velocity, and stream power.

As indicated in Gordon et al. (1992) summary, straight streams seldom occur in a natural setting unless being contained by well-defined stable banks; for instance channelized streams. Even though the stream appears to be straight, meandering thalweg still exists with sinuosity index between 1.0 and 1.5.

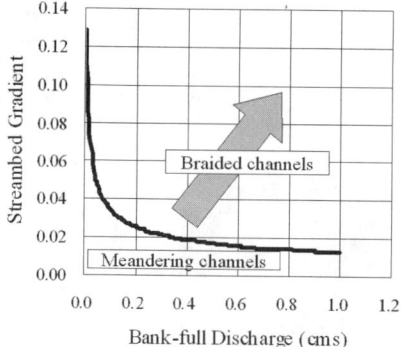

Figure 2. Relation between streambed gradient, bank-full discharge and river morphology

Selby (1985) further refined the river pattern classification from the viewpoint of sediment transport. Variables considered by Selby included sediment supply, dominant textures of floodplain sediment, ratio of bed load to total sediment load, streambed gradient, sinuosity, and river stability. In Selby's classification, confined meander was addressed for streams contained by stream banks or revetments.

Literature reviews all indicate the occurrence of meandering in confined streams. As the implementation of low drop groundsills, further constrain exists at the channel bed that may affect the process of sediment transport. Therefore, providing openings on the groundsills and adjusting opening offset in the subsequent groundsill structure may be a reasonable means to encourage the formation of meandering.

EXPERIMENTAL APPARATUS AND MATERIAL

Hydraulic flume, open-slot groundsill scaled models, and experiment setup

This research was conducted in a semi-circulated flume in Steepslope Hydraulics Laboratory at National Pingtung University of Science and Technology. The flume measured 4 m in length, 0.91 m in width, and 0.30 m in depth. A pair of revetments with the face slope of 1:0.3 (V:H) was furnished to imitate the commonly adopted revetments in Taiwan. The top width of the experimental channel that constrained by revetments was set at 0.2 m, and the model scale was selected at 1:50.

Groundsill scaled models were cut from 14-mm thick waterproof plywood. Each model having a 2cm x 2cm slot cut at the designated offset Δe was placed in the flume in the selected interval Li. Top rim of groundsill was set flush with the sand-filled channel bed, and silicone seal was applied around the perimeter of the groundsills to prevent water from seeping out of the test channel. After the seal was set, experiment sand was filled and carefully leveled. The end product of the entire experiment setup is shown in Figure 3, and the experiment variables as well as arrangement are arranged in Table 1.

Experiment sand and velocity measurement

Sand originated from river sediment was purchased from local factory. They came with three distinct sizes (d_{50}), which were 0.505, 0.745, and 1.370 mm imitating 25.25, 37.25, 68.50 mm in prototype respectively.

Velocity measurements consisted of surface velocity and velocity profile measurements. Surface velocity measurements were taken for each experiment run; whereas, velocity profile measurements were taken only for pre-selected conditions. Floats with an average area of 4 mm^2 and thickness of 0.5 mm were cut from high density Styrofoam. Total of 20 surface velocity measurements were taken in a 1.8 m reach centered at the experiment channel to eliminate the effects from inlet and outlet control sections. As flow became rough due to the occurrence of scour at the channel bed, floats became less suitable for velocity measurement then dye method was used instead.

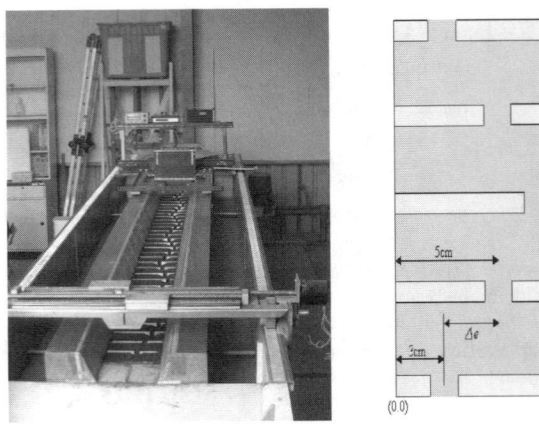

Figure 3. Flume and experiment setup

Table 1. Experiment variables and arrangement

Variables		Range		Units	
Flow rate		0.0000165~0.0012100		cms	
Channel slope		2, 4, 6		%	
Sediment mean diameter		0.505, 0.745, 1.370		mm	
Groundsill interval		0.1, 0.2, 0.3		m	
Location w.r.t. left bank (cm)	Groundsill interval Li (cm)	Offset-interval ratio ($\Delta e/Li$)	Location w.r.t. left bank (cm)	Groundsill interval Li (cm)	Offset-interval ratio ($\Delta e/Li$)
3	10	0.20	4	20	0.25
5	10	0.20	9	20	0.25
7	10	0.20	14	20	0.25
9	10	0.20	9	20	0.25
7	10	0.20	4	20	0.25
5	10	0.20	4	30	0.30
3	10	0.20	14	30	0.30
			4	30	0.30

As for velocity profile measurement, a 1.6-mm OD Pitot tube connecting to a differential pressure transducer, A/D converter, and a personal computer was used for measurements and data acquisition. Velocity profile measurements in the flow depth direction were carefully executed. Signals were captured at 1-second interval but monitored continuously. They were recorded as long as needed for each measurement point so that at least 60~70 acceptable data were stored for later analysis.

Flow- and channel-surface profile measurements

A digital point gauge with the accuracy of 0.1 mm was used for flow- and channel-surface profile measurements. A series of waterproof rules with the accuracy of 1 mm was glued to the sidewall of the revetment for quick reference. Measurements were taken for all groundsill reaches within the preceding 1.8 m reach, and total of five repetitions was gauged during the course of a run.

Outlet of the experiment channel was first blocked, followed by instant cut off of water supply and slow drain in the channel at the end of each run before the channel bed profile measurement was taken. Rill meter that often employed in soil erosion research was used with the assistance of a digital camera to capture the images of final channel bed profiles. The digital images were then manually digitized with respect to a reference point on the images to obtain channel bed profiles' elevations.

RESULTS AND DISCUSSIONS

The Formation of Meandering

Even with the installation of open-slot groundsills, overtopping streamflow still maintains its inertia to creates scour at the toe of the groundsill. The scouring process that occurring evenly across the toe of the groundsill continues until the streamflow finds its way in the course of flow that subjecting to least resistance. In the case of serial open-slot groundsills, the channel bed in the vicinity of open slot is the place where streamflow assembles. When streamflow begins to assemble, higher stream power is gained and the thalweg begins to develop.

In the setting of serial groundsills without openings, the interval between groundsills and the gradient of the channel bed control the flow field which may alter the overtopping streamflow from plunging flow to pseudo-plunging flow or submerged hydraulic jump (Wu et al., 2002). In the situation of serial open-slot groundsills, not only the groundsill interval but also opening offset (Δe) controls the development of thalweg.

Small amount of fluorescent dye was added to the flow during the experiment to highlight the flow pattern as shown in Figure 4(A) so that the contribution of open slots in generating winding flow pattern could be easily visualized. Selective sediment transport as well as deposition at the outskirts of thalweg and floodplain was observed in this study, which resulted the formation of alternate bars as shown in Figure 4 (B).

Development of thalweg and alternate bars was monitored in this study. Experiment results indicate that size of the sediment on the channel bed, flow discharge, and channel gradient control the development. Coarser the sediment gets, shallower and narrower the thalweg achieves provided that channel gradient and flow discharge remain constant. Similarly, finer the sediment exists in the channel, wider the alternate bars extend. However, as channel gradient steepens; such as the case of 6% channel gradient in Figure 5, the diversity of the alternate bars and floodplain reduces and eventually the confined meander disappears.

Increase of open-slot groundsill interval releases the constraint of meander development, which provides additional freedom for overtopping streamflow to mow its corridor with the tradeoff in time for thalweg to be fully developed. However, increasing the groundsill interval may not guarantee the formation of meandering. Observations conducted in this research indicate that sediment diameter, flow discharge, and channel gradient dominate the meandering process even at long groundsill intervals. Under the condition of long groundsill interval, the mobility of the sediment controls the sediment deposition process, which in turns affect the development of alternate bars and the direction of incoming streamflow.

Carrying out the nature of least energy spending while the flow-sediment interaction in the driver's seat, incoming streamflow may either assemble in the thalweg or detour around the bedforms. For streamflow successfully assembles in the conduit, thalweg eventually develops; otherwise, thalweg may either never develop or cultivate the channel bed transversely with large amplitude.

One of the concerns initiated at the design stage of the experiment is the deterioration of scour downstream the openings and that along the thalweg caused by concentrated streamflow. Channel bed surface profile measurements of those experiment settings that having thalweg successfully developed indicate that the width of the scour downstream the opening remains at 2 to 4 times the width of the opening; whereas, the depth of the scour is limited to the elevation of the opening. Hence, proper arrangement of serial open-slot groundsills suggests the possibility of meander formation while individual groundsill maintains its function of channel / gully control.

Figure 4. Meandering of flow path (A) and channel bed (B) (Arrow indicates flow direction)

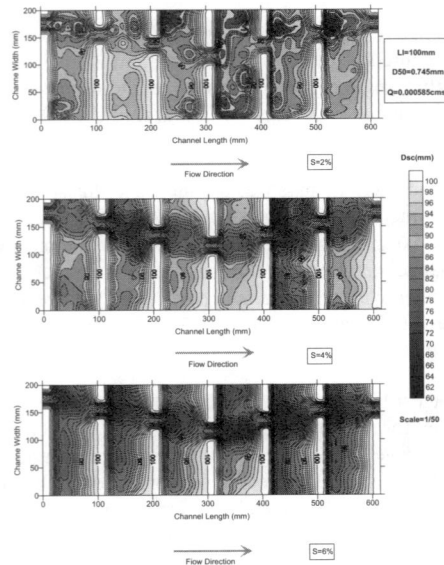

Figure 5. Partial contours of the channel bed at different channel gradients

Sinuosity Index

Sinuosity indexes calculated from channel bed surface profile measurements at different flow discharges, sediment sizes, and channel gradients are shown in Figure 6. The findings can be summarized as follows: (1). Sinuosity index decreases as channel gradient increases, in which case flow field dominates the transport of sediment and the development of meanders. In the case of high offset-interval ratio ($\Delta e/Li$), meanders with the lowest sinuosity occur. (2). Sinuosity index increases as sediment diameter decreases, in which case the mobility of sediment controls the formation of alternate bars that results in higher sinuosity. However, this relationship collapses as channel gradient increases. (3). Sinuosity index decreases as flow discharge increases, in which case flow field controls the transport capacity that results in smaller feature of bedforms.

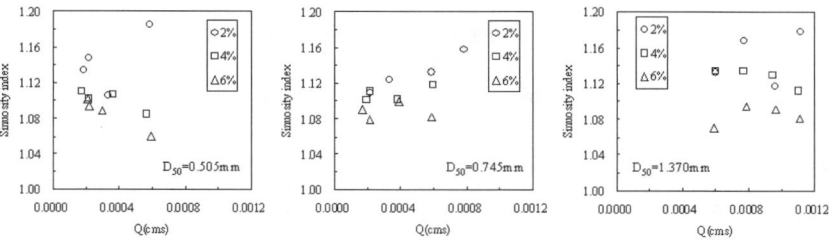

Figure 6. Relations between sinuosity index and flow discharge

Dimensional analysis is further carried out in this study. Variables considered in the analysis include sinuosity index (Sn), channel gradient (S), offset-interval ratio ($\Delta e/Li$), submerged specific density of the sediment (ρ_s/ρ_w-1), relative specific energy (E/h), and shear Reynolds number ($Re^* = u_* d_{50}/v$); in which, ρ_s is the density of the sediment, ρ_w is the density of the fluid, E is the specific energy in the channel reach, h is the mean flow depth in the channel reach.

Multiple regression analysis is then conducted, and the result is shown in Figure 7, Eq. [2], and Table 2. By combining dimensionless terms, Eq. [2] can be written as Eq. [3]; in which, d_{50} is the sediment mean diameter [m], Q is the designed flow discharge [cms], B is the designed top width of the channel [m], S is the channel gradient, and h is the designed normal depth [m]. From the practicability point of view, Eq. [3] can be used to determine the open-slot offset provided that field conditions fall within the variables' ranges; i.e., channel gradient of 2~6%, designed flow discharge of 8.839 ~ 42.426 cms, and sediment mean diameter of 25.25, 37.25, 68.50 mm.

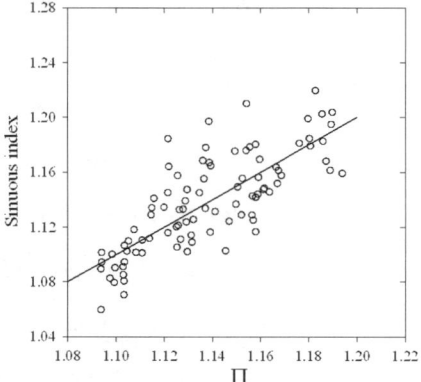

Figure 7. Relation between sinuosity index and dimensionless variable

189

$$Sn = f(\Pi) = 10^{0.0392} \ S^{-0.0261} \left(\frac{\rho_s - \rho_w}{\rho_w}\right)^{0.0554} \left(\frac{E}{h}\right)^{0.0192} \left(\frac{\Delta e}{Li}\right)^{0.0894} \left(R_e^*\right)^{0.0085} \quad ; \quad R^2 = 0.6198$$

[2]

$$Sn = 1.3099 \ \frac{D_{50}^{0.0085} \ Q^{0.0077}}{B^{0.0077} \ S^{0.0218} \ h^{0.0192}} \left(\frac{\Delta e}{Li}\right)^{0.0894}$$

[3]

Table 2. Summary of regression analysis

	Coefficient	Standard error	t-test	P-value	Confidence interval	
					Upper limit	Lower limit
Constant	0.03923	32327.4890	1.206E-06	1.00000	-64275.709	64275.787
S	-0.02611	0.0069	-3.775	0.00030	-0.040	-0.012
(ρ_s/ρ_w-1)	0.08943	0.0101	8.865	0.00000	0.069	0.110
E/h	0.05541	147311.943	3.850E-07	1.00000	-292895.729	292895.842
$\Delta e/Li$	0.01924	0.0166	1.164	0.24787	-0.014	0.052
Re^*	0.00852	0.0052	1.650	0.10258	-0.002	0.019

CONCLUSION

Revetments and low drop grade-control structures like groundsills constrain the amplitude of winding and the development of thalweg in a trained channel, which often destroys the diversity of flow field. Results of this research demonstrate the possibility of meander restoration and alternate-bar reinstatement using open-slot groundsills in trained-channel condition without severe scour on channel bed. Since open slots provide passages for streamflow and transported sediment, the width-depth ratio of the opening may affect the flow field as well as sediment transport as streamflow enters and leaves the open slot, which is not yet covered in this research. Therefore, future research in the aspect of width-depth ratio of the opening is needed to extend the results of this study.

Acknowledgements

The support from Council of Agricultural, Taiwan, R.O.C. through Grant #91-AS-1.3.1-SB-S1 is gratefully acknowledged.

REFERENCES

1. Church, M. 1992. Channel morphology and typology. In *The Rivers Handbook – Hydrological and Ecological Principles*, 126-143. London: Blackwell Scientific Publications.
2. Gordon, N.D., T.A. McMahon, and B.L. Finlayson. 1992. Patterns in Shifting Sands. In *Stream Hydrology, An Introduction for Ecologists*, 288-345. New York, N.Y.: John Wiley and Sons.
3. Leopold, L.B. and M.G. Wolman. 1957. River channel patterns – braided, meandering and straight. Professional Paper 282B. USGS.
4. Leopold, L.B., M.G. Wolman, and J.P. Miller. 1964. *Fluvial Processes in Geomorphology*. San Francisco: Freeman.
5. Morisawa, M.E. 1985. *Rivers, Form and Process, Geomorphology Texts 7*. London: Longman.
6. Rust, B.R. 1978. A classification of alluvial channel systems. In *Fluvial Sedimentology*, Memoir No. 5, 187-198. Calgary: Canadian Society of Petroleum Geologists.
7. Selby, M.J. 1985. *Earth's Changing Surface, An Introduction to Geomorphology*. Oxford: Oxford University Press.
8. Wu, C.-C., C.-H. Hou, C.-N. Chen, and C.-Y. Shih. 2002. Determination of groundsills interval for stream training. In *Proc. River Flow 2002 - International Conference on Fluvial Hydraulics*, 1109-1116. D. Bousmar and Y. Zech, eds. Louvain-la-Neuve, Belgium: Balkema Publishers.

Stream and Riparian Assessment System

Chris S. Mammoliti, Aquatic Ecologist

Tetra Tech EM Inc. 1200 S.W. Executive Dr., Topeka, KS 66615

Scott Schulte, Environmental Planner

Tetra Tech EM Inc., 8030 Flint St., Lenexa, KS 66214

Abstract. *Rapid assessments are now used widely to determine areas for stream and riparian protection, conservation, and restoration. Johnson County, Kansas selected Tetra Tech EM Inc. (Tetra Tech) to develop a Stream Asset Inventory (SAI) that provides information on stream stability, aquatic and terrestrial habitat quality, and water quality. Tetra Tech designed the (SAI) for use by county staff and individuals minimally experienced in stream assessment. The scoring system Tetra Tech developed allows quick classification of stream quality within surveyed reaches. Initial field trial indicates that the SAI is relatively easy to use, and that individuals with a strong background in stream assessment can implement it quickly. Developing a "How-To" manual and training workshop will be necessary for those without formal education in stream assessment. Further testing is needed to determine the regional applicability of the SAI.*

Keywords. Aquatic, assessment, benthic, biological, habitat, macroinvertebrate, monitoring, sampling, stream, streambank, streamflow, substrate, vegetation

Introduction

Johnson County, Kansas is a rapidly developing suburban area of the greater Kansas City metropolitan region. The County faces significant decisions concerning management and conservation of stream and riparian resources within its jurisdiction. To gain a thorough understanding of their stream resources, the County contracted Tetra Tech EM Inc. (Tetra Tech) to inventory the Upper Blue River Watershed using a rapid stream assessment protocol. Incorporating the best elements of a number of accepted stream and habitat assessment methodologies and local research, Tetra Tech designed an approach that provides quickly identified and scientifically defensible indicators of water quality, stream stability, and habitat conditions.

The purpose of this project was to provide a basic field method to evaluate stream condition. Tetra Tech had developed earlier versions of the Stream Asset Inventory (SAI) protocol for use by a wide variety of individuals, including those with little training in biology, hydrology, or environmental chemistry. Developed for the Kansas City region, the SAI protocol drew on past experience and local and national research. This current SAI benefits from lessons learned from two previous SAI versions and the most up-to-date stream quality information. The procedure is primarily based on physical habitat conditions but also integrates elementary biological components. The scoring criteria blend subjective judgment and easy quantitative measure to allow a relatively comprehensive yet rapid evaluation of a stream segment. Of course, this protocol may not detect resource problems influenced by situations outside the study reach—as with any assessment limited to a specific segment of stream. To help overcome this problem, multiple sites were surveyed within the watershed.

Background

Stream ecosystems result from dynamic interactions among water, climate, geology, vegetation, and human influences (Armantrout, 1998). In both urban and rural settings, human activities impact stream resources by altering one or more of five basic characteristics—water quality, habitat structure (including channel dimension, pattern and profile), flow regime, energy source, and biotic interactions (Karr and Chu 1999). The growing demands for multiple-uses of streams and their associated riparian corridors has increased the importance of resource inventories to understand the overall effect of these uses on the biological, chemical and physical features of the ecosystem.

Most resource inventories have focused on physical habitat quality and quantity, based on knowledge that physical habitat complexity influences biological diversity, and that stream systems exhibit consistent patterns that are relatively easy to define and measure. Historically, the objective of these studies was to evaluate suitability of habitat for a species, a group of species, or a species' life stage (Hamilton and Bergesen, 1984). More recently, habitat has become a primary basis for most impact assessments, resource inventories, species management plans, mitigation planning, and environmental regulation (Bain and Stevenson, 1999). Water quality assessment is increasingly important because of rising public concern and various regulations such as the National Pollutant Discharge Elimination System (NPDES) Phase I and II regulations. Finally, increased understanding of stream morphology has made it easier to assess stream stability and protect infrastructure, habitat, and water quality. The following sections describe considerations for survey site selection and scientific rationale for each component of the SAI.

Site Selection

Selecting the survey site was important in planning the Johnson County SAI. Surveying the entire lengths of all streams in a watershed is impossible. Therefore, we selected 20 sites to represent the range of conditions in the Kansas portion of the Blue River Watershed. To ensure that the selected survey sites would provide an accurate baseline condition of the watershed, we stratified separate reaches of the watershed by stream size and evenly distributed survey sites throughout the identified reaches. This allowed an overview of the stream system as a continuum from headwater to confluence. Personnel selecting specific SAI survey sites considered stream channel patterns and representative habitat types, tributary influences, adjacent land uses, spatial distribution, channel disturbances, and accessibility.

United States Geological Survey (USGS) topographic maps, digital orthophotography, and available remote sensing data were used to identify candidate sites. Additionally, information gathered from stakeholders—important natural resource areas, land use patterns, and historic data—was solicited to encompass a variety of environmental settings and comprehensively represent watershed condition. Finally, field reconnaissance of candidate sites verified that each represented the watershed or stream reach.

Assessment Components

The SAI procedure rapidly evaluates site-specific stream conditions via a series of components that provide basic understanding of physical, biological, and water quality characteristics in the Kansas City region. The procedure has four major categories—stream stability, aquatic habitat quality, terrestrial habitat quality, and water quality—each with five scoring components. Each component has a maximum score of 10 for a possible total of 200. By dividing the total score by 20 (the number of components), the assessment provides a qualitative numerical score ranging from 0.0 to 10.0. A score of 10.0 would be considered optimal stream conditions while 0.0 would indicate poor stream conditions. The following sections discuss the assessment categories and the scientific rationale for each component.

Stream Stability

Channel erosion and meanders result from natural river processes. Stream channel stability is typically characterized by equilibrium between erosion and deposition (Leopold et al. 1992). A stable channel may not maintain a constant position in the landscape but maintains a constant cross-section dimension by erosion on one bank and deposition on the opposite. However, the rate of erosion by lateral meander and incision can accelerate in response to floodplain development, channel modification, riparian clearing, or other activities that alter hydrology patterns within the watershed. Designers of the SAI selected the following five components to obtain basic knowledge of stream stability at each site.

The Bank Cut Depth component supplies information on streambed erosion. A streambed's erosion (equivalent terminology is "downcut" or "incision") results from increased runoff volume or velocity. As the streambed downcuts, runoff that previously spread over the floodplain is now confined to the stream channel, increasing volume and velocity, and further eroding the channel bottom (Balch 2001). The greater the bank cut depth, the less stable the stream channel. The bank cut depth is measured at several points to determine an average for the site.

The Root Depth component provides information on the stream bank's ability to resist erosion. Stream channel erosion is influenced by the amount and type of vegetation on the bank. Healthy stream banks are covered by deep, densely rooted vegetation that binds soils and creates friction which slows water velocity (Montana Department of Environmental Quality 1995). The Root Depth component is the value obtained by arithmetically dividing the root depth of streamside vegetation by the height of the bank. The smaller the ratio of root depth to bank height, the less stable the stream channel. Root depth is calculated at the same locations of bank cut depth to determine an average for the site.

The Bank Erosion component also furnishes information on the stream bank's ability to resist erosion. Bank composition importantly influences stream stability because different materials have differing resistances to erosion. Large, solid rock, or tightly packed soils (silts and clays) are less likely to erode than smaller or non-cohesive particles (sands and gravels). Scoring reflects the predominant type of bank material within the site.

The Bed Composition affects the ability of the streambed to resist erosion. The types and sizes of substrate material influence the rate of erosion and streambed degradation. Solid rock, boulders, and tightly packed clays resist erosion more effectively than do smaller particles of unconsolidated cobble, gravel, sand, and soil. Scoring reflects the predominant type of bed material within the site.

Identifying the Types of Erosion present within the reach can indicate in-stream and watershed influences affecting overall stream stability. Presence of different erosion types along a reach indicates multiple factors contributing to stream instability. Scoring is based on the number of erosion types observed within the survey reach.

Aquatic Habitat Quality

Aquatic habitat quality is a function of the stream system's physical and chemical characteristics. This category focuses on five basic physical characteristics that influence life history requirements (food, shelter, reproduction) for fish and macroinvertebrates. Though the amount of available habitat may fluctuate due to seasonal differences in flow, these features remain relatively constant. This category incorporates the assumption that streams with a variety of physical features provide more microhabitats for aquatic organisms, contributing to higher species diversity (Huggins and Moffett 1988).

Flow characteristics are categorized as perennial, intermittent, or ephemeral. Though ephemeral and intermittent drainages are important to overall system health and productivity (American Rivers 2003), the SIA scores them lower because they generally do not provide year-round habitat. Perennial, spring-fed, or intermittent streams with permanent pools typically provide year-round habitat; they host more diverse and abundant fish and macroinvertebrates. Therefore, they receive higher scores. Assessing flow characteristics solely from field observation is difficult. Therefore, before conducting the field assessment, the assessor should acquire historical information from other sources to include when subsequently determining flow characteristics.

The Substrate component considers the quality of the bed material as habitat for benthic fish and macroinvertebrates. The channel substrate can vary significantly within a stream reach and fluctuate frequently in response to flow conditions. Dependence of benthic organisms on substrate particle size has been well established (Waters 1995). In general, abundance is greatest with diverse mixture of pebbles, gravel, and cobble (Minshall 1984) that provides more microhabitat (niches) through greater surface area and interstitial spaces. Benthic abundance decreases in substrates of sand or silt and in large boulders and bedrock. Scoring reflects an average numerical value for the two predominant types of substrate material within the site.

The Macrohabitat Types component considers basic habitat characteristics (pool, riffle, run) related to current speed and depth. Pools are typically bowl-shaped features in the streambed with greater width and depth but very slow current. Riffles and runs are areas with greater current speed. Riffles are shallow areas exhibiting surface turbulence as the water moves over rocks or uneven bedrock bottom. Runs may be deeper and show turbulent flow over an uneven bottom. A stream with all three types of habitat will typically contain a more diverse biological community than a stream with only one or two types. Scoring reflects whether or not all three types are present.

Instream Fish Cover accounts for the types of fish cover available within the survey reach. This influences a variety of factors: shelter, food, reproduction, distribution, abundance, competition, and predation. Cover types include deep pools and backwater areas, riffles, vegetated shallows, logs or large woody debris, overhanging vegetation or banks, boulders, and cobbles. In general, fish diversity and abundance increases when more cover types are available. Scoring reflects the number of cover types observed within the survey reach.

Instream Fish Cover, Instream Macroinvertebrate Cover accounts for the features available as refuge for macroinvertebrates. A wide variety of submerged structures provide a greater number of niches, increasing potential abundance and diversity of macroinvertebrates. Macroinvertebrate cover types include fine woody debris or submerged logs, submerged tree roots and bank vegetation, macrophyte beds, algal mats and leaf packs, and coarse gravel and cobble. Scoring reflects the number of cover types observed within the survey reach.

Terrestrial Habitat Quality

A stream is the product of its watershed, and its health is greatly affected by activities on the land upstream and within its associated floodplain and riparian corridor. High quality terrestrial habitat adjacent to the stream channel is crucial to overall maintenance of stream health. This category considers values of various terrestrial features for evaluating existing stream condition.

The Vegetation Width component provides information regarding the amount of vegetation available to slow and filter overland flows. Riparian vegetation is an important element in the health and condition of stream ecosystems. A wide, well-vegetated riparian corridor benefits the stream physically and chemically—stabilizing banks, filtering excess nutrients and sediment from surface runoff, optimizing sunlight and temperature conditions, and providing organic debris that functions as food and cover (Welsch 1991). The overall component score is an average calculated from the two numerical values obtained by dividing the width of vegetation on each bank by the active water channel width.

Adjacent Land Uses can greatly affect the quantity and quality of runoff water entering the stream. Generally, uses that mimic native habitat conditions are less harmful than uses that significantly alter historic floodplain conditions. Human-made uses exert greater harmful impacts, with impervious surfaces the most detrimental. The overall component score is an average of the numerical values determined for each bank.

The Woodland Richness component enumerates tree, shrub, and woody vine species present in the riparian corridor. Due to increased moisture supplied by the adjacent stream, riparian areas tend to be more structurally diverse and more productive in plant biomass than adjacent upland areas (Brinson et al. 1981). Higher species richness is assumed indicative of better habitat quality and scored accordingly.

The Grassland Richness component enumerates grass and forb species observed within the riparian corridor. As with the previous component, higher species richness assumedly indicates better habitat quality and is scored accordingly. Note that in habitats with high tree and shrub

195

density, canopy shade might decrease grassland richness. Higher grassland scores are expected in more open habitats with less or non-existent canopy cover.

The Undesirable Vegetation component indicates impact of invasive or undesirable plant species on native riparian habitats. Invasive species indicate prior disturbance from activities such as road building, residential development, riparian clearing, agriculture, grazing, mowing, channelization, and fire prevention. A higher percentage of undesirable species reflects an impacted riparian habitat. Score is calculated by dividing the total number undesirable species by the total number of plant species observed in assessing the two previous components.

Water Quality

Water quality conditions affect the stream's ability to support aquatic organisms. This category evaluates five components that indicate the water quality of the stream. Though no water chemistry parameters are measured, these components indirectly reveal poor water quality and conditions that limit aquatic life.

Silt Cover estimates the amount of silt deposition in the survey reach. Presence of sediment is natural and characteristic of most streams. However, excessive sediment deposition and silt cover indicate an altered watershed that can greatly impair stream health by increasing levels of turbidity, nitrogen, and phosphorus in the water (Waters 1995). Sedimentation also eliminates benthic habitat by filling interstitial spaces that fish and macroinvertebrates can occupy. This component is scored by estimating the amount of silt covering the natural bed material within the survey site.

The Undesirable Conditions component identifies general existing conditions that indicate water quality problems. These conditions reflect negative impact(s) on a stream's health and its biologic community—they include odors or foam, various wastes, debris, excessive algae, and channel modifications. Scoring occurs by applying a numerical value to the number of observed conditions.

The Aquatic Organisms component considers general abundance of three animal groups that depend on good water quality conditions—fish, mollusks, and amphibians. Much variability observed in the aquatic biological community results from the system's water quality characteristics. No sampling is required for these components. However, conducting short seine hauls may be valuable for determining fish presence in turbid conditions. Mollusks may be observed in shallow riffles or by searching for recently dead shells along the stream banks and bars. Shells that are old (white and chalky) should not be considered in scoring determinations. Amphibians may be observed directly along the banks, in the water, under rocks and logs—or indirectly through their calls. Large numbers of exotic species such as common carp or Asiatic clam are not considered in scoring determinations. A numerical value is applied for each animal group based on subjective assessment of the group's observed abundance.

Macroinvertebrate Sampling

Presence of various macroinvertebrate species furnishes direct evidence of habitat quality and allows an indirect measure of water quality. Some macroinvertebrate species tolerate poor water quality better than others. The stream assessor may obtain more detailed information on stream resources by sampling macroinvertebrates and observing both the number of organisms and their species. This sampling procedure is optional and requires greater knowledge or training than the basic components identified above. Detailed sampling instructions accompany the field key, and the investigator receives a visual guide to macroinvertebrate organisms, grouped by water quality tolerance.

Conclusion

The SAI scoring system allows a rapid classification of stream sites, enabling SAI assessors to identify reaches suitable for possible conservation, restoration, or recreation. Initial field trial of the SAI indicates that individuals with strong backgrounds in stream assessment can apply the system with relative ease and can implement it quickly. Developing a "How-To" manual and training workshop on the SAI will be required for those without formal education in stream assessment. Further testing is needed to determine the regional applicability of the SAI.

References

American Rivers. 2003. Where Rivers Are Born: The Scientific Imperative For Defending Small Streams And Wetlands. 24 pp.

Armantrout, N.B. 1998. *Aquatic Habitat Inventory Terminology: Glossary*. American Fisheries Society, Bethesda, Maryland. 136 pp.

Bain, M.B. and N.J. Stevenson. 1999. *Aquatic Habitat Assessment: Common Methods*. American Fisheries Society, Bethesda, Maryland. 216 pp.

Balch, P.B. 2001. *Kansas River And Stream Corridor Management Guide*. Kansas State Conservation Commission, Topeka, Kansas. 43 pp.

Brinson, M.M., B.L. Swift, R.C. Plantico and J.S. Barclay. 1981. *Riparian Ecosystems: Their Ecology And Status*. FWS/OBS-81/17. United States Fish & Wildlife Service, Kearneysville, West Virginia. 154 pp.

Hamilton, K. and E.P. Bergersen. 1984. *Methods To Estimate Aquatic Habitat Variables*. Colorado Cooperative Fishery Research Unit, Colorado State University, Ft. Collins. 307 pp.

Huggins, D.G. and M.F. Moffett. 1999. *Proposed Biotic And Habitat Indices For Use In Kansas Streams*. Report No. 35 of The Kansas Biological Survey, University of Kansas, Lawrence. 128 pp.

Karr, J.R. and E.W. Chu. 1999. *Restoring Life In Running Waters: Better Biological Monitoring*. Island Press, Washington, D.C. 206 pp.

Leopold, L.B., M.G. Wolman, and J.P. Miller. 1992. *Fluvial Processes In Geomorphology*. Dover Publications, New York. 522 pp.

Minshall, G.W. 1984. Aquatic Insect – Substratum Relationships. Pages 358–400. *In*: V.H. Resh and D.M. Rosenberg, eds., *The Ecology of Insects*. Praeger Publishers, New York.

Montana Department of Environmental Quality. 1995. *Montana Stream Management Guide*. 30pp.

Waters, T. F. 1995. *Sediment In Streams: Sources, Biological Effects And Control*. American Fisheries Society Monograph 7, Bethesda, Maryland. 251 pp.

Welsch, D.J. 1991. *Riparian Forest Buffers: Function and Design for Protection and Enhancement of Water Resources*. USDA Forest Service, Radnor, Pennsylvania. 24 pp.

EVAPOTRANSPIRATION CHANGES AT THE GLACIAL RIDGE PRAIRIE RESTORATION PROJECT: A REMOTE SENSING PERSPECTIVE

J. W. Oberg[1] and A. M. Melesse[2]

[1] Department of Civil Engineering. [2] Department of Earth Systems Science and Policy
University of North Dakota, Grand Forks, ND 58202

ABSTRACT

A remote sensing approach with the Surface Energy Balance Algorithm for Land (SEBAL) for estimating evapotranspiration (ET) at a prairie and wetland restoration site in northwestern Minnesota is presented. Temporal scales of monthly, inter-seasonal, and seasonal ET rates were compared to three different adjacent land-uses from June to September during 2000 to 2003: (1) a prairie and wetland preserve; (2) a treated site; and (3) a non-treated site subject to future improvements. Comparing ET behavior at the preserve, results suggest restoration efforts have affected monthly and seasonal ET rates within the treated site. From the beginning of restoration efforts in 2001 until the conclusion in 2003, average standard deviations of the seasonal ET within the preserved, treated, and non-treated sites give 47.3, 75.7, and 109.9 mm, respectively, suggesting hydrologic stabilization within the treated site. Seasonal ET averages for the preserved, treated, and non-treated sites show 606.4, 579.7, and 532.2mm, respectively, suggesting a relatively higher ET within the treated site, compared to the non-treated area, possibly caused by the restoration of native grasses and wetlands to natural water levels.

(Key terms: evapotranspiration, remote sensing, SEBAL, Glacial Ridge)

INTRODUCTION

Water management continues to be a topic of major concern for natural resource managers and society at large. With increasing demands and limited availability in many locations, methods for identifying, monitoring, and analyzing the water balance at various scales will be important in order to preserve and manage water resources.

It is estimated that evapotranspiration (ET) accounts for an average of 70 percent of the consumption of annual precipitation in the United States, including up to 95 percent in some arid climates (Gay, 1993). It is known that changes in ET directly affects available water yield. Water yield affects the amount of storage space in a watershed. The amount of storage space in a watershed can determine the amount of streamflow caused by precipitation, and reductions in streamflow during extreme hydrologic events can alleviate losses to life and property. Therefore, ET is a major component of the water balance.

The use of remotely sensed imagery for earth science analyses has increased in the past 20 years due to advancements in research, sensors, software, and increased environmental awareness (Jensen, 2000). Specifically, sensors capable of capturing images in the mid to near infrared (IR) and thermal IR bands of the electromagnetic (EM) spectrum allow for inferences to be made of the water budget at both fine and regional scales. Advantages of using remotely sensed imagery to estimate ET includes high spatial coverage, a reduction in laborious field measurements, and ease of visual interpretation, compared to traditional methods.

The Surface Energy Balance Algorithm for Land (SEBAL) (Bastiaanssen, 1995) model was employed in this study to estimate spatial ET. SEBAL uses remotely sensed imagery with mid, near, and thermal IR, and visible spectrums for estimating ET. The theory that underpins SEBAL is described in Bastiaanssen (1995). The model has been validated in the form of energy and water cycle analysis in mountainous arid climates (Bastiaanssen et al., 2002), irrigation effects on the partitioning of energy fluxes and crop water stress in Mediterranean climates (Bastiaanssen, 2000), crop water deficit monitoring in tropical regions (Hafeez et al., 2002), and precision weighing lysimeter comparison in Idaho (Allen et al., 2001). However, the literature is presently void of studies where SEBAL was used to quantify ET rates for a native prairie recreation project in a northern climate, such as at Glacial Ridge.

SEBAL was used to evaluate ET changes between three unique land-uses at the Glacial Ridge prairie restoration project undergoing in northwestern Minnesota. The three adjacent land-uses studied include (1) a prairie wetland area preserved for the past 30 years; (2) a treated area subjected to restoration efforts from years 2000-2003; and (3) a non-treated area that is a candidate for restoration in the near future.

OBJECTIVES

Specific objectives of this study are to

- better understand how hydrologic cycles are affected by the recreation of land-uses to pre-settlement conditions

- validate the SEBAL model as a hydrologic tool for monitoring watershed recreation in a northern prairie region of the United States, and

- qualify and quantify changes in ET at a prairie restoration in progress.

STUDY SITE

In August 2000, the Minnesota Chapter of The Nature Conservancy (TNC) purchased 24,270 acres of land approximately 16 km southeast of Crookston, in northwestern MN, with the intent to recreate the site's drained wetlands and ecosystems to pre-settlement conditions (TNC, 2003). The restoration objectives include the revival of pre-settlement water levels, drained wetlands, and native prairie vegetation. The methods of restoration include vegetative burning, planting of native grasses, and ditch filling. The area of purchase is commonly referred to as the Glacial Ridge, due to its location at what used to be the eastern beach ridges of the glacial Lake Agassiz. Figure 1 shows the general location of the Glacial Ridge project.

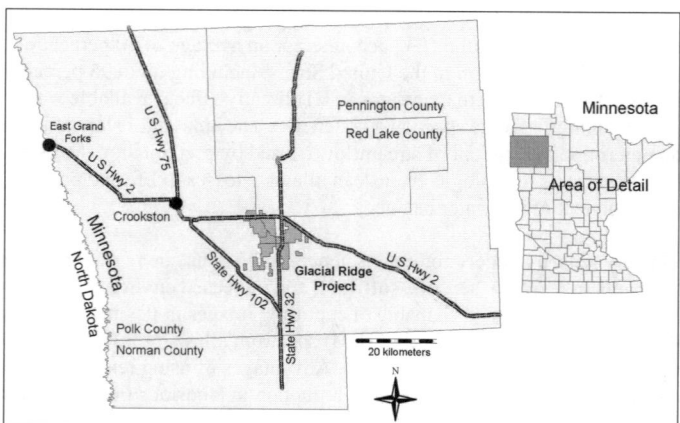

Figure 1. Glacial Ridge Vicinity.

200

METHODOLOGY

PAIRED LAND-USE

A paired land-use method was used to compare changing ET rates. In Figure 2 the control hydrologic area is situated within the Pembina Trail preserve that was restored approximately thirty years ago. This was used as the control site and its ET rates were compared to that of treated and non–treated sites for the months of June, July, August, and September, during 2000, 2001, 2002, and 2003. The non–treated area for this study is an area that as of May 2004 has not received restoration treatment; but as an area administered by the TNC, will likely received treatment in the future. The treated and non–treated sites are shown in Figure 2.

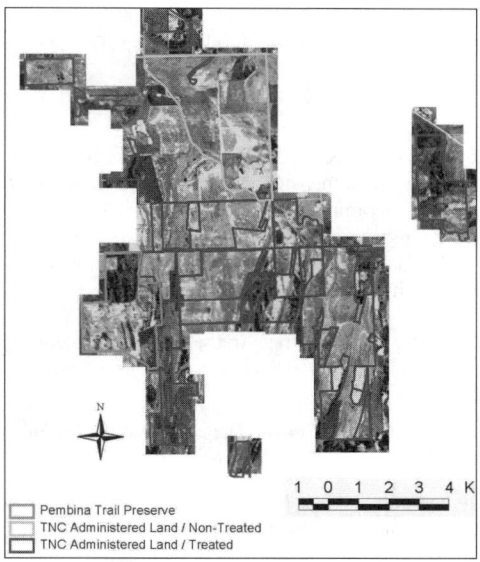

Pembina Trail Preserve
TNC Administered Land / Non-Treated
TNC Administered Land / Treated

Figure 2. Study Sites at Glacial Ridge.

DATASETS

A combined hydrologic and geographic information system (GIS) methodology was followed to derive the datasets used in this study. The datasets include areas affected by the TNC Glacial Ridge restoration efforts. Types of datasets used in this study were in the form of shapefiles, raster images, and images from Landsat Thematic Mapper (TM) and Enhanced Thematic Mapper Plus (ETM+) sensors.

The spatial progress of each of the restoration processes listed above was monitored using Global Positioning System instrumentation and converted into GIS spatial data formats. Using the datasets, each area of interest in the study was clipped from each image date in a GIS module. The clipped images were thereafter converted into grid format to allow for statistical analyses. The common areas of interest defined by the GIS datasets allowed for temporal ET change comparisons to follow. Temporal ET changes in this study were evaluated from monthly, inter-seasonal, and seasonal scales.

For each year, four Landsat images were sampled between June and September. In most cases, each image was sampled to represent the average surface reflectance for each month. Seasonal images for each year were sampled around late July to early August to approximate the average seasonal reflectance.

SEBAL METHOD

SEBAL estimates ET by solving the terms of the surface energy balance derived from the visible, near-IR, mid-IR, and thermal-IR bands of the EM spectrum. By the law of conservation of energy, the equation for the inputs, losses, and storage of energy for a surface on earth is expressed in Campbell and Norman (1998) as

$$R_n = G + H + \lambda ET \ \ (\text{W/m}^2) \tag{1}$$

where R_n is the net radiation flux at the surface, G is the soil heat flux, H is the sensible heat flux to the air, and λET is the latent heat flux.

Net radiation is estimated based on the following relationship (Bastiaanssen et al., 1998),

$$R_n = R_{s\downarrow}(1-\alpha) + R_{L\downarrow} - R_{L\uparrow} - R_{L\downarrow}(1-\varepsilon_o) \ \ (\text{W/m}^2) \tag{2}$$

where $R_{S\downarrow}$ (W/m^2) is the incoming direct and diffuse shortwave solar radiation that reaches the surface; α is the surface albedo, the dimensionless ratio of reflected radiation to the incident shortwave radiation; $R_{L\downarrow}$ is the incoming longwave thermal radiation flux from the atmosphere (W/m^2); $R_{L\uparrow}$ is the outgoing longwave thermal radiation flux emitted from the surface to the atmosphere (W/m^2), ε_o is the surface emissivity, the (dimensionless) ratio of the radiant emittance from a greybody to the emittance of a blackbody.

The soil heat flux (G) is the rate of heat storage to the ground from conduction. Studying irrigated agricultural regions in Turkey, Bastiaanssen (2000) suggested an empirical relationship for G given as:

$$G = R_n \frac{T_s}{\alpha}(0.0038\alpha + 0.0074\alpha^2)(1 - 0.98NDVI^4) \ \ (\text{W/m}^2) \tag{3}$$

where T_s is the surface temperature (°C), α is the surface albedo (dimensionless), and *NDVI* is the normalized difference vegetation index (dimensionless).

Sensible heat flux (H) is the rate of heat loss to the air by convection and conduction due to a temperature difference. Using the equation for heat transport, sensible heat flux can be calculated by (Campbell and Norman, 1998):

$$H = \frac{\rho C_p dT}{r_{ah}} \ \ (\text{W/m}^2) \tag{4}$$

where ρ is the density of air (kg/m^3), c_p is the specific heat of air (1004 J/kg/K), dT is the difference in temperature between the surface and the air (K), and r_{ah} is the aerodynamic resistance (s/m).

With R_n, G, and H known, the latent heat flux is the remaining component of the surface energy balance to be calculated by SEBAL. Rearranging Eqn. (1) gives the latent heat flux where:

$$\lambda ET = R_n - G - H \ \ (\text{W/m}^2) \tag{5}$$

Dividing Eqn. (5) by the latent heat of vaporization (λ) allows a solution for ET.

QUALITATIVE METHOD

A major advantage of using remote sensing to estimate ET over traditional methods is the ability to visually interpret the spatial phenomena being studied. The resulting ET images from remote

202

sensors allowed coverage of much larger areas of interest than possible with costly, elaborate, and time consuming point location studies, such as with the Bowen ratio and eddy correlation techniques. Also, the Landsat images allowed for visual interpretation of the changing ET behavior between two or more datasets

QUANTITATIVE METHOD

The statistical methods used in this study included analysis of means, standard deviations, and correlations of ET rates within and between each level of restoration. To clarify the effectiveness of the restoration treatments, each of the three levels of restoration (preserved, treated, and non–treated) were compared both separately and together with respect to their changing ET rates. The temporal scales used for comparison included monthly, inter-seasonal, and seasonal ranges.

RESULTS AND DISCUSSION

SEASONAL ET

As an overview of changing ET rates spatially, seasonal ET was evaluated at Glacial Ridge from 2000-2003 (Figure 3). For each year, a single image was selected around mid-season in order to represent the average reflectance and emittance from June to September. From Figure 3 it is apparent that an overall increasing trend in seasonal ET has occurred from 2000 to 2003. Particularly, by inspection of the southwest portion of the Pembina Preserve, a wetland area, the seasonal ET appears to have successfully increased from year to year. For treated areas, slight decreases are evident in 2001. From inspection of the non-treated area it is noted that any trend is somewhat uncertain, possibly due to the variations in agricultural land-uses during this time in the form of varying crop types and pasture grazing practices.

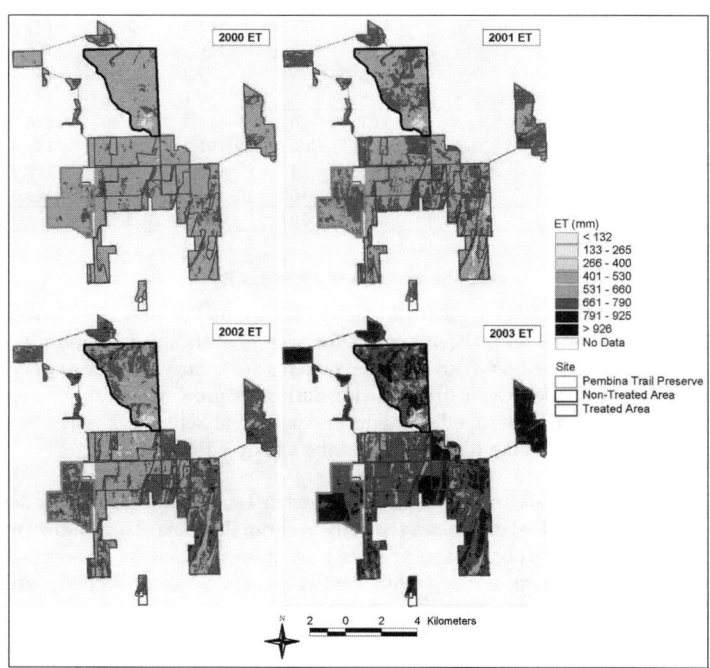

Figure 3. Seasonal ET at Glacial Ridge 2000-2003.

To quantify the seasonal changes statistical comparisons were made between seasonal averages and standard deviations between each area of interest, as given in Table 1. Inspection of the standard deviations in Table 1 show, relative to the Pembina Trail preserve, increasing variability when compared to the non-treated area. Of interest is that the treated area shows agreeable variability with Pembina during 2000 before full restoration efforts and increasing variability until 2003 where the data suggests stabilization of the treated area in its the ET relationship with Pembina. A possible cause of the stronger agreement between the Pembina and treated areas in 2003 could be due to the proliferation of the native grasses establishing permanent vegetation, and the restored wetlands gradually stabilizing the local water tables.

Table 1. Seasonal Average ET (mm) at Glacial Ridge 2000–2003.

| Year | Seasonal ET and Standard Deviation June-Sept. (mm) | | | | | | |
	Pembina	STD	Treated	STD	Non-Treated	STD	ET Avg.
2000	591.5	45.3	584.8	46.3	539.4	91.0	**571.9**
2001	570.4	42.9	537.0	80.2	488.6	106.9	**532.0**
2002	602.2	39.5	567.4	97.0	500.8	116.7	**556.8**
2003	661.4	61.6	629.6	79.3	600.2	125.0	**630.4**
Avg.	**606.4**	**47.3**	**579.7**	**75.7**	**532.2**	**109.9**	

Table 2. Monthly Average ET (mm) and Standard Deviations (mm) 2000–2003.

| Layer | | Pembina Preserve | | Treated Areas | | Non-Treated Areas | | |
		Mean	STD	Mean	STD	Mean	STD	Avg.
2000	June	138.9	13.9	133.2	29.8	112.3	23.1	**128.1**
	July	176.5	13.2	172.1	18.3	157.3	26.6	**168.6**
	Aug	157.4	11.9	161.5	18.7	157.8	24.6	**158.9**
	Sept	118.8	6.3	117.9	8.2	112.0	14.2	**116.2**
2001	June	148.5	12.5	143.6	21.7	118.8	25.6	**137.0**
	July	125.6	12.5	113.6	25.1	103.2	27.1	**114.1**
	Aug	183.7	11.7	174.9	21.9	167.8	29.2	**175.5**
	Sept	112.6	7.7	104.8	17.3	98.9	15.8	**105.4**
2002	June	171.0	17.6	149.6	27.3	102.5	27.3	**141.0**
	July	170.1	19.1	168.8	29.0	159.0	25.2	**166.0**
	Aug	146.7	8.5	137.3	20.9	129.2	25.1	**137.7**
	Sept	114.4	5.0	111.8	10.1	110.1	15.8	**112.1**
2003	June	146.4	18.5	143.7	24.2	118.9	26.0	**136.3**
	July	204.5	16.2	199.6	20.8	191.2	33.0	**198.4**
	Aug	191.1	14.1	172.4	25.3	173.9	38.6	**179.2**
	Sept	119.5	6.2	113.9	10.0	116.2	15.3	**116.5**
Avg.		**151.6**	**12.2**	**144.9**	**20.6**	**133.1**	**24.5**	

INTER-SEASONAL AND MONTHLY ET

From table 2 we note relatively consistent variation of ET within Pembina. This is what we would expect since the preserve has been in place for over 30 years and has a mostly homogenous wetland composition. Also, seasonal patterns are somewhat evident in Pembina with greater standard deviations typically occurring during the growing months of June and July, and lower variations occurring during the months of August and September, when transpiration may have a lessening effect on the ET due to maturing and dying vegetation.

Comparing the means and standard deviations presented in Tables 1 and 2, we note the highest ET averages and lower standard deviations typically occur in Pembina. This follows what we may expect from a hydrologic control area. Tables 1 and 2 suggest that the non-treated area has the highest variation in ET in most cases, while the treated area has comparatively less variation yet higher ET.

To determine more precise relationships for the temporal changes within each area of interest, inter-seasonal and monthly covariance and correlations were calculated for each area of interest, though not presented in this paper. It was found in most cases that the highest variation in ET occurs within the non-treatment area. Comparing inter-seasonal relationships within the non-treatment and Pembina sites, a wide variation in agreement of monthly ET occurred during 2000 ($r \approx -0.1$ to 0.6), compared to Pembina ($r \approx 0.5$ to 0.7). Comparing all three areas of interest, little agreement within the non-treated area occurred during 2002 ($r \approx 0.0$ to 0.5), compared to Pembina ($r \approx 0.5$ to 0.8) and the treatment ($r \approx 0.3$ to 0.7). Similar comparisons within 2003 showed a relatively weak effect within the non-treatment area ($r \approx 0.3$ to 0.9), between months, compared to Pembina ($r \approx 0.8$ to 0.9) and the treatment ($r \approx 0.5$ to 0.9).

CONCLUSIONS AND FUTURE IMPROVEMENTS

The above findings suggest that remotely sensed satellite imagery with SEBAL is a valid and effective method to qualify and quantify ET changes at multi-time scales for a prairie restoration project in a northern climate, such as at Glacial Ridge. Comparing the three areas of interest, results suggest an increasing standard deviation in the seasonal ET (80.2, 97.0mm) during restoration years 2001, 2002, respectively, and a decrease in variation during 2003 (79.3mm). Inter-seasonal correlations suggested converging correlational agreements between changes within the Pembina and treated areas. As expected, the data suggested relatively higher seasonal and inter-seasonal ET variability between both Pembina and treatment and the non-treated site.

Results also suggest that since the monthly ET correlations within the treated areas during 2002 correspond more closely to those within Pembina, compared to the non-treatment, then during 2002 the effects of the restoration may be impacting the ET within the treated areas. Likewise, during 2003, the correlations within the treated area compare more favorably to Pembina, than do those of the non-treatment. Since the treatment correlations appeared to be converging toward those of the Pembina more so than those of the non-treatment, then it appears that restoration is affecting the ET within the treated areas.

Future improvements include a progression toward a more comprehensive surface water balance estimation approach as restoration activities extend into the remaining sites within the Glacial Ridge restoration boundaries. It is noted that that six surface water gage stations are operational within or near the Glacial Ridge site. A more comprehensive water balance approach might include a paired watershed analysis to monitor the effects of restoration activities to surface water runoff. Since the Glacial Ridge area lies within the upper reaches of the greater Red River Valley watershed, a comprehensive study of the water cycle could lead to a better understanding of the effects and feasibility of restoration efforts to pre-settlement conditions to an area historically afflicted with extreme hydrologic events.

ACKNOWLEDGEMENTS

The authors acknowledge members of Upper Midwest Aerospace Consortium (UMAC) for their help. The authors extend their appreciation to Tim Cowdery of USGS of Minnesota, The Nature Conservancy, especially to Chrisie Hura and Jason Ekstein, USGS of North Dakota and North Dakota State Water Commission for providing some of their data. The research was funded by UMAC.

REFERENCES

Allen, R. G., A. Morse, M. Tasumi, W. G. M. Bastiaanssen, W. Kramber, and H. Anderson. 2001. Evapotranspiration from Landsat (SEBAL) for water rights management and compliance with multi-state water compacts. *International Geoscience and Remote Sensing Symposium (IGARSS)*. 2, 830-833

Bastiaanssen, W. G. M. 1995. Regionalization of surface flux densities and moisture indicators in composite terrain. Thesis published by DLO Winand Staring Centre for Integrated Land, Soil and Water Research, Wageningen, the Netherlands. 273 p.

Bastiaanssen, W. G. M., M. Menenti, R. A. Feddes, and A. A. M. Holtsag. 1998. A remote sensing surface energy balance algorithm for land (SEBAL). 1 Formulation. *Journal of Hydrology.* 212-213, p 198-212.

Bastiaanssen, W. G. M. 2000. SEBAL-based sensible and latent heat fluxes in the irrigated Gediz Basin, Turkey. *Journal of Hydrology* 229:87-100.

Bastiaanssen, W. G. M., M. M. D. Ahmad and Y. Chemin, 2002. Satellite surveillance of evaporative depletion across the Indus Basin. *Water Resources Research*, 38 (12): 1273-1282

Campbell, G. S. and J. M. Norman. 1998. *An Introduction to Environmental Biophysics.* 2nd Ed. New York: Springer–Verlag.

Gay, L. W. 1993. Evaporation measurements from catchment scale water balances. In *Proceedings of the First International Seminar of Watershed Management*, ed. J. Castillo Gurrola, M. Tiscareno Lopez, and I. Sanchez Cohen, p 68-86. Hermosia, Sonora, Mexico: Universidad de Sonora.

Hafeez, M. M., Y. Chemin, N. Van De Giesen, B. A. M. Bouman. 2002. Estimation of crop water deficit through remote sensing in Central Luzon, Philippines. *International Geoscience and Remote Sensing Symposium (IGARSS)*, 5, 2778-2780

Jensen, J. R. 2000. *Remote Sensing of the Environment: An Earth Resource Perspective.* Upper Saddle River, NJ: Prentice Hall.

TNC. 2003. Glacial Ridge Project. The Nature Conservancy of Minnesota: Phone Correspondence. 31077 State HWY 32 S Mentor, MN 56736. Ph: 218-637-2146

Monitoring North Carolina Stream Restoration Projects

David Bidelspach, EI
Daniel Clinton, PE
Greg Jennings, PhD, PE
Chris Bass, PE
North Carolina State University
David_Bidelspach@ncsu.edu

Stream stabilization and restoration employs many different tools and techniques. Many of these techniques have not been monitored over an extended period of time. Numerous stream restoration projects have been built in North Carolina since the late 1990s to improve natural stream functions impaired by watershed land use changes. Many of these projects are intended to mitigate off-site impacts to streams from highway construction or other development. We initiated a long-term monitoring project in 2003 to evaluate the success of these projects in meeting restoration goals of stream stability and habitat improvement. Monitoring components include surveys of stream morphology, structure assessment, streambed monitoring, riparian vegetation assessment, and benthic macroinvertebrate sampling on selected stream projects. Results indicate a wide range of successes depending on watershed land uses, design/construction techniques, rainfall patterns, and vegetation management.

Monitoring and Modeling Sediment Transport and Geomorphology Changes Associated with Dam Removal

Fang Cheng, Tim Granata

Department of Civil and Environmental Engineering & Geodetic Science
The Ohio State University, Columbus, OH 43210
(614) 292-0542 (phone), (614) 292-3780 (fax)

The flushing of reservoir sediments downstream is the primary concern associated with dam removal. Sediment release can cause changes in river morphology, flow, aquatic habitats and river ecosystem processes both downstream and upstream. Knowing the dynamics of reservoir sediment transport is the key to understanding the influences of dam removal on river ecosystems. Combining GPS, acoustic methodology and traditional survey techniques, we quantitatively monitored and analyzed changes in river topography and sediment distribution resulting from the removal of St. Johns' dam, Sandusky River, Ohio. River and floodplain elevations were surveyed using Trimble ® 5700 GPS receiver and base station. Bedload transport rates were measured using bedload traps. Mean velocities and discharge at cross sections with bedload traps were measured using a handheld Sontek® FlowTracker acoustic Doppler velocimeter. YSI Sonde® water quality sensors, including turbidity probes, were installed both downstream and upstream capturing the time series changes in suspended sediment concentration within the first week after dam removal. A one-dimensional sediment transport model is being developed to simulate the hydraulic changes using MIKE 11 software (DHI Water & Environment).

Introduction

Dam removal has been proposed as a strategy for river restoration. However, the efficiency of this approach is largely uncertain. The release of dammed sediments downstream has considerable influence on river channel and bed topography, and is one of the major issues associated with dam removal (Pizzuto, 2002). Flushed sediments cause channel erosion upstream and, when deposited downstream, reduce pool volume. In addition, fine sediment may fill the pores between gravels and cobbles damaging the downstream aquatic community (Wohl and Cenderelli, 2000). For example, fine sediment deposition in cobbles during periods when eggs of aquatic organisms hatch could reduce permeability and cause eggs die-off from lack of mixing and transport of dissolved oxygen.

Over 500 low-head dams have been removed during the past decades, and an increasing number of dams are being considered for removal in the near future. Despite of the interest and enthusiasm toward dam removal, nearly no pre- and post- dam removal has been extensively monitored and analyzed, with an exception of one study by Doyle et al. (2003). To accurately estimate the effects of dam removal on river geomorphology and fish habitat and to minimize the negative influences, we need to fully understand the associated dynamics of sediment transport. While some research has described channel responses to sediment wave propagation and movement (Madej and Ozaki, 1996), not enough field data have been used to specifically describe dam removal on the scale of a natural river system. Another method for exploring sediment transport is hydraulic modeling. However, existing models usually assume equilibrium conditions for sediment transport and thus are not appropriate for simulating sediment transport associated with dam removal because of disequilibria condition immediately after the removal. Monitoring and recording the change in channel geomorphology with dam removal would be necessary to determine the sediment transport dynamics and provide the base for model simulations. A complete understanding of the effects of dam removal will help to overcome the simplistic view of removing dams and, thus enhance the dam removal decision making process.

Site Description

The study was performed on the Sandusky River located in NE Ohio. The Sandusky drains an area of 3637 km² and flows northeast into Sandusky Bay and Lake Erie (Figures 1a, b). The St. Johns' dam, a concrete dam, 46 meter long and 2.2 meter high, was located on Sandusky River at river mile (RM) 50. The reservoir area impounded by the dam extends 22.4 km upstream with a slight slope of 0.012%. The total reservoir area was approximately 0.59 km², and the storage was about 0.56×10^6 m³. Sediments accumulated in the reservoir were composed of coarse material (gravel and cobble) and fine sands. Destruction of the dam began on March 18th, 2003, when the west side was breached with a 1.5 m wide, 1.2 m height notch to gradually reduce water levels behind the dam (Figures 2a, b). The whole structure was removed on at 7 AM on November17th, 2003 (Figures 2c, d).

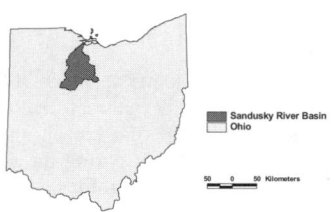

Figure 1a   Location of Sandusky

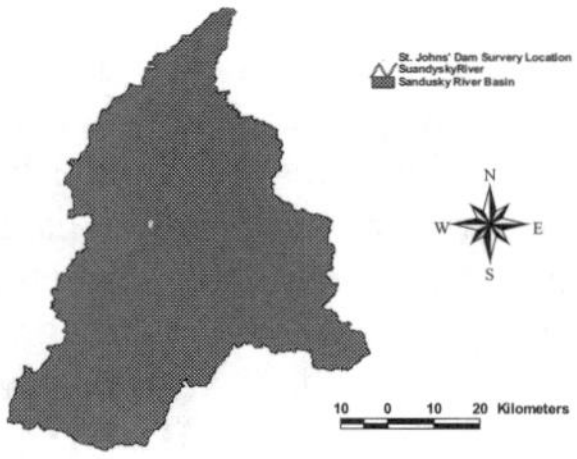

Figure 1b   Sandusky River Watershed

Methodology

River cross sections were surveyed at four stations, two downstream (DN1, DN2) and two upstream (UP1, UP2) of the dam. DN1 and DN2 were located 75 m and 3000m downstream. UP1 and UP2 were located 50m and 12.5km upstream of dam. At each station, cross sections were surveyed as clusters of five closely spaced O(2-3 m) transects over 10 meter in river length. A Trimble® 5700 GPS Receiver was used to map topography of the cross sections, and the raw GPS data were processed using Trimble Geometrics Office® software. Digital elevation models (DEMs) at each section were generated, and changes in river channel width and elevations resulting from dam removal were determined by differencing DEMs of before and after dam removal.

Figure 2a. Breaching West Side of the St. Johns' Dam, March 18th 2003

Figure 2b. St Johns' Dam, a week before removal, Nov 10th, 2003

Figure 2c. Removing St. Johns Dam, Nov 17th 2003

Figure 2d. Two hours later, Nov 17th 2003,

River bed material distributions at the surface were collected using bulk sampling approach. The surface layer was defined by the maximum sediment size in the cross section. Subsurface sediments were collected separately. The pebble count method (Wolman, 1954) was employed for locations with coarse materials (diameter>128mm). These measurements were repeated at the same locations at least twice, both before and after removal situations.

For shallow depths (< 0.6 m), mean velocity was measured using a handheld SonTek® FlowTracker ADV. In deeper sections (depth >0.6 m) a BT-ADP (3 MHz, SonTek Inc.) was used to profile the vertical velocity distribution. Discharge was calculated by summing the products of mean velocity and cross sectional area. A Q-H rating curve was generated by repeating measurements of velocity and depth at different flows.

In situ turbidity was measured using YSI sondes deployed 100 meter upstream of the dam and 200 meter downstream from the dam. The YSI 6600 Sonde® collected real time turbidity every five minutes. Continuous sampling began 2 days before the partial breach and lasted another week after the breach. Turbidity measurements were also taken during the complete removal of the dam. Shear velocities were derived from the velocity profiles at the 95% confidence level. In addition, BT-ADP also provided longitudinal bathymetric profiles of the river within the modeling area, from which were derived high resolution estimates of bed slope.

Three bedload net traps were installed at one station downstream of the dam. The trap consisted of an aluminum frame (0.3 m by 0.2 m) attached with a nylon trailing net, 0.9 m long with 3.5 mm mesh (Bunte, 1998). These traps were suitable for bedload particle sizes of 4 - 90 mm. Pit traps were also installed at stations DN1 and UP2. The pit traps were made from PVC pipes and had a diameter of 0.15 m and 0.3 m height. Each pit trap was buried level with the river bed, and a net with 0.1mm mesh was mounted inside the trap by a C-pipe clamp to collect coarse and fine sediment.

Model Description

One-dimensional sediment transport models have been shown useful to assess aggradation and degradation within river channels (Rathbun and Wohl 2001). Most commonly used channel evolution and sediment transport models include HEC-6, CONCEPTS, and MIKE 11. When equilibrium conditions do not occur for each time step, HEC-6 is not appropriate for simulating a single event, such as flood event.

Conservational Channel Evolution and Pollutant Transport System (CONCEPTS) is a computer model that simulates open-channel hydraulics, sediment transport, and channel morphology. The CONCEPTS model is limited to straight channels or channels of very low sinuosity and homogeneous cohesive bank material. In addition, it only calculates total load for sediment transport.

MIKE 11 is a one-dimensional fully dynamic model for river system and channel. It integrates a hydrodynamic model (HD), which can simulate both subcritical and supercritical flow conditions, with a graded sediment transport model (GST) for non-cohesive sediment transport (NST). In addition, Mike 11 has a water quality model (WQ), and dam break model (DB). All these models can be run simultaneously with the HD module. MIKE 11 GIS links MIKE 11 to ArcView GIS, and can generate digital maps and graphics for data input and output. NST model considers both equilibrium and non-equilibrium conditions, and bed form can also be modeled. Considering the technical advantages of MIKE 11, we use it as our modeling tool.

Preliminary Results

Figures 3 captures the changes on downstream depth and turbidity before, during and after the dam removal. Initially, turbidity from suspended sediments was gradually increasing. During the removal, however, the turbidity increased over 3 folds within 2 hours, and then dramatically decreased (Figure 3a). Thirty-three hours after the removal, turbidity recovered to the pre-removal concentration. A rainstorm increased runoff on the fourth day after the removal (November 20th, 2003) generating another peak in turbidity (Figure 3a). Water depth was gradually dropping then leveled-off immediately before the removal. During removal, water depth increased over 2 fold and then dropped to 1/3 of the level of before the removal following the same trend as the pre-removal time series. Water level exceeded the dam removal period as flow increased due to a rain event on November 20th.

At station DN1 (75m downstream from the dam), bed elevations surveyed within 20m river section. DEMs were generated with 1m grid resolution, where all the surveyed points within a 1m grid were averaged and the mean elevation represents the elevation of that grid. Changes in elevation are plotted in Figure 4. Bed elevation on the left bank (0-5 meter) degradated -0.224 ± 0.034m, while aggradated from the middle to the right bank deposition increased 0.066 ± 0.014m (Table 1).

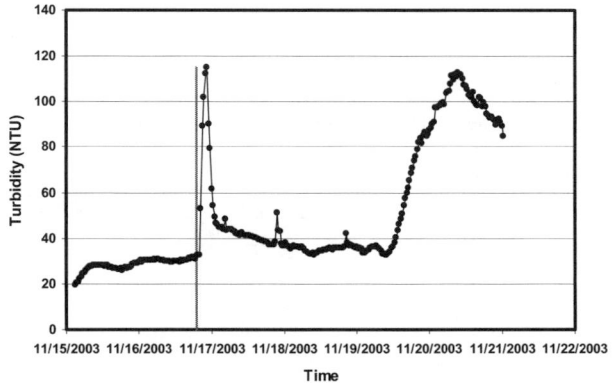

Figure 3a, Turbidity change after dam removal for November 2003. The red line indicates when the removal occurred

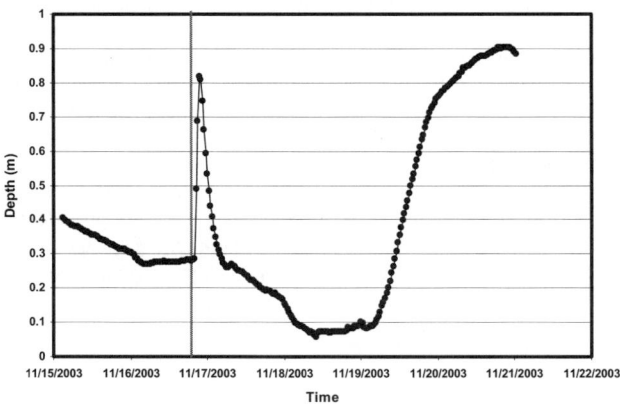

Figure 3b, Depth change after dam removal for November 2003. The red line indicates when the removal occurred.

211

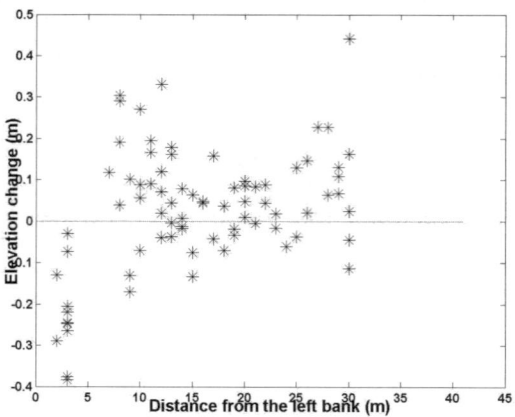

Figure 4, Elevation Change at Downstream 75m, from 11/11/03 to 6/8/04

Table 1. Statistically derived elevation change, 75 m downstream

	DN75m, Left Bank	DN75m, Middle-Right Bank
Mean Change(m)	-0.2238	0.0663
Maximum Change (m)	-0.0302	0.4415
Minimum Change (m)	-0.3830	-0.1695
Median (m)	-0.2452	0.0567
Standard Error (m)	0.0337	0.0141

Conclusions

The methods being developed for documenting changes in sediment transport during episodic events, such as dam removal, are successfully being used to collect the first ever contiguous data set on pre-dam versus post-dam removal conditions. Our preliminary data show significant, but moderate loading of suspended sediments during the dam removal though not more than a bankfull event. Subsequent to removal the channel cross section immediately downstream of the dam began to moderately change in elevation. Continued monitoring and modeling will determine if such adjustments are the result of release of sediments from the reservoir or simply part of the natural processes for this river. Finally, the various methods we used are complimentary in that they provide different pieces of the puzzle for completing the picture of sediment transport. Pebble counts give information on what is in the bed, while sediment traps give information on what is accumulated (or eroded) and at what rates. Velocity measurements link transport to hydraulic variables and topographic DEMs provide an averaged view of the effect of transport on the channel geometry. Once compiled, these data will be used to validate the graded sediment transport model of this river system to predict consequences of the dam removal far into the future.

212

Reference

1. Bunte, K., 1998, Development and field testing of a stationary net-fram bedload sampler for measuring entrainment of pebble and cobble particles, *Report prepared for the Stream Systems Technololgy Center, USDA Forest Service, Rocky Mountain Research Station, Fort Collins, CO,* 74pp

2. Doyle, M.W., Stanley, E.H., and Harbor, J.M., 2003, Channel Adjustments Following Two Dam Removals in Wisconsin, *Water Resources Research,* 39(1), 1011, doi:10.1029/2002WR001714

3. Madej, M.A., and Ozaki, V., 1996, Channel response to sediment wave propagation and movement, Redwood Creek, California, USA. *Earth Surface Processes and Landforms,* 21: 911-927

4. Pizzuto, J. 2002. Effects of dam removal on river form and process, *Bioscience,* 52 (8): 683-691

5. Rathburn, S.L. and Wohl, E.E., 2001, One-Dimensional sediment transport modeling of pool recovery along a mountain channel after a reservoir sediment release, *Regulated Rivers: Research and Management,* 17:251-273

6. Wohl, E.E., and Cenderelli, D.A., 2000, Sediment deposition and transport patterns following a reservoir sediment release. *Water Resources Research,* 36(1): 319-333

7. Wolman, M.G., 1954, A method of sampling coarse river-bed material, *EOS Trans. AGU,* 35(6): 951-956

Targeted Wetland Restoration for Water Quality Improvement: Potential Impact on Nitrate Loads to Mississippi River Subbasins

William G. Crumpton, Associate Professor, Department of Ecology, Evolution, and Organismal Biology, Iowa State University

Non-point source nitrogen loads to surface waters in the Midwest Corn Belt are among the highest in the Mississippi River Basin, and are suspected as a primary source of nitrate contributing to hypoxia in the Gulf of Mexico. Nitrate is transported from crop land primarily in subsurface drainage, especially in extensively tile drained areas like the Corn Belt. As a result, grass buffer strips, woody riparian buffers, and many other practices suited to surface runoff have little opportunity to intercept nitrate loads in these areas. However, wetlands sited to intercept tile drainage have the potential to significantly reduce nitrate loads, and this approach is particularly promising for heavily tile drained areas like the Corn Belt. A performance forecast modeling approach is used to estimate the total nitrate reduction that could be achieved using wetlands as nitrogen sinks in tile-drained regions of the upper Midwest. Not only does the extent of tile drainage vary among the subbasins, but also the efficacy of wetlands intercepting tile drainage varies among the subbasins. This is because several primary determinants of wetland performance vary longitudinally across the upper Midwest, including volume and timing of "runoff", nitrate concentration, and temperature. Our analyses suggest that these would result in a roughly five fold range in mass nitrate removal per acre of wetland restored for different areas of the upper Midwest.

ADDRESSING WETLAND ISSUES – THE KANSAS NRCS APPROACH

L. M. Soffran[1], PE, Member, ASAE, and J. L. McDowell[2]

ABSTRACT

Frequently a wetland in Kansas has been manipulated to improve farming operations or provide a reliable water source for livestock. When a wetland has been converted in violation of wetland provisions, restoration or enhancement of the remaining wetland is the primary approach used to regain the lost wetland acreage and functional capacity units.

In Kansas, the Natural Resources Conservation Service[3] (NRCS) has adopted the Hydrogeomorphic (HGM) Model procedure for assessing wetland functions and values. Four interim HGM models are being used for minimal effects determinations and for mitigations of converted wetlands. The Playa Depression Model is used in the High Plains Playa region of Kansas. A Depression Model is being applied to non-riverine type depressions in the remainder of the state east of the playa region boundary. A Slope Model is used for hillside seeps, and a Riverine Model applies to wooded and herbaceous riverine settings.

Structures are frequently constructed as a component of wetland restoration and creation, and land entered into the Wetland Reserve Program or Conservation Reserve Program. The structural design may include a dike or other type of water impoundment structure. A water level control structure may be included in the design when the objective is shallow water wetlands for wildlife.

KEYWORDS: Hydrogeomorphic, HGM, minimal effects, mitigation, wetlands, wetland functions

INTRODUCTION

The Natural Resources Conservation Service[3] (NRCS) in Kansas is actively providing assistance to customers. Investigative assistance includes performing wetland determinations for Food Security Act (FSA) compliance, as explained in the National Food Security Act Manual (NFSAM), at the request of the landowner for land not under the jurisdiction of the US Army Corps of Engineers (USACE). NRCS also has the responsibility for investigating potential

---

[1] Agricultural Engineer, USDA-NRCS, 760 S. Broadway, Salina, KS 67401-4604, (marty.soffran@ks.usda.gov)

[2] Soil Scientist, USDA-NRCS, 760 S. Broadway, Salina, KS 67401-4604 (jim.mcdowell@ks.usda.gov)

[3] The NRCS and Farm Services Agency are agencies in the United States Department of Agriculture

wetland violations, based on complaints filed with the Farm Services Agency[3].
The determinations are performed by wetland teams in each of the five NRCS administrative areas. These members have received the requisite NFSAM and USACE Regulatory IV – Wetland Identification and Delineation course training.

When a wetland, which meets the NFSAM criteria or USACE jurisdictional wetland definition, has been impacted, the Hydrogeomorphic (HGM) wetland assessment procedure is used in Kansas by the wetland team.

Technical assistance in the form of planning, design and construction inspection is available to individuals on private land. This assistance may be for mitigation of impacted wetlands, or for restoration of wetlands under the Wetland Reserve Program (WRP) and Conservation Reserve Program (CRP). Assistance can also be obtained for non-cost-shared projects to create, restore and enhance wetlands as part of normal operations. Shallow water structures for wildlife is one practice frequently utilized for wetland enhancement.

BACKGROUND

In the 1990s there was interest from various state and federal agencies to find a method of evaluating changes in wetland functions without showing a bias to any particular function which the wetland performs. The HGM model wetland assessment procedure developed by staff at the US Army Corps of Engineers Waterways Experiment Station was one new method developed. HGM models following the HGM Classification for Wetlands (Brinson, 1994) were being developed for some of the wetland types; with expectations that eventually all wetlands would have an HGM model available to address the suite of functions performed by that particular subset. The direction provided in the NFSAM (SCS, March 1996), instructed the individual states to use the HGM procedure to assess the functions that a wetland performs, unless there is another established method for wetland assessment. In Kansas another method did not exist; therefore, this procedure was developed and adopted for use in the state. Different models are needed for assessing wetlands for minimal effects determinations and mitigations; as defined in the NFSAM.

The development of complete HGM models requires the review and approval of the US Army Corps of Engineers Waterways Experiment Station (USACE-WES, Nov. 1997). Sufficient data must be collected to calibrate each function for each model from a statistically sufficient number of sites for a given geographic area. People from federal and state agencies, individuals with expertise in hydrology, biogeochemical, wetland habitat and biology worked together to develop the HGM models needed for wetland types in their geographic area. They identified the various functions that the wetlands performed and the variables that were needed to evaluate the functional effectiveness. A "Measurement or Condition" was defined for each variable to explain the following: (1) pristine conditions would be for that variable (best possible condition that could be found was considered when a truly pristine site did not exist), (2) the worst case situation for the variable, and (3) scenarios between the two extremes. The pristine condition was given an "Index" of 1.0. The worst case condition received a 0.0 Index. Other "Measurement or Condition" scenarios were then developed to address changes in the variable between the two extremes, with an Index somewhere in between. Field tests were performed on sites for a given geographic setting to assess the relative sensitivity of the following: (1) variable "Index" to differing site conditions and (2) function equations resulting from changes in the "Index." Figure1 is an example of a variable with the "Measurement or Condition" and "Index" from the Kansas Playa Interim HGM Model. The models are referred as interim, because sufficient field data has not been collected to develop statistically valid equations and graphs.

KANSAS INTERIM HGM MODELS

Four interim HGM models have been approved for use in Kansas for minimal effects determinations and mitigation.

216

- The Kansas Interim Depression Model is being applied to non-riverine type depressions in the remainder of the state east of the Playa boundary.

- The Kansas Interim Playa Depression Model is used for depressional wetlands in the High Plains Playa region of Kansas.

- The Kansas Interim Riverine Model is being applied to wooded and herbaceous riverine settings.

- The Kansas Interim Slope Model is used for hill side seeps.

Each of the models assesses functions performed by that wetland type. The functions for each model are listed below in Tables 1 through 4.

Figure 1. Example of Model Variable

Model Variable	Ratio of Native to Non-Native Plant Species (V_{pratio}) Measurement or Condition	Index
Definition: The ratio of native perennial plants to annual plants as influenced by the level of disturbance in the wetland. Logic: The presence of a high ratio of native perennial plant species indicates that disturbances that interrupt naturally occurring cycles and other vegetative dynamics are minimal.	All the dominant species in all zones are native perennial species. Impacts from grazing/haying are minimal.	1.0
	Site is abandoned cropland (greater than 5 years). Vegetation consists of a mixture of native perennial species normally found in an undisturbed site and annual plants. Moderate impacts from grazing are present.	0.75
	Severely grazed or cropped site. Vegetation is dominated by FAC or wetter annual plants (i.e., annual smartweed, barnyardgrass and bur ragweed).	0.5
	Site is dominated by annual invasive species which are FACU or upland plants. Examples would include kochia, marestail, bindweed, foxtail, and pigweed.	0.25
	Annually tilled site with no vegetation (excluding crops).	0.0

Table 1. Kansas Interim Depression Model Functions

HYDROLOGY
- Function 1: Maintains Characteristics of Hydrological Regime

BIOGEOCHEMICAL
- Function 2: Maintains elemental cycling (only used in closed basin systems)
- Function 3: Retains particulates (This function has two options: closed basins and throughflow)
- Function 4: Removes dissolved elements and compounds (only used in basins with throughflow systems)
- Function 5: Exports organic carbon and detritus (only used in basins with throughflow systems)

VEGETATION
- Function 6: Maintains characteristic plant community particulates (This function has two options: non-wooded and wooded/forested)

WILDLIFE
- Function 7: Maintain habitat structure within wetland
- Function 8: Maintain food web
- Function 9: Maintains habitat interspersion and connectivity among wetland
- Function 10: Maintains characteristic invertebrate community

Table 2. Kansas Interim Playa Depression Model Functions

HYDROLOGY
- Function 1: Maintains the characteristic of the hydrologic regime
  This function has two options- closed basins and throughflow

BIOGEOCHEMICAL
- Function 2: Maintains elemental cycling (only used in closed basin systems)
- Function 3: Retains particulates (This function has two options: closed basins and throughflow)
- Function 4: Removes dissolved elements and compounds (only used in basins with throughflow systems)
- Function 5: Exports organic carbon and detritus (only used in basins with throughflow systems)

VEGETATION
- Function 6: Maintains characteristic plant community

WILDLIFE
- Function 7: Maintain habitat structure within wetland
- Function 8: Maintain food web
- Function 9: Maintains habitat interspersion and connectivity among wetland

Table 3. Kansas Interim Riverine Model Functions

HYDROLOGY
- Function 1: Dynamic surface water storage (This function has two options: wooded and herbaceous)
- Function 2: Long-term surface water storage for low energy systems (This function only used with no groundwater influence)
- Function 3: Energy dissipation (This function has two options: groundwater influence and no groundwater influence)

BIOGEOCHEMICAL
- Function 4: Elemental cycling (only used in closed basin systems)
- Function 5: Removal of imported elements and compounds (only used in basins with throughflow systems)
- Function 6: Retention of particulates (This function has two options: wooded and herbaceous)
- Function 7: Exports organic carbon and detritus

VEGETATION
- Function 8: Maintains characteristic plant community
- Function 9: Maintains characteristic detrital biomass (This function has two options: wooded and herbaceous)

WILDLIFE
- Function 10: Maintain habitat structure within wetland
- Function 11: Maintain food web
- Function 12: Maintains habitat interspersion and connectivity
- Function 13: Maintains characteristic invertebrate community

Table 4. Kansas Interim Slope Model Functions

HYDROLOGY
- Function 1: Moderation of groundwater flow
- Function 2: Velocity reduction of surface water flow

BIOGEOCHEMICAL
- Function 3: Retention, conversion and release of elements and compounds
- Function 4: Retention of particulates
- Function 5: Organic carbon export

VEGETATION
- Function 6: Maintenance of characteristic plant community

WILDLIFE
- Function 7: Maintenance of habitat interspersion and connectivity among wetland

An example of the Function and its Functional Capacity Index (FCI) Score equation would look like:

Function 3: Retains Particulates
A deposition and retention of inorganic and organic particulates from the water column, primarily through physical processes.

If wetland has no outlet, then FCI Score = $[((V_{sed} + V_{upuse}) / 2) \times V_{mod}]^{1/2}$.

If wetland has an outlet, then FCI Score = $[((V_{sed} + V_{upuse} + V_{wetuse}) / 3) \times V_{mod}]^{1/2}$.

KANSAS INTERIM HGM MODEL USE

The procedure implemented for the interim HGM models used to perform minimal effects determinations is more restrictive than the 2-4-5 Rule found in the NFSAM (USDA-SCS, March 1996). One requirement is there be zero (0) acreage loss. The other requirement is that each Hydrology and Biogeochemical function must stand alone on percent change. These are the only functions that are evaluated for performing minimal effects determinations; though environmental issues that could raise concerns must also be addressed. The allowable percentage impact permitted for minimal effects, based on pre-impacted FCI conditions for each variable, is:

- FCI greater than or equal to 80%, maximum reduction is 20%

- FCI less than 80%, maximum reduction is 40%

Therefore, if any of the Hydrology or Biogeochemical functions exceeds the allowable change, then the activity will not qualify for minimal effects waiver and the activity must be mitigated.

Mitigation requires no reduction in acreage or Functional Capacity Units (FCUs). A FCU is equal to the FCI score multiplied by the acreage. Every function in the HGM model is evaluated for mitigation. If FCU change for each function is positive, such that the function is being replaced with less acreage, the acreage replacement must still be one-to-one (1:1). The maximum acreage requirement is set at 3:1 (3.0 acres replaced for 1.0 acre lost at the impacted site). The calculations can be performed manually for each FCI and FCU equation. A Microsoft® Excel workbook has been developed for each HGM model which performs the FCI and FCU calculations automatically so as to minimize errors.

There are three types of mitigation. The types are as follows: (1) Restoration of the impacted wetland by restoring it to the original size and condition. This will consist of restoring the hydrology to the wetland. (2) Enhancement of an existing site by enlarging the size and providing a biological environment which would successfully establish the wetland characteristics necessary to meet the variable requirements. (3) Creation of a wetland on ground that had never been a wetland before. This type of mitigation is the most difficult to establish, since it requires establishing and maintaining a hydrologic source sufficient to create a hydric soil and support hydrophytic vegetation.

STRUCTURES

Structures have become a key component in many of the wetland restoration and enhancement projects, including WRP and CRP wetlands. They are necessary for the majority of the creation sites. Dikes or embankments are being constructed to contain water to a given depth. The embankment could be for containment of runoff that passes through this area, capture out-of-bank flooding from a channel or nearby stream, or a combination of the two. Normally the existing ground is left in tack and the embankment is constructed of fine textured soil. There is typically a 10-foot top width and the front and back side-slopes are between 5:1 and 10:1. The top of the embankment and both side-slopes are vegetated with perennial grass. The total fill height is normally less than 3-feet. The entire top of the embankment serves as an auxiliary spillway.

A flashboard or stop-log structure is inserted in the embankment for water level control. This permits moist soil management to be practiced when and where appropriate.

CONCLUSION

In Kansas the NRCS is using HGM models for minimal effects and mitigation of impacted wetland concerns. Different models have been developed to address the suite of functions that the wetland performs within a particular landscape setting. The minimal effects criterion takes into consideration the unique wetlands that are found in this state. Mitigation for manipulated wetlands requires replacement of lost acreage and functions performed by the wetland.

REFERENCES

Brinson, M. M., A Hydrogeomorphic Classification for Wetlands, 1994, Wetlands Research Program Technical Report WRP-DE-4, US Army Corps of Engineers Waterways Experiment Station

USDA Soil Conservation Service (SCS), National Food Security Act Manual, Third Edition, March 1996

US Army Corps of Engineers Waterways Experiment Station (USACE-WES) and Environmental Protection Agency Region 6, November 1997, Workshop for Developing Regional Guidebooks Using the Hydrogeomorphic (HGM) Approach

USDA Natural Resources Conservation Service, Kansas interim hydrogeomorphic (HGM) functional assessment model --DEPRESSIONAL WETLANDS, May 5, 2004

USDA Natural Resources Conservation Service, Kansas interim hydrogeomorphic (HGM) functional assessment model – PLAYA DEPRESSIONAL WETLANDS, May 5, 2004

USDA Natural Resources Conservation Service, Kansas interim hydrogeomorphic (HGM) functional assessment model – RIVERINE WETLANDS, May 5, 2004

USDA Natural Resources Conservation Service, Kansas interim hydrogeomorphic (HGM) functional assessment model – SLOPE WETLANDS, May 5, 2004

An Historical Perspective of Hydrologic Changes in Seven Mile Creek Watershed

Kevin J. Kuehner

ABSTRACT

This study documented hydrologic changes, specifically wetland losses, in Seven Mile Creek Watershed. Historical aerial photos along with a Geographic Information System (GIS) were used to assess these changes as it relates to water resources management. The 95.3 km² (36.8 mi.²) study area is a small, agricultural watershed located in south-central Minnesota. More than 130 aerial photographs from seven different periods dating back to 1938 were scanned and rectified for use in a GIS. Wetland areas converted to cropland were then interpreted and digitized. In addition, other land use changes, such as surface and sub-surface drainage modifications and cropping system shifts, were mapped and documented. Results from the study indicate significant hydrologic changes have occurred in the watershed. Analysis of pre-settlement maps and survey notes indicate that about 50% of the watershed was once covered by wetlands. Of those wetlands, it is estimated that 88% of the natural wetlands have been converted to cropland. About 47% of those losses occurred from early settlement (late 1800's) to 1938. From 1938 to 1985, an additional 41% of the wetlands were drained and converted to cropland. This translates to an average annual net wetland loss of 40 hectares (100 ac.) per year. During this same period (1938-1985), 40 km (25 mi.) of drainage ditches were constructed, more than 966 km (600 mi.) of public and private sub-surface drainage systems were installed, and it is estimated that total corn and soybean acreage increased from 30% to 96% within the watershed. The most rapid percent change, a 50% wetland decrease, occurred between 1955 and 1961. The construction of two county drainage ditch systems in 1955 accounts for this change. After 1985 the wetland loss trend has decreased. Wetland increases are a direct result of conservation programs combined with grants from private and state water resource protection programs.

KEYWORDS. wetland, GIS, watershed, water quality, hydrology, drainage water

INTRODUCTION AND BACKGROUND

Research relating to hydrology and water quality has been well documented in the Minnesota River Basin. Southern Minnesota is intensively managed for row crop production (Payne, 1991) and has been extensively drained via subsurface tile (Binstock, in Magner, et al., 2004). Agricultural drainage along with the construction of drainage ditches in southern Minnesota over the last century drained wetlands to improve soil productivity (Leach and Magner, 1992) for crop production. Agricultural drainage is integral to Minnesota's farm economy. However, the cumulative influence of large-scale drainage, particularly the increase in contributing drainage area, has initiated downstream channel instability across southern Minnesota (Magner and Steffen, 2000; Magner, et al., 2004) and accelerated nutrient losses. In particular, studies by Alexander (2000) Magner and others (2004) concluded that elevated nitrate concentrations in area watersheds were largely caused by a combination of increased pathways through subsurface drainage tile systems and minimal contact time with de-nitrifying environments like channel sediment and active floodplains. Gilliam and Skaggs (1986) examined drainage factors that influence nutrient concentrations in tile drained basins. They noted a 10-fold increase in surface water nitrate concentrations when subsurface tile were added to a drainage ditch system. Randall and others (1986) reported tile-drain nitrate concentrations ranged from 16-172 mg/l in shallow groundwater from sub-surface tile drainage systems at the Southern Minnesota Research and Outreach Center, in Waseca, Minnesota.

In addition to direct drainage effects on stream nitrate concentrations, cropping system shifts have been influential in determining the impact of water quality impairments. Cover types such as wetlands, small grains, and perennial crops such as hay, and pastures reduce peak flows. These perennial land covers transpire more in the spring than do corn and soybeans, thereby creating additional room for storage of rainfall in the soil. Compared to a corn-soybean rotation, alfalfa reduced tile drainage loss by 50% in a four-year study at Lamberton experimental station (1990-93) (Randall, 1997). In addition, University of Minnesota studies at Lamberton found nitrate losses in tile drainage from corn and soybean row crop systems to be 30 to 50 times greater than from perennial alfalfa and CRP grass systems. This was due to more water per acre being drained from the row crops and higher concentrations of nitrate in that water (22-28 mg/l) compared to the perennial crops (0.7-1.6 mg/l) (Randall, 1997).

Recognizing the influence hydrologic change can have on water quality, this study helps to quantify the magnitude of these changes in Seven Mile Creek Watershed (SMCW) by examining the historical extent of wetland loss to cropland, engineered surface and sub-surface drainage modifications, and general cropping system shifts. This information will serve as a valuable tool to help educate watershed residents and policy makers about the importance of restoring wetlands for water quality, flood control, and habitat enhancement purposes.

To document and quantify the hydrologic changes in SMCW, historical aerial photos were scanned and rectified for use in a Geographical Information System. More than 130 aerial photos from 1938 through 2003 were used to document the changes in land use and hydrology. In addition, digital elevation models (DEMs), geo-referenced 1854 Public Land Survey Maps, digital soil surveys, and digital ortho-photo-quads (DOQs) were used to estimate pre-settlement land use conditions (see appendix for more information).

STUDY AREA

The 23,551-acre (36.8 sq. mile) study watershed is located in the Minnesota River Basin, within the Middle Minnesota Major Watershed in south-central Minnesota (Figure 1). The watershed is located between the communities of Nicollet and St. Peter in Nicollet County. Flat agricultural fields (0-2% slope) with numerous small depressions dominate the landscape. Dominant soil types include Canisteo-Glencoe complex clay loam, Cordova clay loam, and Canisteo clay loam soils. The landscape transitions to deciduous forest and steep ravines as the creek descends into the Minnesota River Valley. As of 2002, 81% (19,172 acres) of the watershed land use was under cultivation. Cropland in the watershed is dominated by a corn-soybean rotation and accounts for 96% of the cultivated acres. The remaining crops consist of peas, sweet corn, and alfalfa.

Seven Mile Creek is one of Nicollet County's most visible natural resources with a 630-acre county park located at the mouth of the watershed. Since 1985, the Minnesota Department of Natural Resources (DNR) designated the creek as a class 1-D marginal trout stream.

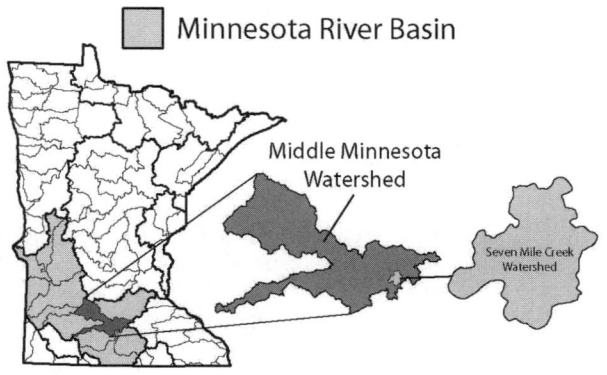

Figure 1. Project Area Location

Methods

Aerial photos from years 1938, 1950, 1955, 1961, 1967, 1978, 1985, 1990, 2002, and 2003 were used in the analysis. Additionally, geo-referenced public land survey maps from 1854, combined with digital elevation models, and soil surveys, provide insight on the pre-settlement conditions within the SMCW (see figure 4 and appendix for more information). College level student interns provided free labor during the rectification process, thereby keeping overall project costs extremely low. A Pentium 4 desktop and a Pentium 3 laptop computer were used to process the images. A rectification software extension for ArcView® called SmartImage® was purchased for $300.00 from the Australian-based company, Mapping and Beyond. The historical photo project is estimated to have taken 500 hours.

The process for documenting the hydrological changes in Seven Mile Creek Watershed can be summarized in five steps:

1. Photo acquisition

2. Scanning

3. Rectification

4. Cropping and creating a mosaic image

5. Aerial photo interpretation, mapping, and analysis

The first step was to acquire the photos. Historical black and white aerial photos from 1950, 1955, 1961, 1967, and 1978 were collected from the U.S. Department of Agriculture, Farm Service Agency (FSA) and Natural Resources Conservation Service (NRCS) offices. All photos were 9x9 prints, except the 1950 and 1978 photos, which were 20x24 and 20x20 inches respectively.

The 1938 photos of Nicollet County were in digital format and obtained from the Minnesota Department of Natural Resources (MDNR). The MDNR acquired the 1938 photographs from the

University of Minnesota, Borchert Map Library. (The Map Library has an extensive set of aerial photos for the state of Minnesota in their collection. Refer to their web site at: http://map.lib.umn.edu/.)

The 1990, 2002, and 2003 were in a digital ortho-photo quadrangle (DOQ) format. A DOQ is a computer-generated digital image of an aerial photograph in which displacements caused by camera orientation and terrain have been removed. The 1990 photo is a black and white United States Geological Survey (USGS) DOQ, and the 2002 and 2003 are color DOQs taken by the FSA Aerial Photography Field Office.

The 1985 photos were taken by the Sidwell Company for county parcel mapping purposes and were in halftone emulsion on mylar format. A piece of white paper was placed on the back of the mylar for scanning purposes. The 1985 photos were obtained from the Nicollet County Environmental Services Department, St. Peter, Minnesota.

The 1938-1978 photos were taken during the summer, while the 1990 and 1985 photographs were taken soon after snowmelt. The 1990 and 1985 photos were especially useful for locating private drainage tile lines.

The second step in the process was to scan the historical photos. The main goal during this process was to maximize photo quality and resolution, while minimizing file size. Most aerial photos were scanned black and white (greyscale) at 400 dots per inch (dpi), and saved as .jpg images using a HPScanJet 5200C series scanner. The 1950, 1978, and 1985 photos were too large for scanning with the HP ScanJet so a Widecom SLC 1036C scanner from the Nicollet County Environmental Services Department was used to convert the photos to digital images. The Widecom scans photos up to 26 inches wide, and digital images were saved at 200 dpi to maintain a smaller file size. Scanning at higher resolution was not necessary with the large format photos.

Once in digital format, the images pertaining to the watershed were geometrically rectified and spatially registered within ArcView. This was the most time consuming of the steps during the historical photo project. Image rectification is the mathematical process of making image data conform to a map projection system. Smart Image version 8.0 (a relatively inexpensive and easy-to-use software extension to ArcView GIS), by Mapping and Beyond, was used to rectify the image. This software program warps the image to match an already registered image like a DOQ. Images are converted to 'real world' ground coordinates by referencing the image to another source that is in the desired map projection (called the 'reference image'). Reference information may be obtained from another image, vector coverages, or map coordinates. In this analysis, the 1990 DOQ was used as the reference image and computer mouse clicks marked the ground control points (GCPs) to depict the same location on both the reference and non-rectified images. The images were rectified using a bilinear or 1.5-order polynomial transformation method with a minimum of 5-10 GCPs to maintain a Root Mean Square error at or below 0.7. The corners of buildings and road intersections were the most commonly used GCPs. Care was taken to spread the GCPs around the entire image.

Note that not all errors were completely eliminated using this rectification process. To correct for all distortion, such as that from relief, a process called orthorecfication must be completed. Orthorecification is the mathematical process of removing the distortion, caused by relief and the camera, within a photograph so that the scale is uniform throughout the output image. However, orthorecification was not considered necessary in this case since all of the cropland areas are located on flat topography ($< 2\%$ slope). Orthorecification software is expensive and, for the small gain in accuracy, was not considered cost-effective for this particular project.

After rectification, the images were cropped down to include just the SMCW. The cropped and rectified photos were then laid on top of each other to create one image, a mosaic. This process creates a single image file out of many separate image files. In essence, the image can be considered a geo-referenced digital map.

The last step in the historical aerial photo project was aerial photo interpretation and mapping within ArcView 3.2. The aerial photos were interpreted for land use changes by comparing 2002 cultivated land to non-cropped land in the historical photographs. Changes in land use, specifically wetland, grassland, pasture, and forest loss were then mapped through a heads-up digitizing process in ArcView. Attributes were then assigned to a polygon attribute table. The 2002 FSA Crop Land Unit was used to compare the cropped versus non-cropped areas in the watershed. For the most part, wetland cover type signatures were relatively easy to delineate. The most common interpretation methods used to distinguish between cropped fields and wetlands were photo tones, textures, and field pattern contrasts. Where there was difficulty determining what was wetland versus cropland, photographs preceding and succeeding the particular photo year were used to validate the polygon's attribute. Wetland areas near farm sites were difficult to categorize and were considered pasture or fallow grassland for this project. For the purpose of this project, a wetland is defined as land area having the water table at, near, or above the land surface or that is saturated for a long enough period to promote wetland or aquatic processes as indicated by hydric soils, and aerial photo-interpreted hydrophytic (water loving plants) vegetation. Open water was also coded as a wetland cover type. In all cases the upland area immediately surrounding the wetland was included as the wetland cover. It was assumed that if the upland area around the wetland was not farmed, then that area was considered hydric and not feasible for farming because of poor drainage during the growing season. These areas could also be interpreted as wet prairie and wet meadow, but for the purpose of this study were uniformly coded as a wetland cover type.

RESULTS AND DISCUSSION

Results from this exercise indicate significant hydrologic changes have occurred in the SMCW. Those changes include conversion of wetland to cropland, surface and subsurface drainage modifications, and cropping system shifts. Tables 1, 2, 3 and Figures 2, 3, 4 summarize the results of the historical photo analysis.

The initial conversion of native prairie and wetland vegetation to cropland was most likely the single greatest hydrologic change within the watershed. The 1851 map in figure 4 estimates the extent of wetlands in the watershed before the area was settled. Pre-settlement vegetation on remaining acres within the watershed was characterized by prairie, big woods, and oak openings. Before the watershed was influenced by European settlement, an estimated 11,000 acres or 47% of the watershed was covered by wetlands. The earliest recorded public water management activity in the watershed began in 1889 when a drainage tile was installed to improve the land for farming. County ditch (CD) number 6, CD 58 and CD 29 were installed in 1889, 1917, and 1918 respectively (figure 3). CD 6 and CD 29 were constructed to drain the northern portion of the watershed into Goose Lake (figure 4). The first public drainage ditch, CD 46, was constructed in 1950 and helped drain the western portion of the watershed. In 1955 two more drainage ditches, CD 24 and CD 13, were constructed to drain the southern and northern portions of the watershed.

Table 1 reports the extent of wetland acres by year within the watershed. Figure 2 graphically illustrates the distribution of wetland acres by year. According to Table 1, the periods with the greatest average annual net change include: 1955-1961, 1950-1955 and 2002-2003. The average annual net change was determined by taking the wetland acreage difference between photo years and dividing that number by the total number of years between photos. For instance, the average annual net loss of 342 acres between 1955-1961 was determined by taking the difference between 4,095 acres and 2,042 acres. The result of 2,053 acres was then divided by 6 years. The change in wetland acreage in SMCW can be organized into three time periods. Those periods include: pre-settlement to 1938, 1938 to 1985 and from 1985 to 2003. During the first period, pre-settlement to 1938, it is estimated that 5,137 wetland acres were converted to cropland, resulting in a 47% loss. This represents an average annual loss of about 61 acres of wetlands per year. From 1938 to 1985 the watershed lost an additional 4,556 acres, resulting in a 41% decrease or a loss of about 97 acres per year. During this period the greatest wetland loss (2,053 acres, 50% change) occurred between the years 1955 to 1961. The construction of CD 13 and CD 24 and subsequent loss of Goose Lake accounts for this significant loss of wetlands. From 1985 to 2003, the wetland loss trend stabilized and decreased. During this period there was an average gain of about 14 acres of wetlands per year. Federal farm bill policies such as Swampbuster, state wetland protection programs such as the 1990 Wetland Conservation Act, and concentrated watershed efforts through a Clean Water Partnership project help explain this reversal.

Table 1. Extent of Wetlands by Year.

Year	Extent of Wetland (acres)	Average Annual Net Change (acres)	% Cumulative Loss	% Change Between Years
1851	~11,000	--	--	--
1938	5,863	-61	47	-47
1950	5,104	-63	54	-13
1955	4,095	-202	63	-20
1961	2,042	-342	81	-50
1968	1,662	-54	85	-19
1978	1,372	-29	88	-17
1985	1,307	-9	88	-5
1990	1,347	8	88	3
2002	1,441	8	87	7
2003	1,561	120	86	8

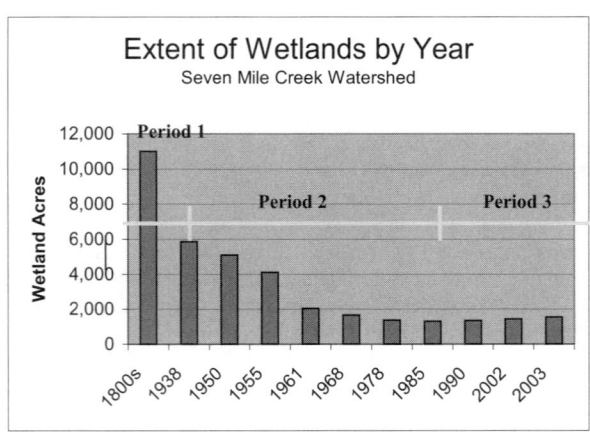

Figure 2. Extent of Wetlands by Year

When compared to wetlands, grassland and forestland conversion was found to be less substantial in the watershed. Table 2 quantifies the extent of forest and grassland conversion in SMCW since 1938. About 374 acres of deciduous forest and 951 acres of grassland and pasture were converted to cropland from 1938 to 1985. From 1985 through 2003 about 200 acres of cropland have been converted back to native grassland. Half of those acres were restored between 2002 and 2003. Restored native grassland serves as a valuable pollutant filter and as a wildlife habitat corridor between cultivated land and surface water within the watershed.

Table 2. Extent of Forest and Grassland by Year

Year	Forest acres	Pasture/Grassland acres
1938	1,852	1,594
1950	1,823	1,251
1955	1,789	1,181
1961	1,639	1,225
1968	1,503	1,022
1978	1,478	672
1985	1,478	643
1990	1,478	683
2002	1,478	733
2003	1,478	823

The general shift away from a wide diversity of hay and small grain crops to corn and soybean production has also produced hydrologic changes over time. Table 3 reports the estimated percentage of crop and wetland acreage within the watershed. Note that no specific crop history data exists for just the watershed. Instead, Nicollet County was assumed to be representative of SMCW, and county averages were used and applied to the watershed preceding 2001. Crop history information was acquired from the United States Agriculture Census data and Farm Services Agency. Using this information, it is estimated that corn and soybean acreage in the watershed has increased from 30% in 1939 to 96% in 2003. An increase in the amount of corn and soybeans grown in the watershed has been at the expense of hay and small grains like wheat, oats, barley, sorghum, flax. From a soil and water conservation perspective, small grain crops require fewer inputs like herbicides and fertilizers and, because they are established earlier than corn and soybeans, tend to transpire more in the spring reducing the potential for surface runoff. Today, 96% of cropland in the SMCW is planted to corn and soybeans with the remaining 4% split between alfalfa, peas and sweet corn. For comparison, 25% of the watershed land use was wetland in 1934 while 7% was in a wetland land cover in 2003.

Table 3. Percentage of Crop and Wetland Acres

Year	Corn	Soybean	Hay/Alfalfa	Small Grain	Other Crop		Wetland (% of watershed)
1872	10	--	3	87			
1934	31	--	24	39	6		26
1939	30	--	22	44	4		25
1950	39	9	14	36	2		22
1954	36	23	13	25	3		17
1964	45	22	13	15	5		9
1969	43	35	9	11	2		7
1974	48	34	6	11	1		6
2002	49	47	3	--	1		6
2003	49	47	3	--	1		7

Figure 3 is a map summarizing the extent of natural versus altered drainage in the watershed. The black, red, and green indicate public and private drainage systems while the blue indicates unaltered drainage. Using rectified historical photos, approximately 620 miles of private tile lines were documented. Since tile line signatures only appear in aerial photography when certain soil moisture and cover conditions are present, it is safe to say there are more private tiles than indicated. Most likely more than 1,000 miles of private tile have been installed in the watershed. Figure 3 illustrates the connectivity of the watershed. From a water quality perspective, this may increase the transport of nutrients like nitrate if nitrogen is applied at rates in excess of recommended agronomic rates. In addition the figure also suggests the contributing drainage area to Seven Mile Creek was significantly smaller and less efficient during pre-settlement conditions compared to current conditions. According to 1854 Public Land Survey Maps, tributaries to Seven Mile Creek totaled about 7.3 miles. Most of these tributaries meandered and extended from large upland, prairie pothole wetland complexes. Many of the prairie potholes that were once scattered throughout the watershed were closed flow intermittent wetland systems, which filled with rain and melting snow and then slowly evaporated or drained through the ground-water system during the later summer months. Under most soil moisture conditions, water was trapped in these potholes and only a small portion entered the creek as runoff. Today there are 25 miles of channelized tributaries that directly connect the watershed to the Creek. This represents a three-fold increase in surface drainage. The cumulative effect of these changes may increase peak flows, resulting in accelerated stream bank instability and water quality impairments (Magner, 2000).

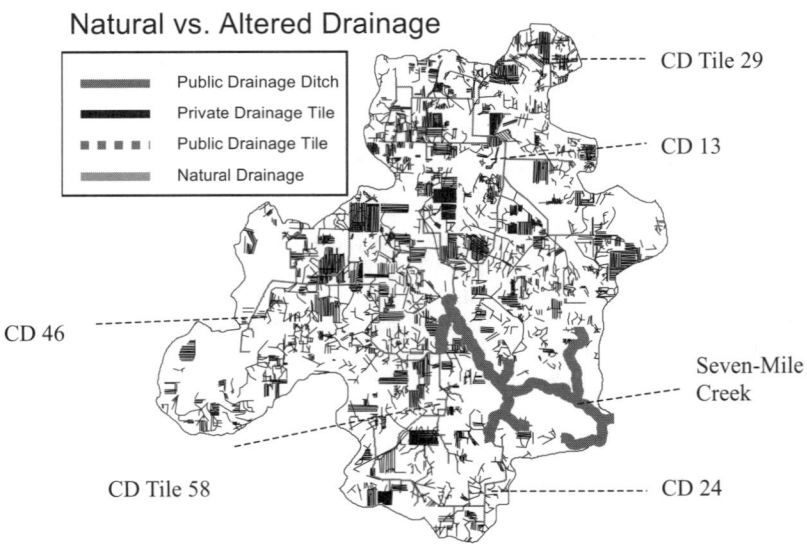

Figure 3. Natural vs. Altered Drainage in SMCW

The use of historical aerial photos and GIS has proved to be effective for quantifying hydrologic changes in the Seven Mile Creek Watershed. The hydrologic changes documented in this exercise also demonstrate the effectiveness of engineered agricultural drainage systems. The benefits of agricultural drainage are numerous. Drainage on wet agricultural soils allows timely field operations, helps plant growth to begin early, and helps achieve improved levels of productivity. In summary, drainage benefits crop production by minimizing risks, improving efficiency, and increasing net income. Although integral to the agribusiness economy, concerns relating to agricultural drainage systems including wetland habitat loss, impaired water quality, and hydraulic overloading has increased.

Recently, the SMCW Clean Water Partnership Project with funding from the McKnight Foundation are exploring ways to restore wetlands while maintaining agricultural productivity and profitability. Innovative solutions include linking wetlands to subsurface drainage systems for the purpose of filtering agricultural runoff before entering streams or ground water. For instance, in one project, a 12-inch public ditch tile (CD 58) draining 200 acres of cropland was petitioned by the Brown-Nicollet-Cottonwood Water Quality Board and Nicollet Soil and Water Conservation District through Minnesota Drainage Law 103e.227 to be routed into a newly restored 50-acre wetland (see map 2003 in figure 4). The wetland is expected to remove 4,400 lbs. of nitrate-N per year and act as downstream flood control by storing up to 55-acre feet of water during a 100-year, 24-hour storm. Additional research to determine how much restoration of this type is needed on a watershed scale to help address downstream environmental concerns will be beneficial for watershed managers. Furthermore, additional research on what the long-term significance of these hydrologic changes mean to the landscape and receiving water bodies in relation to runoff, erosion, nutrient and pesticide losses, turbidity, stream bank erosion, etc. will be useful in developing and implementing realistic water resource protection polices.

Results Summary

- Conversion of 9,693 wetland acres to cropland from 1800s to 1985 (88% loss).

- Surface drainage changes--Between 1938 and 1985, 33 miles of public drainage ditches installed. This represents a 3 fold increase in surface drainage to Seven Mile Creek compared to pre-settlement conditions.

- Sub-surface--40 miles of public drainage tile were added in the watershed. Also, more than 600 miles of private subsurface drainage tile systems were added.

- Corn and soybean acreage in the watershed has increased from 30% in 1938 to 96% in 2003.

- Most of the shift to corn and soybeans has been at the expense of large reductions in wetlands, small grains, and hay.

- 1,325 acres of forest, grassland and pasture are converted to cropland from 1938-1985.

- A total of 5,881 acres of wetlands, forest, pasture, and grassland were converted to cropland from 1938-1985. This represents 32% of the cropland in 2002.

- Four large lakes were drained and converted to cropland: Goose Lake, Fox Lake, Overson Lake, and an unnamed lake (figure 4).

- Top three periods of greatest average annual net wetland change.
    1. 1955-1961: loss
    2. 1950-1955: loss
    3. 2002-2003: gain

- From 1985 to present, federal farm bill policies, state wetland programs, and watershed-based water quality projects helped to reverse the wetland loss trend. Since 1985, 254 acres have been restored back to grassland and wetland.

CONCLUSIONS

The use of rectified historical aerial photos and GIS has proven to be effective for quantifying hydrologic changes in the Seven Mile Creek Watershed. Results from this project indicate that significant changes in watershed hydrology have occurred. Some of those changes include the construction of more than 600 miles of drainage systems and subsequent conversion of 9,693 acres of wetlands to cropland (88 % loss). After 1938, the rate of wetland loss appeared to be the greatest between 1955 and 1961 (50% decrease). The construction of two drainage ditches account for this change. The average annual net loss of wetlands during the 1938 to 1985 period was about 100 acres per year. After 1985, the wetland loss trend decreased. The greatest wetland gain occurred between 2002 and 2003 (8% increase). Federal Conservation Reserve Programs (CRP) combined with grants from private and state water resource programs to help restore wetlands for water quality are primary reasons for this increase.

Acknowledgments

The McKnight Foundation and Minnesota Pollution Control Agency's Clean Water Partnership Program provided funding for this project. In addition, special thanks are extended to the following: Adam Cordes, 2003 Gustavus Adolphus College Fall Semester Intern, for helping scan, rectify, and archive the historical aerial photographs. Crystal Mustain, Minnesota Department of Transportation, for the scanned and rectified Nicollet County 1854 Public Land Survey Maps. Bill Geary of Nicollet County NRCS for help in locating and organizing historical aerial photos. Chris Steffl, Nicollet County Environmental Services, for use and instruction on the large format scanner and for 1985 Sidwell photography. Minnesota State University Mankato-Water Resources Center, for the GIS base layers and website integration. Minnesota Department of Natural Resources for the scanned 1938 photography.

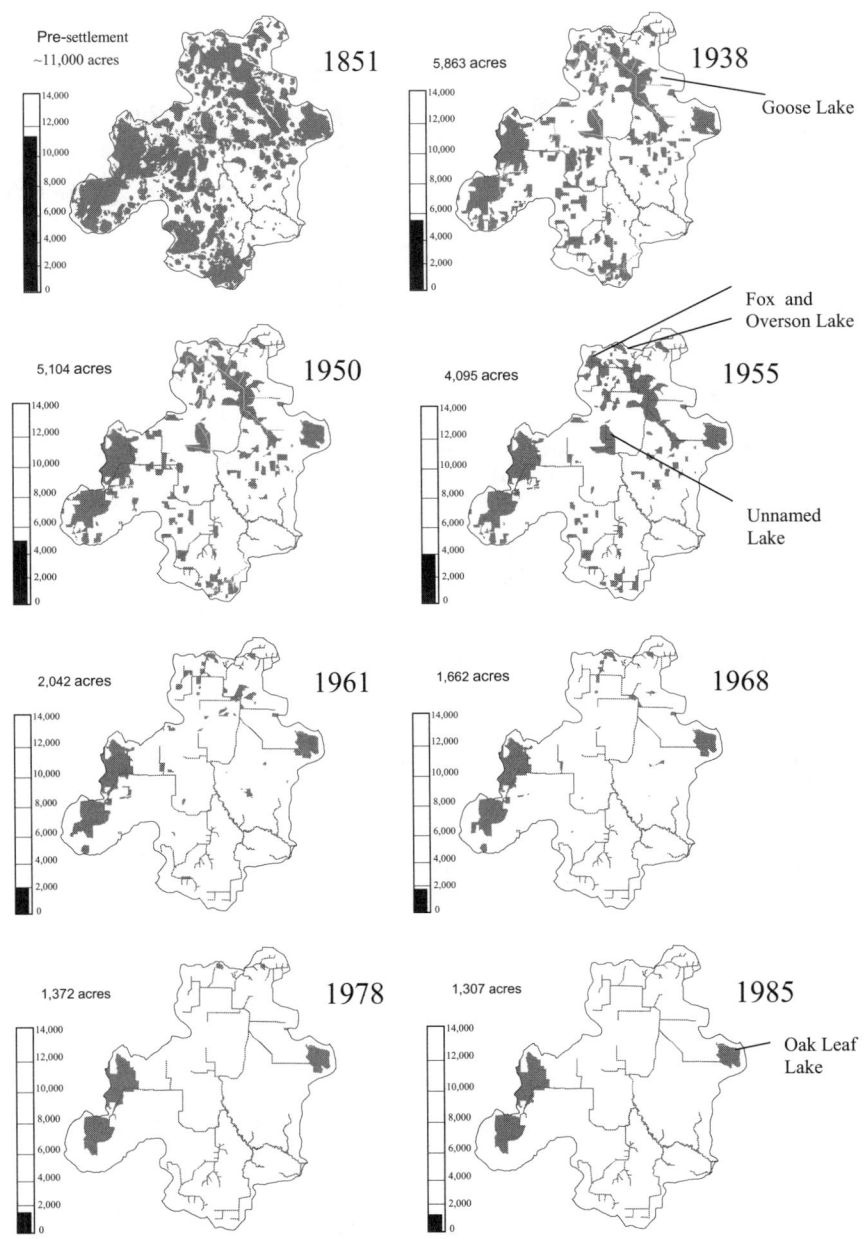

Pre-settlement
~11,000 acres

1851

5,863 acres

1938

Goose Lake

5,104 acres

1950

4,095 acres

1955

Fox and
Overson Lake

Unnamed
Lake

2,042 acres

1961

1,662 acres

1968

1,372 acres

1978

1,307 acres

1985

Oak Leaf
Lake

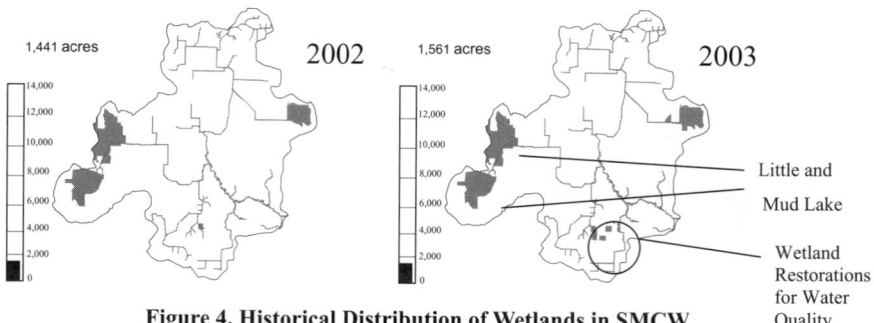

Figure 4. Historical Distribution of Wetlands in SMCW

REFERENCES

1. Alexander R. B., R. A. Smith, and G. E. Schwarz. 2000. Effect of stream channel size on the delivery of nitrogen to the Gulf of Mexico. *Nature* 34:758-761.

2. Gilliam, J. W. and R. W. Skaggs. 1986. Controlled agricultural drainage to maintain water quality. *Irrig. Drain. Engin.* 112(3):3-10.

3. Gresham, W. G. 1916. Nicollet and LeSueur County History. Bowen and Company, Indianapolis, Indiana.

4. Leach, J. and J. A. Magner.1992. Wetland drainage impacts within the Minnesota River Basin. *Currents* 2:3-10.

5. Magner, J. A. and L. J. Steffen. 2000. Stream morphological response to climate and land-use in the Minnesota River Basin. In *Proc. Joint ASCE Water Resources Engineering, Planning & Management Conference.* Reston VA: ASCE.

6. Magner, J. A., G. A. Payne, and L. J. Steffen. 2004. Drainage effects on stream nitrate-N and hydrology in south-central Minnesota. *Environmental Monitoring and Assessment.* 91:183-198.

7. Mustain, C. M. 2004. Historical Surface Hydrology of Nicollet County Minnesota. University of Minnesota.

8. Payne, G. A. 1991. Sediment, nutrients, and oxygen demanding substances in the Minnesota River: selected water quality data, 1989-90. USGS Open-file Report 91-498. Denver, CO: United States Geological Survey.

9. Randall, G. W., G. C. Buzicky, and W. W. Nelson. 1986. Nitrate losses to the environment as affected by nitrogen management. In *Proceedings of the Agricultural Impacts on Groundwater, NWGA.* Dublin, OH.

10. Randall, G. W., D. R. Huggins, M. P. Russelle, D. J. Fuchs, W. W. Nelson, and J. L. Anderson. 1997. Nitrate losses through sub-surface tile drainage in CRP, alfalfa, and row crop systems. *J Environ. Qual.* 26:1240-1247.

11. United States Department of Commerce. 1934-2002. Agriculture statistics for Minnesota and counties: Washington D.C.

Appendix

Creation of the Pre-settlement Wetlands Map

This map (1851 in figure 4) estimates the possible extent of wetlands before the watershed was settled, drained, and converted to cropland. For the purpose of this project, pre-settlement is considered before 1851 or after the signing of the Treaty of Traverse De Sioux. By no means does this represent the absolute extent or distribution of wetlands before European settlement, but represents a logical estimate. Combinations of data layers were used and are listed below. The 1938 aerial photography interpreted wetlands map was used as the base map. For the purpose of this study the blue areas on this map represent all types of prairie pothole wetlands including wet prairie and open water. The remaining land covers in the watershed were considered prairie, oak openings, and big woods forest. Using these data sources it is estimated that about 11,481 acres could have been a wetland land cover in the watershed before extensive settlement and water management of the watershed occurred.

- Rectified 1854 Public Land Survey Maps (PLS)

Original Public Land Survey records were used to assess the spatial extent of hydrologic conditions in the Seven Mile Creek Watershed prior to European settlement. From the late 1700s through the early 1900s, the US General Land Office oversaw the surveying of the unsettled lands of the lower 48 United States. Established by Thomas Jefferson, the Public Land Survey is a legal land reference system set up to ease the inventory and transfer of property.
The original public land surveys in Nicollet County were conducted over 150 years ago by A.D. Anderson and use a section, township and range grid system. In 1854 surveyors traversed Nicollet County to survey and mark section corners. At the same time the surveyors drew maps and notes of the surveyed area. The data derived from land surveyor notes included descriptions of the landscape, vegetation, stream locations, and wetlands found within eyesight of the survey transects. Although only accurate near the section lines and corners, the map and corresponding surveyor notes are the only detailed record of pre-settlement hydrologic conditions within the watershed.

As part of a historic surface hydrology pilot study of Nicollet County to predict archeological sites, geo-referenced PLS maps were acquired for Nicollet County from the Minnesota Department of Transportation in 2003 (Mustain, 2004). The rectified images help indicate the original extent and location of streams, and wetlands within the watershed. The PLS images were clipped to the watershed and wetland and stream features were mapped using ArcView. About 4,000 wetland acres were in the watershed were mapped using this historical information.

Since the survey only included the area immediately adjacent to the section transects, a more accurate reflection of past surface hydrology would be obtained through close exploration of elevation, geomorphology, soils and related data sets in addition to the 1854 Public Land Survey. The following data sets were used in combination with the Public Land Survey.
- MDNR Pre-Settlement Vegetation Map

Information contained in the surveyors' notebooks from the Public Land Survey are a valuable record of the composition and distribution of pre-European vegetation over much of the United States. This map was created to represent the presettlement vegetation of Minnesota based on Marschner's original analysis of Public Land Survey notes and landscape patterns. Marschner compiled his results in map format, which was subsequently captured in digital format by the MDNR.

- 1996 Digital Soil Survey

1996 digital soil survey and soil scientist interpretations were used to delineate additional wetland areas in the watershed. Land capability soils that were classified as hydric and IIIw were used.

Land classification shows, in a general way, the suitability of soils for most kinds of field crops. The soils are grouped according to their limitations for field crops. Capability classes range from 1-8 with 8 being the most severe limitation. A subclass of w, s, or c is also given to describe the limitation:

w= soil limitation because area is too wet.
s= soil limitation because it is shallow, droughty or stony.
c = limited soil potential for growing crops because of climate extremes- too cold or dry.

By selecting IIIw soils, areas with high water holding capacity and low permeability are selected. These areas would indicate a wetland land cover at one time. There are a total of 3,500 acres of IIIw soils in the watershed.

- 1990 National Wetland Inventory (NWI)

The National Wetlands Inventory is a national program sponsored by the US Fish and Wildlife Service (USFWS). The National Wetlands Inventory (NWI) of the U.S. Fish and Wildlife Service produces information on the characteristics, extent, and status of the Nation's wetlands and deepwater habitats. Linear wetland features (including selected streams, ditches, and narrow wetland bodies) are mapped as part of the National Wetlands Inventory. According to this information 1500 acres of wetlands are idendtified in the watershed.

- USGS 30 meter digital elevation model

Digital elevation models (DEM) are data files that contain the elevation of the terrain over a specified area, usually at a fixed grid interval over the surface of the earth. USGS 30 meter DEM, Spatial Analysis and MDNR Hydro Tools Extensions in ArcView were used to identify low elevation signatures. The wetness index calculation was performed on each elevation cell. Low cell values indicate lower surface elevations, and therefore may be indicative of historic wetland basins.

Wetness index $= \ln(A_s/\tan B)$

A_s==contributing catchment area in meters squared
B= slope of cell measured in degrees

- June 2002 Color DOQ

June 2002 color FSA photos were used to delineate wetland basins. In June of 2002 the watershed received over 9 inches of rain, nearly four to five inches over the normal rate. Cropped prairie pothole basins are very distinguishable using photos during this period. Cropped wet areas were then mapped. This represents about 1,015 acres.

NATURAL CHANNEL DESIGN FOR AGRICULTURAL DITCHES IN SW MINNESOTA

W. T. Christner Jr.[1]
J. Magner[2]
E. S. Verry, Ph.D.[3]
K.N. Brooks, Ph.D.[4]

ABSTRACT

Agricultural drainage ditches are a common occurrence throughout the Minnesota River Valley. Current ditch construction utilizes a trapezoidal form engineered to contain both small and large volume flows. However, following high flows, sediment accumulation in the channel bottom necessitates periodic channel cleaning. This (inefficient) design results in annual costs to local governments and private citizens. Joint research by the University of Minnesota (UMN) and Minnesota Pollution Control Agency (MPCA) investigated the use of a "compound" channel design to reduce and/or eliminate the need for periodic ditch maintenance. Compound channels incorporate smaller, self-maintained, "natural" channels within the larger flood channel geometry. An 800-m section of Judicial Ditch #8 (JD #8) in Swift County, MN was over-widened during routine cleanout maintenance. The over-widening allowed a smaller, low-flow channel with an active floodplain to establish within the larger flood channel. Measurements indicate a naturally stable B4 channel has developed within the larger flood channel. The smaller, stable channel allows for higher velocities during low flow conditions that efficiently move water and sediment. The ability of this channel design to transport sediment represents a potential savings in periodic clean-out maintenance. Additional benefits include enhanced fish and lowland bird habitat.

KEYWORDS. Drainage, drainage channels, natural channel design, sediment, aquatic habitat, compound channel, Minnesota.

INTRODUCTION

The use of natural channel design for the restoration of degraded stream channels is based on an understanding of the fluvial geomorphic processes operating within a watershed. While extensive work has been performed on natural channels, less thought has been extended to its application on agricultural ditches. Agricultural ditches are man-made structures designed to move water through a watershed where natural drainage features are lacking. Typical drainage ditch design incorporates a trapezoidal channel shape (Figures 1 and 2) devoid of any natural stream channel morphology. This criteria is contrary to those established for natural channel design and creates a dysfunctional water and sediment transport system.

Natural channel design is based on the fundamental concept that stable stream channel morphology is determined by the channel forming or bankfull discharge (Q_{bkf}). This discharge is associated with a more frequent return interval, generally between 1 to 2 years (Leopold et al., 1964). Agricultural ditch construction does not consider this design criterion. Agricultural drainage ditches are designed; "…to carry the maximum anticipated flow when filled to 80% of

---

[1] Ph.D. Candidate. Department of Water Resources Science, University of Minnesota.

[2] Hydrologist. Minnesota Pollution Control Agency, St. Paul, MN.

[3] Chief Forest Hydrologist (retired), North Central Research Station, Grand Rapids, MN.

[4] Professor. Department of Forest Resources, University of Minnesota.

their depth" (Troeh et al., 1991). Return intervals associated with these discharge values occur less frequently and are typically greater than 50 years. Consequently, agricultural ditches require periodic clean-out maintenance to remove sediment that accumulates due to improper channel design (Figure 3). Chippewa County, MN estimates average annual ditch maintenance costs between $65-$115 per linear foot (Magner, 2001). Ditch cleaning is a common land improvement practice performed in SW Minnesota (Figure 3).

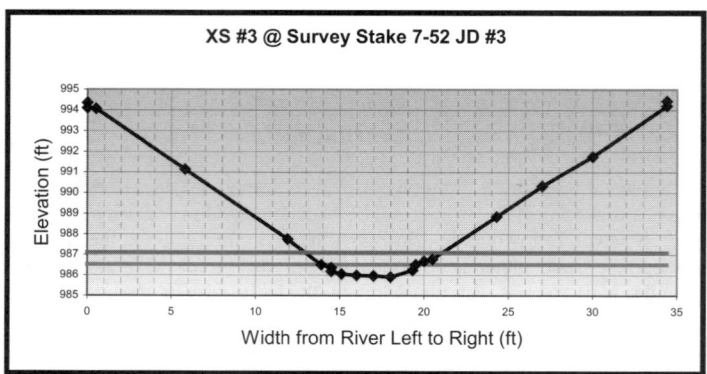

Figure 1. Cross-section of typical trapezoidal channel design. Judicial Ditch #3 (JD #3), Murray County, MN.

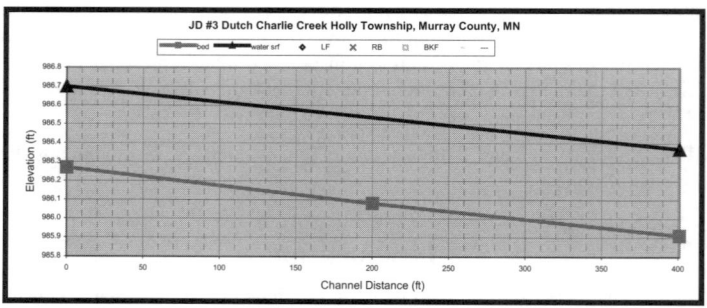

Figure 2. Longitudinal profile of typical drainage ditch design, JD #3 Murray County, MN.

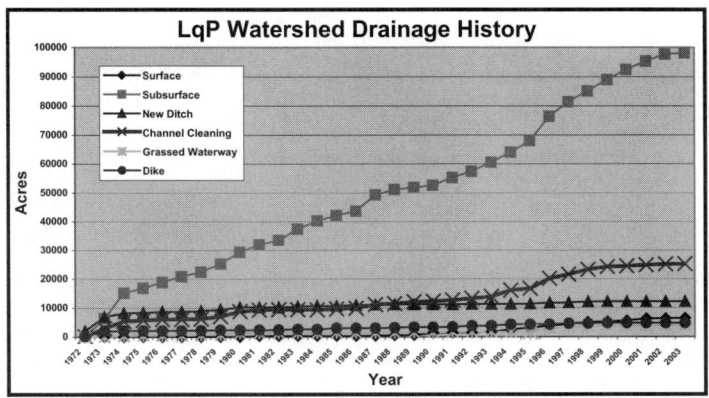

Figure 3. Six most common land improvement practices performed in the Lac qui Parle watershed between 1972-2003 and the number of affected acres.

METHODS

In order to apply natural channel design principles to agricultural ditches, a suitable reference reach must exist. This reference reach provides design data on channel dimensions such as: width, depth, slope, velocity, particle size distribution, sediment load, and discharge necessary to maintain a stable channel or, in this situation, ditch. Since agricultural ditches are man-made structures, no "natural" ditches exist. This makes collecting the required reference data problematic. A solution to this problem was inadvertently constructed in Swift County, MN. A 0.8 km section of Judicial Ditch #8 (JD #8), was over-widened during routine clean-out maintenance. The over-widening produced a wider, U-shaped channel. The new and wider channel allowed the low-flow channel in JD #8 room to meander and establish its own dimension, pattern, and profile along with an active floodplain.

A Topcon self-leveling laser was used to collect longitudinal and cross-sectional data during the summer of 2001. Channel metrics were collected for the natural section of JD #8 and two trapezoidal sections, one immediately upstream, and the other a tributary to JD #8 henceforth referred to as the Middle Fork Trapezoidal. A total of seven (7) cross-sections and three longitudinal profiles were surveyed after methods described by Harrelson and others (Harrelson et al., 1994). Each cross-section was classified to determine the presence of a natural channel design (Rosgen, 1996). Particle size distribution for each cross-section was determined after methods described by Wolman, (Wolman, 1954). Meander geometry data (Figure 4) were directly measured in the field. Radius of curvature (R_c) values were determined according to equation 1.0.

Equation 1.0: $R_c = \left(C^2/_{8M}\right) + \left(M/_2\right)$

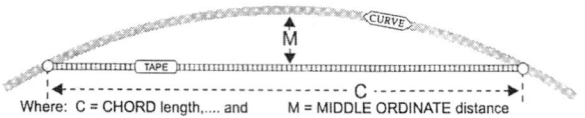

Where: C = CHORD length,.... and   M = MIDDLE ORDINATE distance

Figure 4. Radius of curvature measurements. (after Rosgen, 2001).

RESULTS AND DISCUSSION

Data collected from the cross-sectional and longitudinal surveys confirm the over-widened section of JD #8 has shifted from a trapezoidal channel design to a channel that exhibits natural facets. The channel morphology established in this 0.8 km reach is a direct result of the hydrologic conditions operating within the watershed. Data collected in the natural reach of JD #8 confirm the development of a moderately entrenched B4 channel type (Figure 5) with low sinuosity (Rosgen, 1994). This natural channel is contrasted by an entrenched trapezoidal channel type, with no sinuosity immediately upstream (Rosgen and Silvey, 1996). Four reaches were classified to determine channel morphology: cross-sections 2, 4, and 6 on the natural reach, and the trapezoidal cross-section (Table 1). Differences in channel morphology between the natural and trapezoidal reaches are related to differences in the width of the floodplain (W_{fpa}), channel depth (d_{max}), and sinuosity (Table 1). The natural and trapezoidal reaches are similar only in their bankfull cross-sectional areas (A_{bkf}). All other channel forming variables are distinctly different due to the over-widening of JD #8 that resulted in a larger floodplain. Consequently, the natural reach is less entrenched than the trapezoidal reach with an entrenchment ratio (ER) of 2.5, 2.1, and 2.2 compared to an ER of 1.2 for the trapezoidal reach. The entrenchment ratio is defined as the vertical containment of a stream (Kellerhals et al., 1972). The ER indicates the degree to which a stream is incised in the valley floor and its ability to access the floodplain. Entrenched streams have limited access to their floodplains. Because of its lack of entrenchment the natural channel on JD #8 has created a low-flow channel able to access its floodplain through a greater range of flows. Additionally, the low-flow channel produces higher velocities during base flow conditions than the wider trapezoidal channel. Higher base flow velocities result in better sediment transport. The wider trapezoidal channel design produces lower velocities at base flow, which reduces its ability to transport sediment. This results in sediment deposition and leads to periodic cleanout maintenance. Sediment deposition also has implications for the quality of aquatic habitat.

Table 1. Channel metrics for JD #8.

Location	A_{bkf} (m²)	W_{bkf} (m)	W_{fpa} (m)	d_{max} (m)	d_{g} (m)	Sinuosity	W/D	ER	Rosgen Stream Type
JD#8 Trapezoidal	3.8	8.4	10.1	0.67	0.52	1.0	17	1.2	F5
JD#8 New #1	3.9	7.5	18.6	1.01	0.52	1.2	15	2.5	C4
JD#8 New #2	4.1	9.0	16.2	1.04	0.46	1.2	18	1.8	B4
JD#8 New #3	4.2	7.7	16.3	1.01	0.55	1.2	13	2.1	B4
JD#8 New #4	4.1	9.6	13.7	0.73	0.43	1.2	24	1.4	B4
JD#8 New #5	4.1	7.9	17.1	1.01	0.52	1.2	16	2.2	B4
JD#8 New #6	5.0	10.0	14.5	0.88	0.49	1.2	20	1.5	B4

In addition to the differences in channel dimension and profile, the natural channel exhibits greater sinuosity than the trapezoidal channel with sinuosity values of 1.2 and 1.0 respectively. The meander geometry associated with this sinuosity supports the presence of a stable B4 channel on JD #8. Table 2 contains radius of curvature values collected at nine (9) meander bends on JD #8. Figure 6 illustrates the relationship between channel pattern (radius of curvature) and channel dimension (bankfull: depth, width, and cross-sectional area). The mean R_c value plotted in Figure 6 agrees with the established relationships, indicating a stable channel design for JD #8.   Mean bankfull depth for JD #8 equals 0.5m (1.6 ft), mean bankfull width equals 8.6m (28.2 ft), and mean bankfull cross-sectional area equals 4.4 m^2 (47.4 ft^2).

Table 2. Radius of curvature values for JD #8.

Site	R_c (m)
1	11.1
2	13.2
3	16.5
4	18.1
5	15.8
6	14.0
7	16.8
8	19.3
9	26.1
Average	16.8

All six cross-sections in the new reach of JD #8 have similar bankfull channel areas. However, while average depth of the trapezoidal and natural reaches are similar with values of 0.52 m and 0.46 m respectively, the maximum depth (d_{max}) of the trapezoidal channel is 0.67 m, an increase of 23 %. This is contrasted by a 56 % increase in the depth of the natural reach with a maximum depth of 1.04 m. The longitudinal profile in Figure 7 illustrates the difference in channel bottom variability and residual pool depth.

Residual pool depth is the difference in streambed elevation from the pool top to the pool bottom, and indicates the aquatic habitat available to fish. Streams with greater variability (i.e. deeper pools) provide better habitat and buffer in-stream temperature changes (Harper and Everard, 1998; Shields et al., 2000).

The natural reach on JD #8 offers the best overall fish habitat with the highest mean residual pool depth of 0.37 m, and greatest residual pool depth variability of 0.06 m (Table 3). Mean residual pool depth for the trapezoidal reach varies from 0.24 m in the Middle Fork, to 0.18 feet in the Main Fork. Low residual pool depth limits the ability of a trapezoidal channel to provide aquatic habitat and buffer changes in water temperature, which may result in fish mortality (Kohler and Hubert, 1999).

Table 3. Comparison of residual pool depth data for trapezoidal and natural sections of JD #8 (All measurements in meters).

Feature	Middle Fork Trapezoidal	Main Fork Natural	Main Fork Trapezoidal
Count	20	18	10
Minimum	0.09	0.15	0.06
Maximum	0.46	0.61	0.24
Mean	0.24	0.37	0.18
Variance	0.04	0.06	0.01

Another indicator of aquatic habitat is channel substrate size. Figure 8 illustrates the difference in channel substrate between the natural and trapezoidal reaches on JD #8. Seventy-two percent of the sediment in the trapezoidal reach is 1.0 mm in size or less. The same particle size accounts for only thirty-seven percent in the natural reach. Additionally, the D_{90} particle in the trapezoidal reach is 2 mm, compared to 32 mm in the natural reach. Greater channel substrate variability results in better aquatic habitat (i.e. better riffle quality where quality refers to well-aerated riffle w/o fines clogging interstitial spaces). It also indicates the natural channel is actively transporting finer sediment. This is important from a maintenance perspective. Self-maintaining channels do not require periodic cleanout maintenance. This can translate into substantial savings in county monies.

One of the main drawbacks of agricultural ditches, from a landowner's perspective, is the amount of land they remove from production (Troeh et al., 1991). Any proposed change in ditch design must address the amount of land required to maintain the design. Meander belt width

(W$_{belt}$) describes the lateral extent of river meander, or the amount of land required to maintain the channel design. The average meander belt width of the natural channel on JD #8 is 16.1 m with a maximum value of 18.6 m (Table 4). The meander width ratio (MW$_r$) is the ratio of the meander belt width to the bankfull channel width (W$_{bkf}$) and describes the degree of lateral channel containment. The average meander width ratio for JD #8 is 1.9 with a maximum value of 2.5. This indicates the natural channel on JD #8 requires 16.4–21.6 meters to maintain the channel belt width. Current meander belt width of the trapezoidal channel on JD #8 varies between 20.7-24.4 meters (68-80 ft). This is 2.8 m (9.2 ft) more then the required meander belt width of the natural channel design. This indicates that at this watershed scale, natural channel design can be accomplished without removing any more land from production.

Table 4. Meander width ratio data for JD #8 (All measurements in meters).

Site #	W$_{bkf}$	W$_{belt}$	MW$_r$
1	7.50	18.6	2.5
2	9.02	16.2	1.8
3	7.65	16.3	2.1
4	9.63	13.7	1.4
5	7.86	17.1	2.2
6	9.99	14.4	1.4
Average	8.63	16.1	1.9

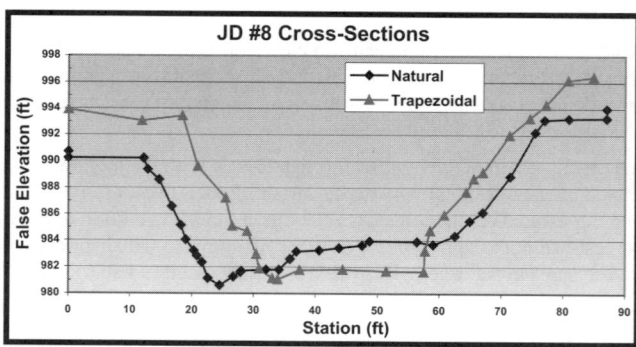

Figure 5. Cross-sections on JD #8 illustrating the differences in channel area of the natural and trapezoidal reaches.

240

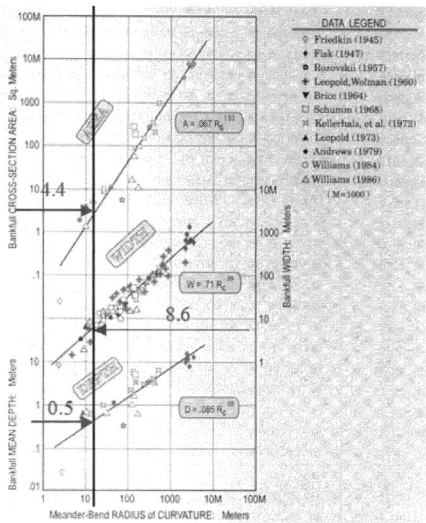

Figure 6. Relation of radius of curvature to channel dimensions for JD #8 (In Rosgen 1996, after Williams 1986).

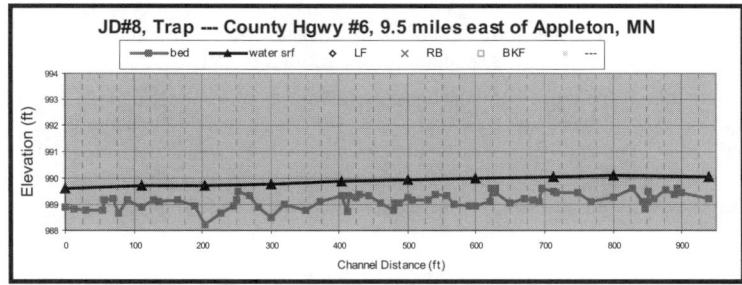

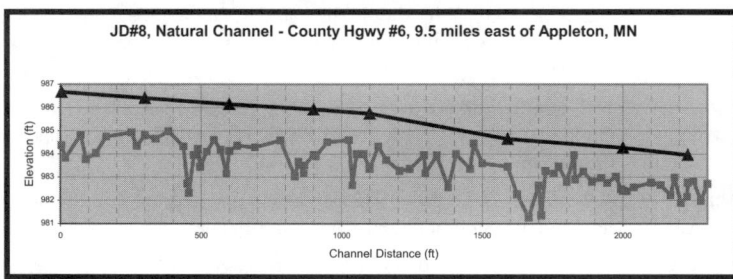

Figure 7. Longitudinal profile of the trapezoidal (upper) and natural (lower) reaches on JD #8. Note the greater channel variability in the natural reach.

241

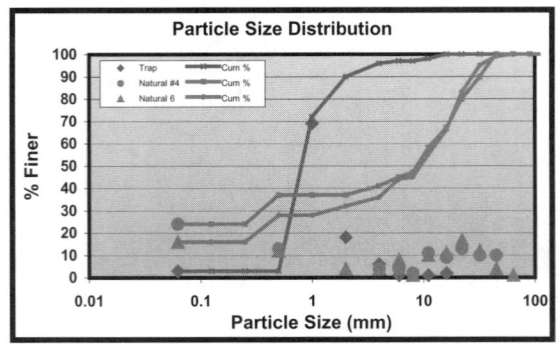

Figure 8. Particle size distribution for JD #8.

CONCLUSIONS

The natural section of JD #8 displays a large variance in channel width and depth throughout the reach indicating the existence of a variety of channel morphology and aquatic habitat. The meandering, natural channel represents a stable channel design or dynamic equilibrium in the morphology of JD #8. Since stable channels develop as a result of the current hydrologic conditions operating within a watershed they are able to efficiently transport the water and sediment delivered to the stream, through the watershed, without aggrading or degrading (Rosgen and Silvey, 1996). The ability to efficiently transport sediment represents a potential savings in public and private monies spent on ditch clean-out maintenance.

There are more than 27,000 miles of open agricultural ditches in Minnesota (Taff, 1998). Chippewa County, MN spends an average of $65-$115/ft on annual ditch maintenance (Magner, 2001). Application of natural channel design to agricultural ditches in Minnesota has the potential to offer substantial savings to public and private entities through reduced ditch maintenance costs without the need to remove any additional land from production.

REFERENCES

1. Harper, D. and Everard, M., 1998. Why should the habitat-level approach underpin holistic river survey and management? Journal of Aquatic Conservation: Marine and Freshwater Ecosystems, 8(4): 395-413.

2. Harrelson, C.C., Rawlins, C.L. and Potyondy, J.P., 1994. Stream channel reference sites: an illustrated guide to field techniques, USDA Forest Service General Technical report RM-245, Fort Collins, CO 80526.

3. Kellerhals, R., C. R. Neill and D. I. Bray, 1972. Hydraulic and geomorphic characteristics of rivers in Alberta., Research Council of Alberta, Alberta, Canada.

4. Kohler, C.C. and Hubert, W.A., 1999. Inland fisheries management in North America. American Fisheries Society, Bethesda, Maryland, 718 pp.

5. Leopold, L.B., Wolman, M.G. and Miller, J.P., 1964. Fluvial process in geomorphology. W. H. Freeman and Company, San Francisco, CA.

6. Magner, J.A., 2001. Hydrologist, Minnesota Pollution Control Agency. In: P. Communication (Editor), Saint Paul.

7. Rosgen, D.L., 1994. A classification of natural streams. Catena, 22(3): 169-199.

8. Rosgen, D.L., 1996. The reference reach, a blueprint for natural channel design. In: D.F. Hayes (Editor), Wetlands Engineering & River Restoration Proceedings, American Society of Civil Engineers, Reston, Virginia. American Society of Civil Engineers, Denver, CO.

9. Rosgen, D.L. and Silvey, H.L., 1996. Applied river morphology. Wildland Hydrology, Pagosa Springs, CO 81147.

10. Shields, F.D.J., Knight, S.S. and Cooper, C.M., 2000. Cyclic perturbation of lowland river channels and ecological response. Regulated Rivers: Research & Management, 16(4): 307-325.

11. Taff, S.J., 1998. Managing Minnesota's drainage system. Minnesota Agricultural Economist, 692(2).

12. Troeh, F.R., J. Arthur Hobbs and Donahue, R.L., 1991. Soil and Water Conservation. Prentice Hall, 530 pp.

13. Wolman, M.G., 1954. A method of sampling coarse river-bed material. Transactions of the American Geophysical Union (EOS), 35: 951-956.

CHARACTERIZATION OF SURFACE WATER POLLUTANTS THROUGH RIVERBED SEDIMENT ANALYSIS: ASSESSMENT OF LEBANESE COASTAL RIVERS

Nadim Farajalla

There have been few studies in Lebanon on the water quality of rivers; however most concentrated on water contaminants and did not address the issue of sediments and riverbed sediments. Further, very few studies relied on long-term water quality monitoring of rivers. Thus their conclusions were based on "instantaneous" assessments of water quality data only. The primary objective of the study is to assess the "health" of perennial rivers in Lebanon through the analysis of their riverbed sediments. This assessment will be used to aid in identifying the various sources of pollution in watersheds of sampled rivers. The main pollutants targeted for analysis are nutrients (nitrates and phosphates), heavy metals, and commonly used pesticides as these are typically correlated with effluent discharges from domestic sewage, industry, and agriculture.

Twelve of Lebanon's 15 perennial coastal rivers were sampled during a period stretching from October through November, 2003. During this period river flows are at their lowest and depth of flow is less than one meter in most of the sampled rivers thereby allowing easier access to riverbed sediment. In nearly all rivers, samples were collected at a distance of 300m to 500m from the sea to avoid tidal inflows and from behind weirs and barrages whenever these were encountered. Sediment was collected using 12cm hand-held bucket augers at depths ranging from 10 – 30cm. Samples were air dried, sieved and stored.

Extraction and analysis are currently being carried out with results expected within the next four to six weeks. As such no results can be reported at this time but will be available by the deadline for manuscript submittal.

A Procedure for Assigning Management Categories to Urbanizing Watersheds

Dipmani Kumar[1]

In the development of a watershed plan, it is useful to classify subwatersheds into a number of management categories based on the potential quality of streams in the subwatersheds. In this paper, a procedure for classifying subwatersheds into different management categories is suggested. The procedure is based on estimating the ultimate imperviousness of subwatersheds and establishing a statistical relationship between imperviousness and the quality of streams. The procedure is illustrated using biological stream quality data from Fairfax County, Virginia. The results of applying the procedure to classify subwatersheds in Fairfax County as *protection*, *critical*, or *restoration*, based on the predicted ultimate imperviousness of the subwatersheds, are also presented.

KEYWORDS: Urban watersheds, management categories, stream biological quality, imperviousness, regression

INTRODUCTION

Fairfax County is located in the northeastern part of the state of Virginia, bordering the Potomac River. The County was almost completely rural until World War II, with dairy farming being the most important single industry. Today, it is highly urbanized, and approaching ultimate build-out conditions, as envisioned in the County's Comprehensive Plan. The total land area of Fairfax County, including incorporated towns is 395 mi^2. It is the most populous jurisdiction in Virginia as well as the Washington D.C. metropolitan area. The 2003 population was estimated to be 1,015,600 with 369,900 households. Most land in the county is devoted to residential, commercial, recreational, and open-land uses, with heavy industry essentially non-existent.

There are two major physiographic provinces in the County. The Coastal Plain province lies to the east, characterized by low gradient topography and sandy soils. The Pieldmont Upland province lies to the west, and is characterized by rolling hills and a more rocky substrate. A sub-region of the Piedmont Upland province, the Triassic Basin, located in the southwest corner of the county consists of areas of somewhat lower relief and large expanses of shale and red sedimentary sandstone. The southwestern part of the County drains into the Occoquan River. Except for two watersheds (Sugarland Run and Horsepen Creek), which drain into adjoining Loudoun County, the northern and southeastern parts of the county drain to the Potomac River.

Fairfax County's Stream Protection Strategy (SPS) program is an ongoing biological monitoring effort with the overall goal of identifying and assessing trends in stream conditions countywide. A baseline study (Fairfax County, 2001), documented current conditions throughout the county based on watershed and stream conditions, and provided a foundation for prioritizing and implementing watershed management strategies.

---

[1] Stormwater Planning Division, Suite 449, 12000 Government Center Parkway, Fairfax, Virginia 22035; dkumar@fairfaxcounty.gov

The SPS baseline study established broad management categories for future watershed protection and restoration efforts, based primarily on overall stream rankings of biological quality and projected development (inferred from predicted ultimate imperviousness). Management areas identified in the SPS baseline study covered relatively large areas, and projected development was determined from zoned land-use, which generally resulted in overestimates of ultimate imperviousness.

Objectives

This project was conceived as a follow-up to the SPS baseline study, with the overall goal of developing criteria for assigning management categories at the subwatershed scale, and applying the criteria to the county's subwatersheds based on predicted ultimate imperviousness. The specific objectives were:

- Delineate county subwatersheds with drainage areas of 1 mi^2 or less.
- Estimate subwatershed ultimate imperviousness based on vacant and underutilized land parcels using both planned and zoned land-use.
- Classify county subwatersheds into different management categories based on the potential quality of streams within the subwatersheds.

METHODS

Subwatershed delineations

To support automated subwatershed delineations, 1:24,000 scale USGS Digital Elevation Models (DEMs) with one-arc-second cell sizes (98.411 feet or approximately 30 meters) were obtained for all 7.5 minute quadrangles within the County. In addition, 1:100,000 scale National Hydrography Data (NHD) were obtained for the two USGS hydrologic cataloging units bounded by the County.

Initial testing with the USGS DEMs revealed acceptable accuracy in subwatershed delineations could be achieved for the planning-level scope of this project for all areas of the County, other than two watersheds in the southeastern part of the county, which are relatively flat. The USGS DEMs for these areas were supplemented by a DEM generated from 120,533 spot elevations available for these watersheds.

A synthetic stream grid, was created at a 0.5 mi^2 threshold and found to closely follow the NHD streams. Subwatershed delineations were completed for the entire County employing automated identification of links (confluences) on the synthetic stream grid.

Subwatershed ultimate imperviousness determination

The following countywide GIS layers, shown for an example subwatershed in Figure 1 were utilized in determining future (ultimate) subwatershed imperviousness:

- Zoning districts
- Comprehensive plan land use designations at the parcel level
- Vacant parcels (1999)
- Underutilized parcels (2000)

Future subwatershed imperviousness estimates were obtained by projecting the imperviousness of vacant and underutilized parcels based on the county's comprehensive plan and zoning district designations. Imperviousness values for the various comprehensive plan land uses and zoning districts were taken from a analysis performed by the County to support its Chesapeake Bay ordinance (Fairfax County, 1992). Other land parcels were assumed to be stable and future imperviousness set equal to current imperviousness, which was determined from a number of countywide built environment planimetric GIS layers, including: building footprints, roads, parking lots, and sidewalks. The higher of the planned land-use and zoned land-use value was then taken to be the best estimator of future imperviousness.

Criteria for assigning subwatershed management categories

The biological quality of streams is generally regarded as a strong indicator of their overall health. Based on observations that macroinvertebrate diversity in urban streams had an inverse relationship with drainage area imperviousness (Jones and Clark, 1985), it is now accepted practice to infer the biological quality of urban streams from indices that measure the diversity, richness, and composition of benthic macroinvertebrates.

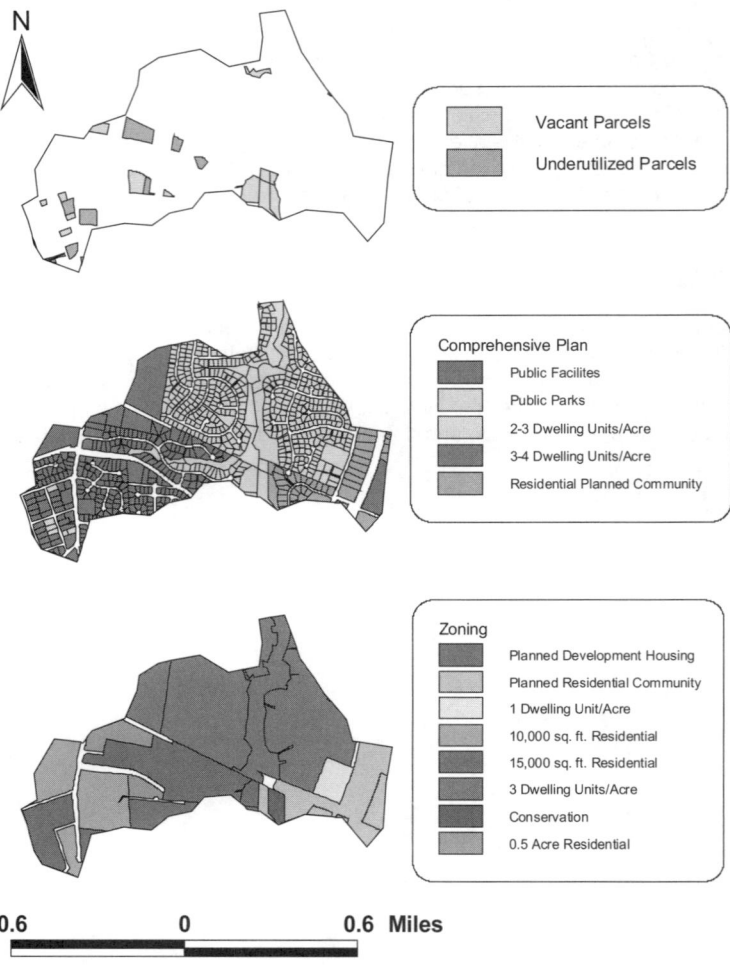

1. GIS layers utilized in estimating future imperviousness (shown for an example subwatershed).

To establish criteria for classifying subwatersheds into management categories, data from the SPS baseline study were employed in a regression analysis to evaluate the relationship between stream biological quality and drainage area imperviousness. In the SPS study, stream biological quality at was inferred at 123 targeted sites throughout Fairfax County from an Index of Macrobenthic Integrity (IMBI) composed of a number of individual metrics for macroinvertebrate diversity, richness, and composition.

A number of linear regression models were evaluated for their utility in predicting IMBI from the <u>current</u> drainage area imperviousness (IMP). A cubic polynomial regression model was found to provide the best fit to the available data. Ordinary least squares (OLS) regression was employed and the major assumptions associated with OLS regression – the response variable follows an approximately normal distribution with constant variance were visually assessed.

As part of the regression analysis, a confidence interval (CI) for the mean regression line was also determined. A CI represents the range of values within which a parameter being estimated is expected to fall with a given confidence. Thus a *p% CI* IMBI values for a given IMP value represent the range within which the <u>mean</u> of a large number of IMBI values obtained for that IMP is expected to fall *p%* of the time.

To determine the range of drainage area imperviousness values representing 'critical' conditions, a desired stream quality level and confidence level have to be defined. The desired stream quality was chosen to be 50% of the reference condition streams (which have IMBI values of about 100) and a 99% confidence level selected. *Critical* subwatersheds were taken to be those with ultimate imperviousness values that contained a predicted IMBI value of 50 within the 99% confidence interval. Subwatersheds with ultimate imperviousness less than the critical range were regarded as *protected*, while those with ultimate imperviousness greater than the critical range were deemed to be *restoration* subwatersheds.

RESULTS

The results of the regression analysis and determination of critical imperviousness range are shown in Figure 2. The range of critical imperviousness values, based on a desired stream quality of 50% of the reference condition and a 99% confidence level was found to be 10.7 to 17%. As would be expected, a higher desired stream quality would result in lower values for the range, while a higher desired degree of confidence would have result in a wider range.

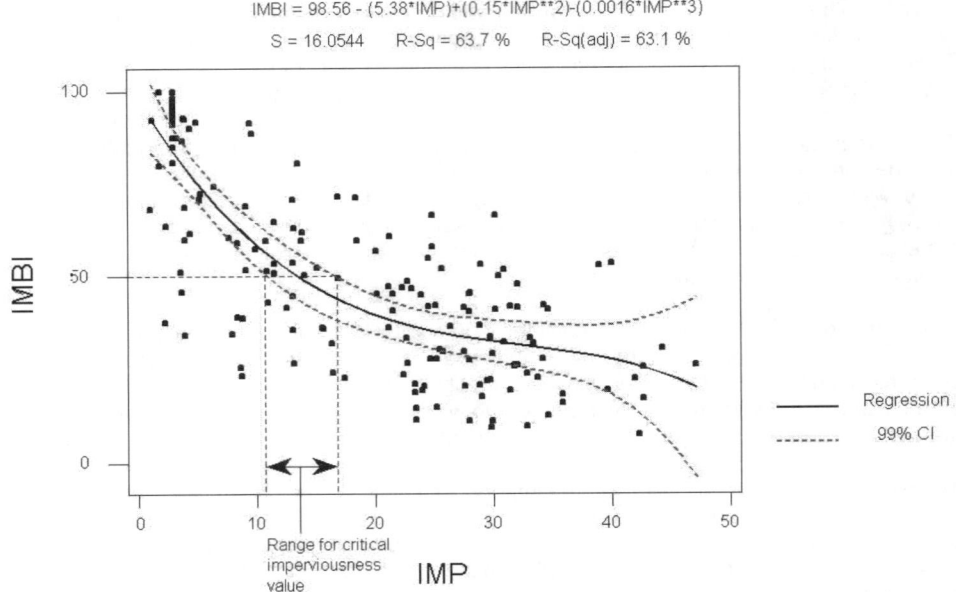

Figure 2. Regression of Index of Macrobenthic Integrity (IMBI) versus % Imperviousness (IMP) and determination of critical imperviousness.

The results of applying the critical imperviousness range to assign management categories to subwatersheds in Fairfax County based on their predicted ultimate imperviousness are shown in Figure 3. Of the 473 subwatersheds, 182 (representing 29.6% of the total county area) were classified as *protected*, 57 (13.4% of the county area) were classified as *critical*, and the remainder (56.7% of the county area) were classified as *restoration*.

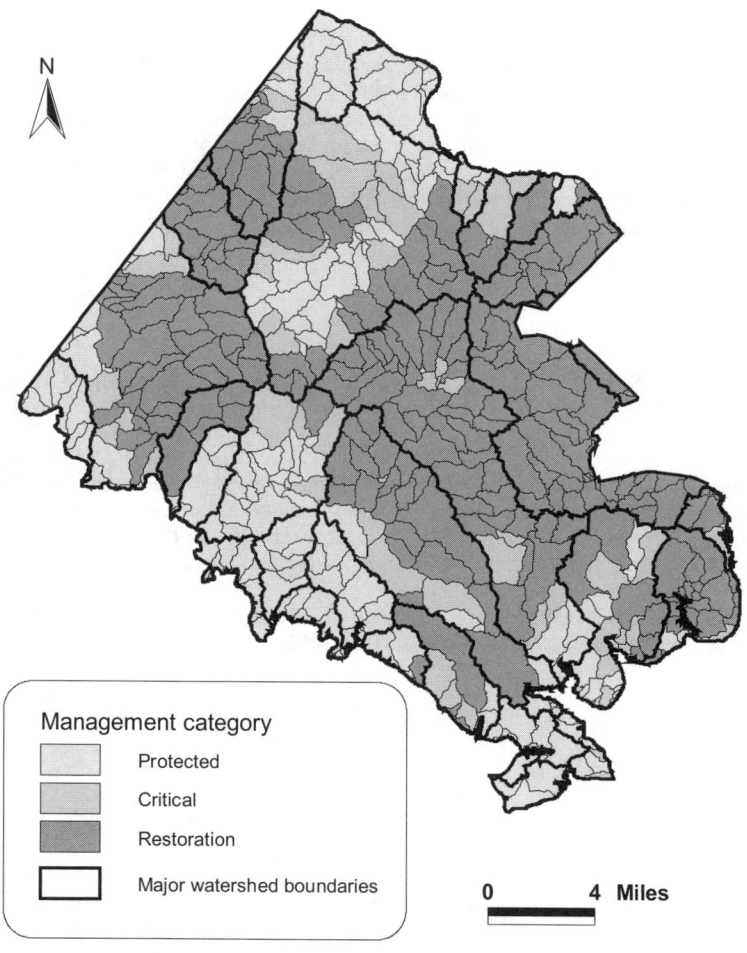

Figure 3. Management categories of Fairfax County subwatersheds based on predicted ultimate imperviousness.

SUMMARY

Fairfax County, located in the northeastern part of the state of Virginia, is highly urbanized, and approaching ultimate build-out conditions, as envisioned in the County's Comprehensive Plan. A procedure for classifying Fairfax County's subwatersheds into different management categories was suggested. The procedure was based on estimating the ultimate imperviousness of subwatersheds and establishing a statistical relationship between imperviousness and the quality of streams. The procedure was illustrated using biological stream quality data obtained from a baseline study completed in 2001.

The range of critical drainage area imperviousness values, based on a desired stream quality of 50% of the reference condition and a 99% confidence level was found to be 10.7 to 17%. This range was used to assign management categories to subwatersheds in Fairfax County based on their predicted ultimate imperviousness. This resulted in 29.6% of the total county area being classified as *protected* subwatersheds, 13.4% of the county area classified as *critical* subwatersheds, while the remainder (56.7% of the county area) were classified as *restoration* subwatersheds.

REFERENCES

Fairfax County, 2001. Stream Protection Strategy Baseline Study. *Stormwater Planning Division, Department of Public Works and Environmental Services.* (http://www.fairfaxcounty.gov/dpwes/environmental/sps_main.htm)

Fairfax County, 1992. Justification of 40% Phosphorus Removal Requirement for New Development in Resource Protection Areas. Paper submitted to the Chesapeake Bay Local Assistance Department to support Fairfax County's proposed designation of Chesapeake Bay Preservaton Areas. *Department of Planning and Zoning.*

Jones, R. C. and C. C. Clark. 1987. Impact of watershed urbanization on stream insect communities. *Water Resources Bulletin* 23(6): 1047-1055.

RESTORATION OF THE UPPER 40 TRIBUTARY, AN URBAN STREAM RESTORTION CASE STUDY

T. M. Evans, ASLA

ABSTRACT

Restoration of the entire two mile Upper 40 tributary of the Chagrin River is currently underway in the eastern suburbs of Cleveland, Ohio. The project represents a unique collaborative partnership between the Villages of Mayfield, Gates Mills, and Cleveland Metroparks and illustrates all of the funding, design, and construction issues common to urban and suburban stream restoration.

Increased runoff from suburban development over the last 50 years had created severe flooding and erosion problems, causing a source of conflict between watershed partners for 20 years. Once the partners reached a consensus on cooperation, they commissioned a restoration masterplan for the tributary. An interdisciplinary consultant team of landscape architects, engineers, ecologists, and permitting specialists diagnosed watershed problems, identified restoration locations, and developed feasible watershed solutions. More critical however, was developing an implementation plan consisting of estimated project costs, funding sources, and a funding plan. The stream restoration masterplan featured natural channel design techniques and restoration of floodplain functions. The project will dramatically increase stormwater storage capacity, restored floodplain functions to 2000 lineal feet of stream, daylight over 2000 lineal feet of formerly culverted stream, construct 2-3 acres of wetlands, and enhance a 40 acre wetland complex.

Phase One of a four phase restoration master plan was constructed in 2004, phases two and three are currently in design. Phase two construction is anticipated in the fall of 2004. Total project costs will exceed over $2 million; the project was awarded over $1.2 million in two Clean Ohio grants.

KEYWORDS. Urban stream restoration, urban watershed restoration, grant funding, natural channel design, landscape architect, multidiscipline consultant team.

INTRODUCTION

A unique collaborative partnership between the Villages of Mayfield, Gates Mills, and Cleveland Metroparks Restoration is currently undertaking the restoration of the entire two mile Upper 40 tributary of the Chagrin River. The project provides a good illustration of all of the funding, design, and construction issues common to urban and suburban stream restoration and represents one prototype for solving urban watershed problems due to urbanization which have occurred over the span of several decades. The project provides a story of conflict resolution, consensus building, innovative restoration techniques, permitting issues, a multidiscipline team approach, as well as creative funding strategies.

A watershed restoration master plan was completed 2001. Initial funding was obtained in 2002 ? Phase One of a four phase restoration master plan was constructed in 2004, phases two and three are in design, with construction anticipated in fall of 2004. Total project costs will exceed $2 million; the project was awarded over $1.2 million in two Clean Ohio grants to fund riparian restoration.

PROBLEM DESCRIPTION

The Upper 40 project is typical of urbanization impacts in the upper reaches of the watershed in one jurisdiction producing impacts downstream in other jurisdictions. Much of the 2000 acre upper watershed was developed in the 1960s and 1970s prior to enactment of stormwater management ordinances. Increased runoff from build out of the suburban watershed of Mayfield Village over the last several decades created severe flooding and erosion problems downstream in the middle reaches of the watershed in Cleveland Metroparks North Chagrin Reservation. The result was severe and chronic flooding and erosion which have eroded away a park road, which has been closed to the public for over 20 years. Erosion produced tons of sedimentation which degraded wetlands, and private property, caused flooding on Chagrin River Road, state route 147 in the lower reaches in Gates Mills. The closing of the Park Road severed access to a 200 acre zone of the 2000 acre North Chagrin Reservation. This created a source of conflict between watershed partners for over 20 years.

The flooding and erosion problems were the subject of 3 previous studies. Litigation had been discussed; earlier rehabilitation strategies using traditional engineering approaches such as constructing an earthen dam across a ravine to form a stormwater storage reservoir were stopped due to safety and permit objections.

A new city administration in Mayfield Village made the commitment to cooperatively pursue watershed restoration with Cleveland Metroparks and Gates Mills. The Upper 40 Tributary is a tributary of the Chagrin River, designated a State Scenic River, which added to the desire for cooperation. Once the partners reached a consensus on cooperation, they commissioned a restoration master plan for the watershed.

An interdisciplinary consultant team of landscape architects, engineers, ecologists, and permitting specialists developed a comprehensive tributary restoration strategy. The master plan diagnosed watershed problems, identified restoration solutions, identified permitting requirements, and identified restoration locations. More importantly the master plan provided a roadmap for implementation including cost estimates, funding sources, and a funding plan.

The stream restoration master plan employed natural channel design techniques, and restoration of floodplain functions. The project illustrates prototype approaches to addressing watershed problems in upper, middle, and lower watershed segments. The proposed solution utilized both storage and conveyance solutions to address erosion and flooding problems common to most urbanized watersheds. Overall project goals are to dramatically increase stormwater storage capacity, as well as to daylight and restore a natural channel to over 2000 lineal feet of formerly culverted stream, restore 3-4 acres of degraded wetlands, and enhance a 40 acre wetland complex.

Upper segments of the watershed include an interstate interchange, large office parks and a cemetery. Such high value land uses currently appraise at $ 200,000 to 300,000 per acre; making any large scale land acquisition cost prohibitive. The master plan therefore primarily looked at available public land as well as smaller parcels of potentially acquirable private land. Opportunity locations for restoration projects were identified in both city and metropolitan park locations. One acquisition property of 1.5 acres was required, in order to enlarge a city park.

FUNDING PLAN

A critical and fundamental aspect of watershed restoration is obtaining funding. Most communities do not regularly budget funds for stormwater projects. Project timing was quite fortuitous; in 2000 Ohio voters passed Issue 1, the Clean Ohio Fund. Eligible purposes include acquisition of open space land, restoration of functioning floodplains, ecologically informed design, and reforestation. Eligible costs are land acquisition, acquisition costs, engineering, permitting, as well as construction.

The consultant team prepared a grant application in the first round of the Clean Ohio program. Grant text was carefully crafted to address specific funding criteria. Cost estimating for stream restoration projects is challenging due to the specialized nature of work, difficult access conditions, and seasonal timing constraints. An added challenge with grant funding is realistically and accurately estimating design and construction costs during the conceptual phase of the project. Phase one total project costs, including land acquisition, engineering services and construction were estimated at $860,000. The project was awarded 75% or $ 570,000; Mayfield Village contributed the remaining 25% or $290,000.

<h1 style="text-align:center">PHASE ONE</h1>

Urbanization of the upper watershed during the 1960s and 1970s resulted in three classic impairments in urban watersheds, much more runoff, loss of depressional storage capacity, and stream channelization. The predictable result was dramatic increases in the volume, rate and velocity of runoff which overwhelmed and destroyed the downstream drainage system.

The primary goal of phase one is to increase stormwater storage capacity, thereby replacing lost headwater storage capacity in the upper watershed, in order to reduce peak flow rates downstream. Floodplain functions are to be restored to the formerly channelized stream through excavation of about six feet of earth to construct a floodplain along the stream in two locations thereby creating 8-10 acre feet of stormwater storage. A grading plan was designed to avoid straight lines, to mimic nature, maximize storage volume. The existing straight stream channel was relocated to form meanders in two locations. The floodplain was graded to trap water and create 1-2 acres of floodplain wetlands.

Geotechnical investigations were conducted to assess materials to be excavated during excavation; The borings verified that soft shale would be encountered in the lower reaches of excavation. Fortunately this did not cause cost escalation.

The location for the phase one storage project was planned for an upper watershed area in an existing city park, requiring displacement of one of two Village ballfields. A second nearby storage location required acquisition of a 1.5 acre parcel, and working around an existing home structure.

Hydraulic modeling was performed to assess the effects of stormwater storage and various control structures in achieving the goal of reducing peak flow rates. A variety of control structures were evaluated in regards to reduction of peak flow rates, cost effectiveness, visual appearance, and to minimize maintenance. Ultimately the simplest, least obtrusive, and most cost effective control structures was found to be steel plates, designed to mounted in front of existing 5 x 9 elliptical culverts. Computer modeling of watershed hydraulics indicated that an 18" opening in the plate would achieve a 25% reduction in peak flow rates during critical design storms.

A landscape plan for the floodplain was devised to include a wide variety of native shrubs and trees. Native rock boulders were carefully placed to form step pools and to naturalistically armor stream bends. City crews removed fencing along stream banks to restore riparian buffers strips

The consultant team obtained a nationwide permit from the Army Corps of Engineers for stream restoration. Nationwide # 27 for stream and wetland restoration is intended to permit stream and wetland restoration activities and pose no limit to lineal footage of stream restoration.

Plans and specifications were prepared for public bidding. Plans were developed to solicit unit prices for all construction items, common in municipal engineering. Although this requires some additional effort during plan development, it provides much greater flexibility to add or subtract quantities in order to make field adjustments during construction. Unit prices also provide a cost database for accurately estimating the next project.

The phase one project was on Mayfield Village property and thus was administered by Mayfield Village. The bidding process resulted in receiving 3 bids. A variety of bid alternates allowed the village the flexibility to tailor an authorized bid amount to the available funding. Due to the receipt of bids lower than estimated, the Owner was able to have additional restoration work performed with available funds. The low bid Contractor had never done a stream restoration project before. Effective project specifications and a generous amount of field supervision were required to direct construction crews to achieve the desired naturalistic effect.

An ancillary benefit was the positive utilization of excavated soil. Excavation of 12,000 cubic yards of earth for the floodplain required the displacement of one of the village's two ballfields. The village badly needed to not only replace a ballfield but to construct additional ballfields. Therefore a soil disposal site was designed to be the site of the future ballfields; only a half mile haul away versus more than 5 miles a Contractor may have had to haul. The soil disposal area was graded to prepare the site for construction of 3 ballfields. Separate grant funds were obtained, and a separate bid package prepared for ballfield construction. Coordination of the construction timing of the two projects was quite challenging but it resulted in considerable savings in grant funds, and provided additional benefits to Mayfield Village.

PHASE TWO

The consultant team prepared a second Clean Ohio grant application for Phase two total project costs, estimated at $1,200,000. Clean Ohio funded 63% or $ 760,000, Mayfield Village and Cleveland Metroparks contributed the remainder.

The middle segment of the watershed, in Cleveland Metroparks North Chagrin Reservation, was the site of extremely severe flooding and erosion which resulted in the destruction of a park road and culvert infrastructure. In the 1940s, a culvert was sized for then current flow rates, placed down the center of the ravine, backfilled, and a park road constructed. The intervening 50 years saw substantial suburban development which intensified with the construction of an interstate interchange in the 1960s. The culvert size, although adequate upon installation, was woefully inadequate for the developed watershed, was overtopped chronically. Former roadside ditches were eroded away to a depth of 6' to native bedrock. Tons of sediment were washed downstream and deposited in delta fans, degrading floodplain wetlands.

The primary goal of phase two is to restore 2000 lineal feet of natural channel in the middle segment of the watershed and restore degraded wetlands in the lower segment of the watershed. Project tasks include removal of the existing road and culvert, and regrading a natural multi stage channel to provide both a base channel and a flood channel. Design of the naturalized channel section was very challenging due to high velocities resulting from the 10% average gradient. The typical cross section will restore a multi stage channel, and an all purposes trail in a constricted and meandering ravine. To accommodate the erosive velocities, plans are developing to construct the channel using native shale as the stream bed and armor the stream banks with native rock and bioengineered plantings. The stream channel will incorporate step pool and rock cascade structures to reduce velocity, dissipate energy, and add habitat value. Existing culverts will be rehabilitated to provide stream crossings for the trail. The grading plan was developed to carefully balance cut/fill, minimize haul, and control earthwork costs. Phase Two is in design now.

Runoff rates were modeled for storm events of 1 year through 100 years and a variety of channel sizes carefully analyzed for capacity and velocities. Modeling carefully assessed stream velocities and identified appropriate armoring techniques.

The consultant team applied for a nationwide permit from the Army Corps of Engineers for stream restoration. Nationwide # 27 for stream and wetland restoration is intended to permit stream and wetland restoration activities with no limit to lineal footage of stream restoration. In the lower segment of the watershed, in the floodplain of the Chagrin River, the project will restore 2-3 acres of floodplain wetlands, which will provide 5-6 acre feet of badly needed stormwater storage capacity to relieve flooding. Enlarged wetlands will enhance amphibian habitat as well. A 40 acres oxbow wetland degraded by sedimentation will be enhanced through removal of accumulated sediment, and installation of a water level control structure.

Stream restoration construction has been timed to start in the fall of 2004. Channel grading is planned to be completed in the fall during periods of lower precipitation. Landscape work is to be completed in the spring months of May and June of 2005. The consultant team anticipates considerable construction direction to direct grading, rock placement, and landscape installation. The multidiscipline consultant team provided a complete suite of services including conceptual watershed restoration master planning, cost estimating, grant assistance, topographic survey, geotechnical investigations, permitting, hydraulic engineering, landscape architecture, bidding and construction administration. Team members included a landscape architect as project manager and included surveyors, geotechnical engineers, ecologists, hydraulic engineers, and construction administration personnel.

CONCLUSION

A number of lessons have been learned from the Upper 40 restoration project.

Historical context. The impacts of urbanization on urban streams are typically quite massive, having been ongoing since European settlement of North America over 150 years ago. Impacts include runoff rates and volumes which are probably 5-10 times the original rates, replacement of headwaters streams with a storm collection system, loss of depressional storage, as well as floodplain filling and stream channelization on a massive scale. Reversing 150 years of watershed impacts is not quick, nor easy, nor inexpensive. Reconstructing thousands of feet of stream where only a channelized ditch or a culvert exists may require extensive construction but which represents a fraction of the cost invested in watershed development.

Real estate. Many watershed segments in urban watersheds are fully built out and restoration is not even physically possible. In communities with high real estate values, restoration is not financially possible. Therefore identifying opportunity locations for restoration along stream corridors is typically the single largest challenge to stream restoration.

Grant funding. The only reason most stream projects are undertaken is to protect or repair infrastructure due to erosion or flooding. The majority of communities typically do not have a dedicated funding for watershed projects; therefore grant funding sources are critical to undertaking many projects. Grant funding, many times requires development of engineering and construction cost estimates at the conceptual design stage, requiring considerable skill and experience.

Storage versus conveyance. Nearly all urban watersheds face the double trouble of increased runoff and reduced storage capacity. Restoration of storage capacity in urban watersheds, through floodplain restoration or constructed wetlands is a cost effective strategy for reducing peak flows, and is overlooked in many restoration plans. Stream restoration is not just about the stream channel, but the more about the riparian corridor.

Naturalistic design. Designing stream geometry, rock structures and landscape plantings which avoid rigid geometric patterns and which appears both naturalistic, attractive, or even random is far more difficult than it seems. Too many restoration projects rely on engineered geometries, look artificial, and lose sight of the larger goal of stream restoration.

Unit Price Contracts. We have found administering construction contracts with unit prices provides significantly more flexibility to address unforeseen field conditions, add or subtract items of work to adapt to bid pricing.

Construction supervision. Stream restoration typically requires a generous amount of field direction during construction in order to properly adapt the project to field conditions and make minor design adaptations.

The Upper 40 Tributary Restoration is significant in that it represents the first attempt to restore an entire urbanized tributary in Northeast Ohio and perhaps in Ohio. Lessons learned from this project are currently being applied to more cost effectively restore other watersheds in Northeast Ohio.

Demonstration of Greenway Development to Protect Ecological Services in Small Urban Streams

ROBERT A. MORGAN[1], MARTY MATLOCK[2], LUANNE DIFFIN[3], MARC NELSON[4]

ABSTRACT

The city of Rogers, Arkansas is part of the rapidly growing Fayetteville-Springdale-Rogers metropolitan area of Northwest Arkansas. The natural resources of the area, particularly water resouces have been stressed as the cities struggle to develop infrastructure needed for the rapid growth. Increased wastewater effluent, nonpoint source pollution and loss of riparian vegetation have degraded the water resources. In 2002, the US EPA and the Arkansas Soil and Water Conservation Commission awarded a grant to the city to demonstrate protection of ecological services through the use of greenways in their stormwater system. The demonstration stream is a second order urban stream draining roughly 16 km². The watershed is rapidly developing as commercial, industrial, multi and single-family residential property. In the initial phase of the demonstration, an evaluation of the stream was made including evaluation of; habitat, biodiversity, trees, land use and land use change, and hydrologic/geomorphologic conditions in the stream to analyze the ecological services being provided by the stream. The project is unique in its combination of environmental sciences and engineering in the initial planning stages. A local stakeholders committee consisting of the landowners, local educators, and interested citizens has been created to direct the project. As a result of this project, a local Greenways and Trails Committee has been established to extend the greenway concept to the rest of the city. The city has changed its development approval process to require dedication of riparian areas for greenway trails development.

KEYWORDS. Stream restoration, ecological services, riparian vegetation, stormwater management.

INTRODUCTION

The City of Rogers, Arkansas is a rapidly growing community in Northwest Arkansas. The rapid growth is fueled by the growth of Wal-Mart, the nation's largest retailer, and Tyson Foods, the nation's largest supplier of meat products both of which are located in the Metropolitan Statistical Area (MSA) consisting of Fayetteville, Springdale, Rogers and Bentonville, and the surrounding non-incorporated areas. During 2002, the population of the MSA grew by an annual rate of 9.5 percent. The population of the metropolitan area in 1999 was 285,017 (US Bureau of Census, 1999). This figure is expected to double or triple by 2020.

The rapid growth of the area has put the areas infrastructure and natural resources under stress. Particularly stressed are the water resources of the area because of increased imperviousness of the watersheds, physical changes to the streams and increases in discharge of wastewater from municipal treatment plants. Out of necessity, standard stormwater drainage practice in the area

---

[1] Room 203 Engineering Hall, University of Arkansas, Fayetteville AR 72701

[2] Room 203 Engineering Hall, University of Arkansas, Fayetteville AR 72701

[3] Rogers Water Pollution Control Facility, 4300 S. Rainbow Rd, Rogers AR 72756

[4] Chem 101, University of Arkansas, Fayetteville AR 72701

has been to channelize streams to increase the hydraulic capacity without regard for the ecological services provided by those streams. In many areas, streams have been straightened to conform to property lines. Earthen trapezoidal channels are often built to replace the natural stream, and then as peak flows continue to increase the lining is replaced with concrete. Detention storage is usually installed in new developments to attenuate peak discharges. Major concerns in the watersheds draining this area are elevated nutrients, chronic turbidity and siltation (ADEQ 1997). The identified sources of the pollutants were point and nonpoint sources of nutrients, construction, and stream bank erosion.

The Environmental Protection Agency (EPA 2002) proposed reestablishment of ground water recharge, natural stream restoration, and preservation, enhancement or establishment of riparian buffers as management measures for the control of urban nonpoint source pollution. The National Park Service (NPS 1995) identified greenway parks and trails as a measure to protect the intrinsic values of streams while at the same time increasing real property values, expenditures by residents and tourism. A project team consisting of the University of Arkansas Biological Engineering Department, the Arkansas Water Resource Center, and the Rogers Water Utilities applied for and received a section 319(h) grant from the Arkansas Soil and Water Conservation Commission and the US EPA to demonstrate these values of greenways and the ability of greenways to protect or enhance ecological services provided by the stream corridor. The project proposal was submitted to the Arkansas Soil and Water Conservation Commission (ASWCC) and subsequently selected for implementation starting in early 2003. Funding for the project came through Region VI of the Environmental Protection Agency (EPA). $490,000 federal dollars are being provided which are being matched by $378,000 from the City and the U of A. The specific objective of this project was to demonstrate methods and technologies for protecting critical ecological services in urban streams. Tasks to be performed were:

- Monitor water quality and quantity through the installation of an automatic monitoring station on the demonstration reach and collection of rainfall data from a gage in the watershed.
- Develop and implement educational curricula for elementary, secondary, and graduate schools
- Collect and analyze hydrologic, geomorphologic and ecological data
- Develop and implement greenway design working with local stakeholders, public officials and engineers
- Outreach and technology transfers through the organization of a local greenways and trails committee, development of a short course for engineers and developers and holding of conferences on greenway development.

The underlying goal of the project was the adoption of more environmentally sound and sustainable drainage practices by the city, local developers, and engineers.

To develop support from the City for the project, technology transfer tours were organized to similar cities where greenways have been used successfully in the drainage plan. Denver, Longmont, and Fort Collins, Colorado were selected for the tour. Longmont and Fort Collins are both very similar to Rogers in that they are medium sized cities with rapidly growing populations. Denver is noted for its development of the South Platte River through a rundown part of town as a recreational trail. The public works and parks departments in each of these cities were gracious in their support of the tour with personnel and information. Attending the tour were representatives of the City Council, Planning Commission, Parks Department, Public Works Department, Water Utilities, Arkansas Department of Environmental Quality and Arkansas Soil and Water Conservation Commission. Shortly afterward, a second trip was organized to Springfield, Missouri, the only town in the Ozarks that has incorporated greenways into its development plan at this time. On the trips, the City officials questioned the local officials about technical aspects of the drainage system, public support for the greenways, funding, and administrative needs to support the greenways system.

259

A 1,490 m (4,900 ft.) reach of the Blossom Branch of Osage Creek in south Rogers was selected for the demonstration because it was typical of undeveloped streams in the area. And, it could be monitored and observed as development approached the creek from all directions. In addition, the Rogers Public Schools owned half of the reach, and the other two property owners along the reach were supportive of the effort as well. The drainage area of Blossom Branch is 9.3 sq. kilometer at the upper end and 15.5 sq. kilometers at the lower end of the project area. The average slope of the watershed is 2.8%. Maximum width of the watershed is 4.5 kilometers; the maximum length is 5.6 kilometers. Land cover is currently 8.1% farmstead, 9.7 % roads and pavement, 8.5 % commercial, 15.2 % industrial, 57.7 % residential and 0.8% water. The last of the farmland within the watershed has been planned for development within the next five years. For this project, the stream was divided into three reaches, an upper reach of 380 meters that had been channelized through pasture, a middle 550 meters flowing through an intact riparian zone, and a lower 450 meters flowing through mixed pasture and forest.

METHODS

Habitat Assessment. Habitat assessments were conducted at eight sites on the stream in the project area. EPA's Rapid Bioassessment Protocol for Streams and Wadeable Rivers (Barbour et al. 1999) was used in conducting the assessments. Three different teams of assessors from the Biological Engineering's Ecological Engineering class rated the stream on ten different parameters at each site. The results from the three teams were averaged for the final rating.

Riparian Zone Evaluation. To evaluate the condition of the riparian zone in the project area, a survey was made documenting the size and type of each tree within the 100-year floodplain. This data on the type of tree was entered into a GIS database using the CityGreen (American Forests Washington DC) extension. CityGreen was then used to evaluate the value of carbon sequestration provided by the remaining trees within the riparian zone compared to the historic value. This analysis allowed evaluation of the loss of the pollution attenuation provided by the riparian forest.

Hydrology. The watershed was delineated using ArcView GIS software (ESRI, Redlands, CA) and 1:24,000 scale digital ortho quarter quads (DOQQs). The watershed area was determined at the entrance to the project and at the lower end of the project. Landuse/landcover was determined using ArcView and the DOQQ's also. Soils were delineated from ARKSYS (Center for Advanced Spatial Technologies, University of Arkansas). ArcView was then used to determine the average watershed slope and stream length. The Natural Resources Conservation Service's curve number method was then used to develop synthetic hydrographs for the project area for the 1, 2, 5, 10, 50 and 100 year storms during the estimated pre-development period, at present, and at the estimated full development condition.

Stream cross sections were surveyed through the project area at 60 meter intervals along the stream. The Army Corps of Engineer's HEC-RAS hydraulic model was used to compute the water surface elevation through the project area, and this information was then compared to the FEMA flood study base flood elevations. Manning's n was adjusted to calibrate the HEC-RAS model to the FEMA base flood elevations. HEC-RAS was also used after completion of the restoration design to document no rise from the base flood elevations.

Geomorphology. Channel slope and cross sectional area were taken from the surveyed stream cross sections. The bankfull elevation and width was determined in the field from observation of

260

channel scour, depositional features and change of grade. The flood prone area was determined by the width of the channel at twice the bankfull depth. Entrenchment was then determined as the ratio of the flood prone area width to the bankfull width. Channel sinuosity was determined by dividing the channel slope by the valley slope through each reach of the project. Wolman pebble counts were conducted from bankfull to bankfull in each reach to determine D_{16}, D_{50} and D_{84}. Analysis was then made of stream power in the design reach to determine the competence of the stream to move the material in the system. Ackers and White's formula (Simoes, F.J. M. 2001) was used to model the sediment transported by the stream. Sediment supplied by the watershed was modeled by the modified universal soil loss equation (Ward and Elliot 1995).

Technology Transfer. The main objective of the demonstration project was for the City administration to adopt more sustainable and environmentally sound drainage practices. Specifically, it was desired that the city adopt a greenway system to protect the ecological services provided by the stream. Reluctance by the city to modify their practices was based on perceived need to maintain clean ditches for hydraulic conveyance of floodwater and on the cost of alternative practices. To overcome this reluctance, technology transfer tours were conducted where city officials were given opportunity to visit similar municipalities that had adopted the desired practices and to discuss with local officials about the outcome of those practices. Longmont, Fort Collins, and Denver, Colorado and Springfield, Missouri were visited on these tours. Representatives of the Mayor's office, the Planning Commission, an influential local Consulting Engineer and the directors of the City's Planning Department, Parks and Recreation Department and Water Utilities were also involved in the tours. At each stop on the tour, local officials were present to discuss with the tour any questions that might arise.

RESULTS AND DISCUSSION

Habitat Assessment. The average habitat assessment score for the eight sites assessed was 82.3 out of a possible 200 points indicating that there was great potential for improving the aquatic habitat along the stream corridor. Budgetary constraints dictated that restoration be restricted to reach 1 at this time. In reach one, the assessment score 96.5 was slightly higher than the average for the entire project area. However, the stream was still very low compared to the potential. Individual parameter scores (Figure 1) indicated that there was especially good potential for improving pool variability, channel flow status and restoring vegetation in both the right and left banks.

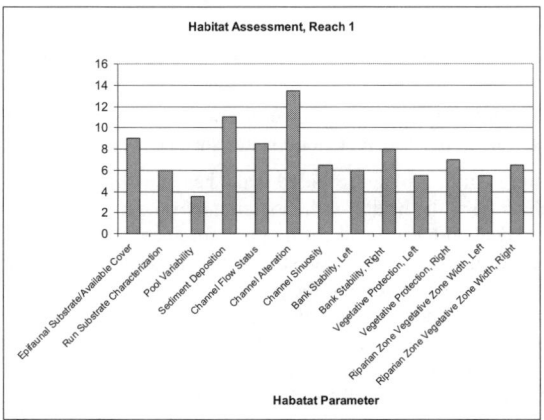

Figure 1. Reach 1 habitat assessment scores

Riparian Zone Evaluation. The analysis performed by CityGreen indicated that the city has lost almost 90% of the economic value of the ecological services provided by the riparian forest of Blossom Way Creek (Table 1). This figure was based on the value of carbon sequestered and stored in the timber in pre-developed conditions versus the amount of timber left today.

Table 1. Ecological Services From Forests in Blossom Branch Creek Under Three Conditions

Land Use Condition	Air Pollution Removal		Carbon Storage (T)	
	Lbs removed	Value ($)	Stored	Sequestered
Pre development	290,000	690,000	135,000	1,050
Current	31,000	73,000	18,000	51
Fully developed	500	1,200	240	0.5

Hydrology. Urbanization of the watershed has caused an increase in flow ranging from 6.5 fold at the 1 year storm to 1.8 fold at the 100 year storm. (Figure 2). Because of this increase in flow, the design stream has been repeatedly channelized and enlarged. The restoration design would have to consider the increased flow in sizing of the channel. In the project area, there was ample room to allow overbank flooding, so it was decided to keep the channel small so that low flows could be confined to a narrower channel with adequate depth to provide some pool and channel variability.

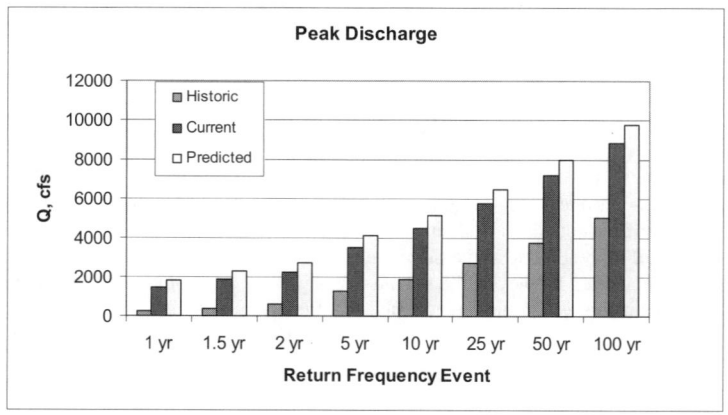

Figure 2. Historic, Current and Predicted Peak Discharges

The results of the HEC-RAS model of the project for pre-restoration and post-restoration conditions showed no rise in the 100 year flood water surface elevation as a result of the project.

Geomorphology. Survey of the stream revealed that a head cut was working its way upstream from earlier channelization at the bridge at the lower end of the project. Any restoration work in reach 1 would have to deal with this head cut first and foremost. It was decided to build an artificial riffle above the cut as a stabilization structure. The geomorphic characteristics (Table 2) of the stream indicated that reach one of the project area was generating more power because of slope and bank height than the material in the stream was able to resist. As a result, the stream was eroding by downcutting into the bed. The sediment yield from the watershed as computed by the modified universal soil loss equation for a bank full storm (650 cfs) was 1040 tons. The Ackers and White model predicted that the sediment transport in the stream at bankfull conditions would be slightly higher at 1080 tons per day. This small difference was not considered to be significant except that it seemed to reinforce the data showing that the stream was degrading in this reach. Visual inspection of the stream banks and bed also reinforced this idea. In reaches two and three, the stream had already cut down to bedrock, so it was actively

262

eroding its banks and widening out. To correct these problems, the design for reach 1 called for increasing the sinuosity of the stream to reduce slope and also to restore the flood plain at a lower bank height. The combination of reduced slope and bank height provided for stream power equal to the d_{50} particle and should be a stable channel. Form riffles were provided in the lengthened stream to provide for consistent grade control, scouring of pools and variability of flow to improve instream habitat.

Table 2. Geomorphic characteristics of Blossom Branch Creek

Reach #	BF w, m	BF d, m.	s m/m	Q, cms	W/D	Sinuosity	Tractive Force kg.m^2	D_{50} cm
1	7	1.3	0.0041	11.8	5.2	1.26	5.43	5
2	10	1.2	0.0022	33.7	8.4	1.08	2.6	2.2
3	11	1.15	0.0061	49.8	9.6	1.19	7.0	3

Technology Transfer. After completion of the greenway technology transfer tours, the city established a local greenways and trails committee through the Rogers-Lowell Chamber of Commerce. This committee started meeting in the fall of 2003. Later that fall, the City Planning Commission adopted a 45 kilometer greenway trail system into its comprehensive master plan for development. Since adoption of the master plan, developers along the two major streams through the city have been required to provide easement for the greenway, or to donate the land within the FEMA floodway to the city for drainage and recreational purposes. Most developers have elected the latter option as it reduces their property tax, and also eliminates the perceived liability of having the trail on their property. Because of dedication of the greenway area, the stream can now be maintained in a more natural condition than previous practices allowed.

Restoration Design. The overall design (Figure 3) for the restoration of reach one called for control of the active head cut with an artificial riffle, lengthening the stream and lowering the banks to reduce stream power and also to reduce bed erosion, and restoration of riparian vegetation to provide shade for the stream and to protect the new banks from erosion. Artificial riffles along the new alignment of the stream also were provided to increase the variability of flow within the channel and improve pool/riffle structure as the habitat assessment indicated was needed. Finally, to provide more sustained flow, and also to provide thermal refuge in the stream, the abandoned channel was filled with filter gravel and then covered with filter fabric and soil creating subsurface water retention. Drainage from the parking lot of the adjacent public school is to be redirected to this subsurface retention cell. Summer thunderstorms on this impervious area will then provide recharge water to the stream.

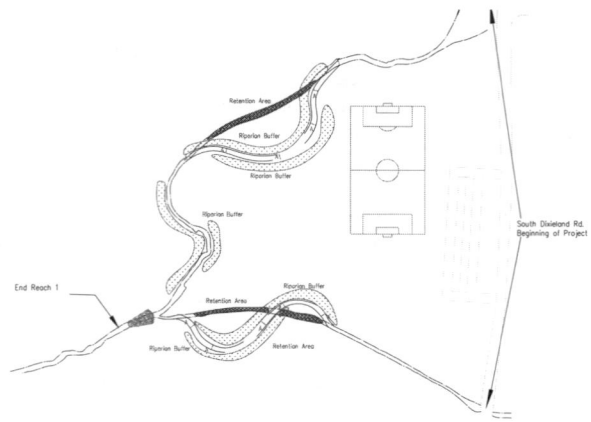

Figure 3. Reach 1 restoration design

CONCLUSION

This project has demonstrated that analysis of ecological services can provide rationale for the design of restoration projects in the stream. Also, the analysis provided a means to communicate to the public officials the need to complete reconstruction projects. In Rogers, the initial resistance of the city administration modification of the standard drainage practice was over come by an effective technology transfer program. As a result, the city has adopted greenways as a component of their master planning process.

Further technology transfer programs and demonstrations need to be conducted in the area to convince local developers and consulting engineers of the need to preserve greenways. Of particular interest in Northwest Arkansas will be study of how property values of developments with greenways compare to those that do not cooperate with the program. Research should be conducted to determine the mental models that the developers are working under and to develop technology transfer and demonstration projects that will result in modification of those mental models.

REFERENCES

1. ADEQ 1997. Assessment Report, Prepared Pursuant to Section 319(a)1(A,B,C,D) of the Federal Water Pollution Control Act. Little Rock Arkansas. Arkansas Department of Pollution Control and Ecology.

2. Barbour et al. 1999. Rapid Bioassessment Protocols for Use in Streams and Wadeable Rivers. Washingon DC. USEPA Office of Water.

3. EPA 2002. National Management Measures to Control Nonpoint Source Pollution from Urban Areas. Washington, DC. USEPA Office of Wetlands, Oceans, and Watersheds

4. NPS 1995. *Economic Impacts of Protecting Rivers, Trails, and Greenway Corridors.* San Francisco, CA. National Park Service, Western Region.

5. NPS 1995. *Economic Impacts of Protecting Rivers, Trails, and Greenway Corridors.* San Francisco, CA. National Park Service, Western Region

6. Simoes, F.J. M. 2001. *GSTARS 2.1 GUI Tutorial.*, Denver, CO. US Bureau of Reclamation

7. US Bureau of Census, 1999. Metropolitan Area Population Estimates for July 1, 1998 and July 1, 1999 and Population Change for July 1, 1998 to July 1, 1999. MA-99-2. Available at: http://eire.census.gov/popest/archives/metro/ma99-02.txt. Accessed 18 Nov. 2003

8. Ward, A. D. and W. J. Elliot, 1995. Environmental Hydrology. Boca Raton, Fl: Lewis Publishers.

Preliminary Results of a Dam-Removal Analysis on Brewster Creek Near St. Charles, Illinois, 2002-2004

K. M. Kosky[1], T. D. Straub[2], D. P. Roseboom[3], and G. P. Johnson[4].

Abstract

The benefits of gradually removing a dam (through multiple notches) are to reduce the total project cost and reduce possible environmental effects by allowing the impounded sediment to slowly move downstream, and a stable stream and revegetated floodplain to form upstream. Notching, in this study of a dam on Brewster Creek, near St. Charles, Illinois, involves cutting a given height (in five 12–18 inch notches over approximately a 9 month period) across the length (or some portion of the length) of the dam. Brewster Creek is a tributary of the Fox River in northeastern, Illinois. Sediment, dissolved oxygen, and geomorphic response are being monitored before, during, and after a gradual (notching) removal of the dam. The study area includes the creek reach immediately below the dam and above the lake. Preliminary data analysis indicate that during and after the removal, the relation between the sediment transported to the study area from upstream and the sediment transported out of the study area remained relatively stable. This preliminary result indicates that the notching system created a fairly slow and predictable sediment transport response to storms, when compared to known upstream sediment loads. This result corresponds to the slow geomorphic response at the site since inception of the notching sequence in 2003. The creek responded to the five notches removed over the course of 9 months by gradually cutting through the former lakebed sediment to establish a meandering channel. Notchings did not appreciably affect dissolved oxygen concentrations in Brewster Creek.

KEYWORDS: sediment, transport, automated sampling, dam, removal, calculation, costs, analysis

Introduction

The Illinois Department of Natural Resources (IDNR) declared the dam on Brewster Creek near St. Charles, Illinois a Class I structure, having a high probability of causing loss of life and/or substantial economic loss in the event of a catastrophic failure. The dam was in disrepair and, therefore, needed to be repaired or removed. Costs were prohibitive to repair the dam; therefore, the owners decided to remove the dam.

Costs and possible environmental effects of removing dams, and managing or removing the sediment impounded behind them also can be substantial. In Illinois, State regulations require complete sediment containment during any construction project. Project engineers developed a

---

[1] Kane County Department of Environmental Management. 719 Batavia Avenue, Geneva, IL 60134. koskykaren@co.kane.il.us.

[2] USGS, Illinois District, Water Resources Division. 221 N. Broadway Ave., Urbana, IL 61801. tdstraub@usgs.gov.

[3] USGS, Illinois District, Water Resources Division. 221 N. Broadway Ave., Urbana, IL 61801. roseboom@usgs.gov.

[4] USGS, Illinois District, Water Resources Division. 221 N. Broadway Ave., Urbana, IL 61801. gjohnson@usgs.gov.

first design for the total removal of Brewster Creek dam that complied with these regulations. The projected construction, oversight and restoration costs of the dam removal under this first design were $1.17 million, which exceeded available funds.

The Illinois Environmental Protection Agency (IEPA) granted permission for a pilot dam removal project, consisting of cutting five 12–18 inch notches, over approximately a 9-month period, across the length (or some portion of the length) of the dam. The projected construction, oversight, and restoration cost for the gradual removal was $330,000.

The benefits of gradually removing a dam at this site using multiple notches included reducing the projected removal project cost and reducing possible environmental effects by allowing the impounded sediment to slowly move downstream and a stable stream and revegetated floodplain to form upstream.

PURPOSE AND SCOPE

The purpose of this paper is to present preliminary analysis of the sediment data collection (pre-notching, during notchings, and post-notching) and to document some geomorphic changes with selected project photos. Dissolved oxygen data, collected during notchings, also will be summarized. A final report is planned describing the entire study after the channel and floodplain have stabilized. Note that all data presented are provisional and subject to revision.

METHODS

Sediment concentrations and load are being monitored at two gaging stations (one upstream and one downstream of the dam) (Figure 1), by the U.S. Geological Survey (USGS), in cooperation with Kane County Department of Environmental Management (KCDEM), and the IEPA. From June 2002 through March 2004, more than 600 sediment samples were collected at the two gaging stations. Samples are obtained at various stream stages throughout the year, including all five dam notchings from June 2003 through February 2004. The USGS also monitored dissolved oxygen concentrations during each notching event.

Biologists from the IDNR and the Shedd Aquarium in Chicago are monitoring stream biota. Engineers and surveyors from the IDNR, Office of Water Resources and USGS are surveying geomorphic changes. Lastly, KCDEM is maintaining a Web site showing the progression of geomorphic changes by way of photographs taken during the course of the study.

Notching criteria will be developed based on watershed flow and sediment released in relation to the amount of stored sediment behind the dam. The notching criteria developed through this pilot project may be expanded to other watersheds so that the IEPA can develop criteria for other possible dam removals in Illinois.

267

Figure 1. Location of two USGS gaging stations near the dam on Brewster Creek near St. Charles, Illinois.

RESULTS

The following discussion summarizes preliminary results based on data collected from June 2002 through March 2004. Note that all data presented are provisional and subject to revision. A summary of sediment loads from storm events and notchings are given in Table 1 and Figure 2, along with peak flow and rainfall totals (Table 1).

The percent difference in sediment loads (downstream minus upstream) before any notchings was near zero or negative for storm events (Table 1 and Figure 3). This result means that the impounded water behind the dam trapped sediment. Also, as expected, after the notchings started, there was a shift in the percent difference in sediment load from negative to positive (Table 1 and Figure 3). At this point in time, the positive percent difference in sediment load has remained relatively stable, with a median value of 65 percent.

Data from the first storm after notch 5 yielded the highest sediment load (163 tons), but the difference as compared to the upstream gaging station is 66 percent. This result means that the sediment load only increased by 66 percent between the upstream and downstream gaging station, similar to the results of the other notchings and storms between notchings. Storms occurred between all notchings (except between notch 1 and 2) allowing for transport of sediment (given the new base elevation near the dam), although the peak flows during some storms are relatively small (Table 1). A series of photos through time illustrate the gradual formation of a stream channel in response to the five notchings (Figure 4). Dissolved oxygen concentration during the notchings never dropped below 7 milligrams/liter (mg/L) at the downstream gaging station.

Table 1. Sediment loads at gaging stations upstream and downstream of a dam on Brewster Creek near St. Charles, Illinois, along with peak flow and total rain. [ft³/s, cubic feet per second]

Total Sediment Load	Percent Difference	Peak	Total

268

Time Period	Downstream (tons)	Upstream (tons)	(Downstream minus Upstream)	Flow (ft³/s)	Rain (inches)	Event Type
07/08-07/15/2002	6.5	6.8	-4	20	1.41	Storm
08/03-08/10/2002	19.0	21.6	-12	27	1.08	Storm
08/12-08/16/2002	19.2	20.4	-6	58	3.33	Storm
08/21-08/29/2002	33.7	36.3	-7	81	3.34	Storm
04/29-05/03/2003	36.6	76.4	-52	100	3.67	Storm
05/04-05/12/2003	51.5	49.1	5	87	3.59	Storm
05/13-05/22/2003	14.4	13.7	5	42	0.58	Storm
06/19-06/22/2003	1.8	1.1	63	23	0.01	Notch 1
07/14-07/19/2003	12.5	9.1	37	19	0.00	Notch 2
07/30-08/07/2003	2.5	1.5	65	9	1.45	Storm
08/19-08/24/2003	2.9	1.1	165	4	0.00	Notch 3
11/01-11/08/2003	23.4	10.3	127	21	3.09	Storm
11/17-11/28/2003	30.6	19.09	60	35	1.92	Notch 4 and Storm
12/08-12/19/2003*	135.0	73.6	83	41	2.16	Storm
02/18-02/28/2004	8.9	7.0	27	16	0.52	Notch 5
03/03-03/09/2004	162.9	97.9	66	90	1.83	Storm

*estimated downstream sediment load due to sampler malfunction

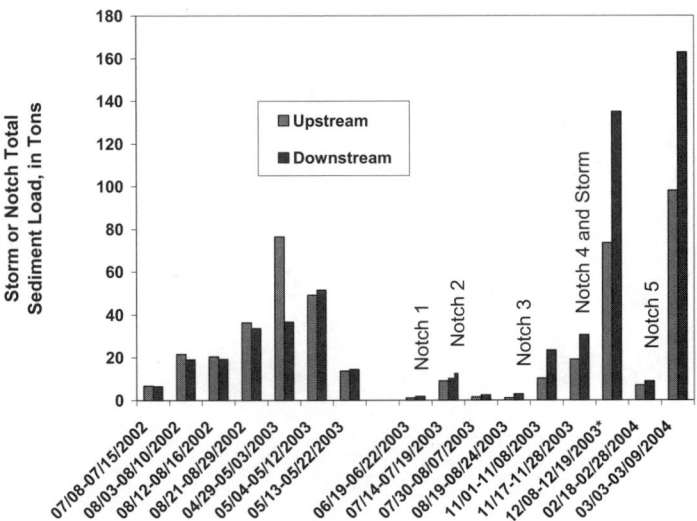

Figure 2. Total sediment load at the downstream and upstream gaging stations for storms and notches of a dam on Brewster Creek near St. Charles, Illinois.

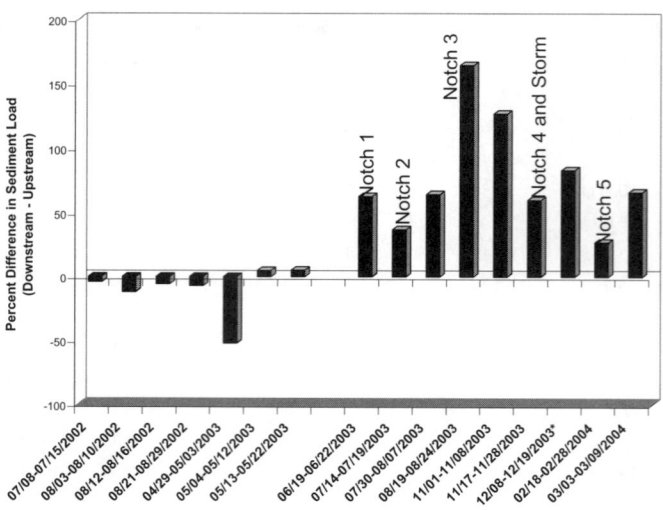

Figure 3. Percent difference in sediment load (downstream minus upstream) for storms and notches of a dam on Brewster Creek near St. Charles, Illinois.

Brewster Creek/ Willow Lake, pre-dam removal

June 2003

August 2003

March 2004

June 2004 (8 weeks after spring burn)

Figure 4. Geomorphic and vegetative response of Brewster Creek to gradual dam removal (photos taken looking upstream from bridge that is located 15 feet downstream of dam)

271

CONCLUSIONS

The benefits of gradually removing a dam on Brewster Creek, near St. Charles, Illinois, are to reduce the total project costs and reduce possible environmental effects by allowing the impounded sediment to slowly move downstream, and a stable stream and revegetated floodplain to form upstream. Through the monitoring of sediment movement, dissolved oxygen concentrations, and geomorphic changes, notching criteria can be developed based on flow and sediment released in relation to the amount of sediment stored behind the dam.

The costs to remove the dam by using a gradual (notching) removal method were approximately one-quarter ($330,000) the costs projected to remove the dam using traditional sediment-containment techniques ($1.17 million).

As expected, the dam trapped sediment before removal. During and after removal, the impounded sediments are transported downstream. Also, during and after removal, the relation between the sediment transported to the study area from upstream and the sediment transported out of the study area remained relatively stable. This preliminary result indicates that the notching system created fairly slow and predictable sediment transport response to storms when compared to the expected upstream sediment load. This result corresponds to the slow geomorphic response at the site since inception of the notching sequence in June 2003. Based upon regional estimates of annual sediment yield (0.3 tons per acre per year), the current sediment loads of all events analyzed is less than would be expected for a watershed of this size.

Brewster Creek gradually cut through the former lakebed sediment to establish a meandering channel. As expected, the channel formation was most active during and after major storm events. The channel emerged after the first notching and gradually developed with the slow rate of notchings. After the fifth notch, the channel widened to the point that small meanders emerged within the channel banks, with gravel and cobble point bars on the inside bends.

Results indicate that the notchings did not appreciably affect dissolved oxygen concentrations in Brewster Creek. Before removal of the dam, the upstream lake was almost full of sediment, with maximum water depths of approximately 2 ft. The lack of any appreciable effects because of notchings could indicate a well-mixed lake with no areas of low dissolved oxygen concentrations given the low water depths in the lake.

Acknowledgements

Special thanks to the following agencies who provided funding and support for this project: the Illinois Environmental Protection Agency, the Illinois Department of Natural Resources, the U.S. Geological Survey, the U.S. Fish and Wildlife Service, the Kane County Riverboat Grant Fund, the U.S. Department of Agriculture Natural Resource Conservation Service, the Fox River Ecosystem Partnership, and the Northeast Illinois Planning Commission.

The Restoration of Coldwater Fork
Using the Reference Reach Approach

J. George Athanasakes, PE and
Michael F. Adams, Jr., PE[1]

ABSTRACT

A coal slurry release from the Big Branch Slurry Impoundment on October 11, 2000 resulted in the flow of approximately 300 million gallons of slurry into the Coldwater Fork watershed. Emergency response to this spill required pumping of slurry and the use of excavators. Efforts were made to maintain the pre-spill channel configuration of Coldwater Fork during the clean up; however, portions of the creek were realigned. Since the slurry release, significant portions of Coldwater Fork have exhibited recovery including the reestablishment of pools and riffles, gravel substrate and macroinvertebrates. At the downstream limits of the project, a series of head-cuts formed which threatened to destabilize recovering portions of the creek. A head-cut also formed near the confluence with Lynn Bark Branch.

The stream restoration design focused on minimizing reworking of the stream in areas exhibiting good natural recovery, and restoring destabilized areas using natural channel design techniques. Channel dimensions developed from reference reach data were used to design two areas where the channel alignment and geometry were altered during the slurry cleanup. The appropriate use of the reference reach was validated by simulating a design on the reference reach to verify sediment transport competency equations would accurately predict the mean depth required to pass the sediment delivered to the stream. This paper will discuss the design and construction of this project with emphasis on use of the reference reach approach and sediment transport competency calculations for design.

KEYWORDS. Natural channel design, dimensionless ratios, reference reach, sediment transport competency equations.

INTRODUCTION

An uncontrolled slurry release caused by the breaching of the Big Branch Slurry Impoundment on October 11, 2000, resulted in the release of approximately 300 million gallons of slurry into the Coldwater Fork watershed. The slurry release, which is considered by EPA to be one of the worst coal mining disasters in the eastern United States, impacted over 60 miles of stream. These impacts ranged from significant accumulation of fine coal slurry within streams and adjacent floodplains, to significant turbidity in extreme downstream reaches. The most severely impacted reach included a mile-long section of Coldwater Fork immediately below the impoundment, where up to 6 feet of slurry spilled out into the floodplain (Figure 1). The emergency response to this spill required pumping of slurry out of the area as well as excavation with heavy equipment as shown in Figure 2. Cleanup efforts were initiated and were conducted around the clock to remove the slurry as quickly as possible. During the cleanup, the focus was on minimizing impacts to existing streams; however, the upper portion of Coldwater Fork was realigned as debris and slurry were removed from the Coldwater Fork valley, resulting in a stream with a nearly trapezoidal cross section and boulder armoring.

---

[1] J. George Athanasakes, PE, is an Associate with Fuller, Mossbarger, Scott and May Engineers, Inc., Louisville, KY 40223. Michael F. Adams, Jr., PE, is a Senior Project Engineer with Fuller, Mossbarger, Scott and May Engineers, Inc.

Figure 1: View of Coldwater Fork Valley Following the Slurry Spill.

Figure 2: Excavation of Coal Slurry.

Since the slurry release occurred, significant segments of Coldwater Fork have exhibited a remarkable recovery. Portions of Coldwater Fork are re-establishing stable riffle-pool sequences and riffle armor is beginning to accumulate in the upstream portion of the stream. The remaining sections of Coldwater Fork lack appropriate riffle/pool sequences and/or exhibit head-cuts or substantial bank erosion. Banks largely consist of clay and sand and the channel bottom is primarily comprised of sand and gravel. At the down-stream limits of the project near the confluence of Walnut Fork, a series of head-cuts (Figure 3) formed, which threatened to destabilize the recovering portions of the creek. In addition, a head-cut also formed near the confluence with Lynn Bark Branch (Athanasakes, 2004).

Figure 3: View of Head-Cut Near Confluence of Walnut Fork.

DESIGN METHODOLOGY

A natural channel design was prepared for approximately 6,000 feet of stream along Coldwater Fork. The stream restoration design for this project focused on minimizing impacts to the stream in areas exhibiting good natural recovery, and restoration of destabilized areas using natural channel design techniques to achieve proper dimension, pattern and profile, especially within areas exhibiting head-cuts. The new channel was designed using dimensionless ratios developed from a reference reach utilizing the RIVERMorph© Stream Restoration software.

Reference Reach

A reference reach was used to develop pattern, profile and cross sectional dimensionless ratios (Rosgen, 1998) for the restoration design. The data collected from the reference reach included pool-to-pool spacing, sinuosity, meander wavelength, belt width, pool cross sectional properties, hydraulic slopes, width/depth ratios, entrenchment ratios, etc. The measured data from the reference reach was divided by either the bankfull width of the reference reach or the bankfull depth to establish the dimensionless ratios. These dimensionless ratios were then multiplied by the bankfull properties of the proposed design reach to establish the new channel's geomorphic properties.

The reference reach used for the project was located in a nearby watershed in close proximity to the Coldwater Fork site (Figure 4). The reference reach chosen had a similar slope to the proposed design reach and was transporting similar size material. In order to evaluate the suitability for the reference reach for use in the design, an assessment of the reference reach was performed utilizing the Pfankuch methodology (Pfankuch, 1975). The reference reach had a Pfankuch rating of 80 (good rating) and classified as a C4 stream type according to the Rosgen classification system (Rosgen, 1996). The reference reach had bedrock control immediately upstream.

Figure 4: View of Reference Reach.

Debate continues to occur within the stream restoration profession regarding the validity of reference reaches for design and the use of sediment transport competency equations in spite of documented success with the methodology. The use of sediment transport competency equations will be discussed in more detail in the following paragraphs. For this particular project, the authors decided to complete design calculations on the reference reach to predict the depth required to pass the largest particles transporting through the stream. This calculated depth was then compared to the actual measured bankfull depth at the site to help judge the suitability of the selected stream as a reference reach and to verify that sediment transport competency equations were an applicable predictor of sediment transport for this particular stream.

Impacted Reach Design

The project was divided into five reaches (based on drainage area and sub-catchment) of similar bed material, bank composition and bankfull slope. Three reaches were on Coldwater Fork and two were on Lynn Bark Branch and Walnut Fork, tributaries to Coldwater Fork. Lynn Bark

Branch and Walnut Fork were considerably steeper than Coldwater Fork following the emergency cleanup.

The approach to determining bankfull properties for Coldwater Fork and its tributaries differed from the typical methodology. Following the emergency cleanup, re-alignment and placement of boulders along the banks of Coldwater Fork and its tributaries, there was a relative absence of bankfull indicators due to bank armoring and recent excavation of the channel. Reliable bankfull indicators from Wolf Creek were used as support for the few bankfull indicators present along Coldwater Fork in the following fashion. First, the ratio of the bankfull cross sectional area for the Wolf Creek reach drainage area to the predicted bankfull cross sectional area shown on the Eastern United States regional curve (Rosgen, 1996) for the same drainage area was derived. Next, this ratio was applied to the predicted bankfull cross sectional areas on the Eastern United States Regional Curve for the drainage areas of the respective reaches of Coldwater Fork and its tributaries. The estimated bankfull cross sectional area was then compared to possible bankfull indicators observed during the geomorphic survey. Indicators corresponding to these predicted bankfull properties were used to determine bankfull elevation throughout the reach (Athanasakes, 2004).

The design consisted of a combination of stream relocation; bankfull bench construction augmented with in-stream structures; and placement of in-stream structures without significant alteration to the cross section. In the relocated sections, particular attention was given to improving the stream pattern and cross section based on the reference reach. Structures to maintain grade, concentrate flows away from the banks, and improve habitat were included in the form of cross vanes and j-hooks/log vanes. Riffle armor was also included for the relocated sections. The relocation and stabilization design increased the length of Coldwater Fork from approximately 4,995 to 5,385 linear feet of stream by increasing the sinuosity in the relocated reaches. An additional 70 linear feet and 155 linear feet of stream were added to Lynn Bark Branch and Walnut Fork, respectively. As a result of the increased channel length, the bankfull slope was decreased, which resulted in a decrease in shear stress.

For the bankfull bench construction areas, a 20- to 30-foot flat area was excavated on the left bank at the bankfull elevation. The bench was designed to reduce shear stresses during flows above bankfull. Cross vanes and j-hooks/log vanes were also used to provide grade control and bank protection, respectively.

Sediment Transport Competency Calculations

In order to verify that the designed reach can efficiently transport sediment without excessive aggregation or degradation, entrainment calculations were performed to determine the sediment transport competency. Critical dimensionless shear stress was calculated utilizing Andrew's equations (Andrews and Erman, 1986; Andrews and Nankervis, 1995). These equations were modified by Rosgen (2002), whereby sediment data from bar samples are used instead of sub-pavement data:

$$\tau c_i = 0.0384 \left(\frac{d_{i-Bar}}{d_{50-Bed}} \right)^{-0.887} \qquad \text{(Eqn. 1)}$$

where τc_i = critical dimensionless shear stress

d_{i-Bar} = largest particle in bar sample (mm)

d_{50-Bed} = median particle size from riffle pebble count (mm)

The bankfull mean depth required to mobilize the largest particle encountered in a bar sample was determined using Equation 2:

$$d = \frac{\left(\tau c_i \times 1.65 \times d_i / 304.8 \right)}{S} \qquad \text{(Eqn. 2)}$$

where d = required mean depth (ft)

d_i = largest particle in bar sample (mm)

S = riffle slope at bankfull

Using Equations 1 and 2, the depth required for sediment transport competency was compared to the design depth. If they differed, the dimensions were changed in an iterative process using the RIVERMorph© Stream Restoration software until they were within approximately 10 percent of each other (Athanasakes, 2004). Table 1 summarizes the design parameters.

Table 1. Natural Channel Design Parameters

	Reach 1	Reach 2	Reach 3	Lynn Bark Branch	Walnut Fork
Designed Valley Slope	0.00595	0.00464	0.0046	0.00584	0.00459
Cross Sectional Area	34 ft²	62 ft²	73 ft²	39 ft²	30 ft²
Riffle Bed D_{84}	60.38 mm	51.69 mm	73.07 mm	51.69 mm	57.17 mm
Riffle Bed D_{50}	32.72 mm	29.96 mm	31.66 mm	29.96 mm	21.94 mm
Bar D_i (Max Particle Size)	55 mm	90 mm	90 mm	90 mm	65 mm
Bar D_{50}	8.11 mm	12.71 mm	12.71 mm	12.71 mm	8.38 mm
Avg. Meander Wavelength	183 ft	247 ft	300 ft	196 ft	172 ft
Sinuosity	1.22	1.2	1.19	1.2	1.24
Bankfull Slope	0.00486	0.00386	0.00387	0.00487	0.0037
Width to Depth Ratio	17.5	17.5	17.5	18.0	16.5
Bankfull Width (W_{BKF})	24.4 ft	32.9 ft	35.7 ft	26.5 ft	22.4 ft
Pool Width	30.6 ft	41.3 ft	44.8 ft	33.2 ft	28 ft
Bankfull Depth (D_{BKF})	1.4 ft	1.9 ft	2.0 ft	1.5 ft	1.4 ft
Riffle Max Depth	2.1	2.8	3.0	2.3	2.0
Pool Max Depth	3.7	4.9	5.2	4.1	3.6
Classification	C4	C4	C4	C4	C4
Sediment Transport Required Mean Depth	1.48 ft	1.83 ft	1.97 ft	1.45 ft	1.39 ft

A similar methodology was used for the reference reach to establish the suitability of the proposed stream as a reference reach and to validate that sediment transport equations are accurate predictors of depth required to pass sediment for the reference reach stream. For the reference reach, the measured bankfull mean depth of the channel was 1.77 feet. The calculated depth required to pass the largest representative particle encountered at the reference reach site was 1.69 feet, which is within 5 percent of the actual measured depth.

CONSTRUCTION

Construction was completed using a design/build approach with the intent of making the reconstructed stream look as natural as possible. A set of construction drawings were prepared by Fuller, Mossbarger, Scott and May Engineers, Inc. (FMSM) to guide the construction process. Construction was completed by Enviro-Pro, Inc., working with FMSM field engineers. Construction commenced in December 2003 and was completed in May 2004.

In areas where the stream was relocated, the new floodplain for the stream was first excavated at the proposed bankfull elevation. The amount of excavation necessary to achieve this elevation varied throughout the project. The floodplain was excavated as wide as lateral constraints would permit, and in constrained areas, the width of the floodplain was constructed such that the entrenchment ratio of the stream exceeded a value of 2.2. Final grading along the floodplain varied to create some occasional depression areas to support wetlands and to minimize the uniform appearance of the floodplain.

Following the floodplain excavation, the channel was excavated to the appropriate pattern and profile indicated in the design plans (Figure 5). The new profile of the stream was constructed by measuring down from the excavated floodplain using a range of riffle and pool depths established using dimensionless ratios derived from the reference reach. Riffles were first excavated and shaped according to the design drawings. Afterwards, pool areas were subjected to additional excavation. Due to field conditions encountered during construction, changes were made to the alignment in an effort to make the newly constructed channel appear as natural as possible. A view of the completed channel alignment in the upstream relocation area is shown in Figure 6.

Figure 5: Floodplain and Channel Excavation along Lynn Bark Branch.

Figure 6: Typical View of Lynn Bark Meander Pattern.

Structures (J-hooks, log vanes and cross vanes) were installed at strategic locations throughout the restored reaches (Figures 7 and 8). Rocks and logs used for hydraulic structures were readily available from the mining operation located near the site and were stockpiled prior to construction. One of the goals of the project was a natural-looking stream; thus, the use of structures was minimized and wood was used where feasible. Additionally, structure features were varied where possible, such as step heights and general vane shape, so that the structures looked slightly different throughout the site. Structures were built "in the dry" to minimize sedimentation in the downstream reaches of Coldwater Fork during construction. After flow was established over the structures, minor adjustments were often needed to achieve the intended flow patterns over the structures. The purpose of the various structures used throughout the project were to provide grade control, maintain deep scour pools and provide protection against bank erosion. All boulder structures were built with footing boulders that were at least the same size and often larger than the boulders visible at the surface.

Figure 7: Cross Vane on Lynn Bark Branch.

Figure 8: Log Vane on Coldwater Fork.

A major goal of the project was to restore habitat features depleted during the slurry removal process. Accordingly, the stream profile was designed to include a variety of deep pools and riffles. Riffles were constructed using cobbles and gravel from the abandoned channel as well as material brought in from off-site (Figure 9). At several riffle locations, boulder clusters were placed in the stream to create converging and diverging flows and provide holding cover for fish.

Figure 9: Close Up of Constructed Riffle along Coldwater Fork.

Erosion control blanket was installed on both banks up to the bankfull elevation. The blanket consisted of coir fabric that was staked near the water surface and buried in a trench at the top of the bank near the bankfull elevation. After initial erosion control blanket placement, live stakes were installed through out the project area to provide quick revegetation and to further secure the blanket (Athanasakes, 2004).

Monitoring

As a condition of the 401 and 404 permit, monitoring will be required at the site for a minimum of three years. Monitoring will include a longitudinal profile, cross sections and photo documentation taken on an annual basis. Photo documentation will be completed in two phases each year, one during the growing season to document the success of revegetation efforts and one during dormancy to visually document the overall stream pattern. In addition, a series of pebble counts and bar samples will be obtained to measure the gradation of bed materials and note any changes in particle size distribution that may be occurring at the site. Monitoring reports will be submitted annually.

CONCLUSIONS

A natural channel design was prepared for approximately 6,000 feet of stream along Coldwater Fork. The stream restoration design for this project focused on minimizing reworking of the stream in areas exhibiting good natural recovery, and restoration of destabilized areas using natural channel design techniques to achieve proper dimension, pattern and profile, especially within areas exhibiting head-cuts. A reference reach was used to develop pattern, profile and cross sectional dimensionless ratios for the restoration design, which was located in a watershed close to the Coldwater Fork site. The reference reach had a similar slope to the proposed design reach and was transporting similar size material. In order to evaluate the suitability for the reference reach for use in the design, sediment transport competency equations were performed on the reference reach to predict the required depth to pass the largest particles transporting through the stream. The calculated depth based on sediment transport competency equations was found to be within 5 percent of the measured mean bankfull depth of the reference reach. This methodology is recommended by the authors to confirm the applicability of a reference reach for use in a natural channel design.

REFERENCES

Andrews, E.D. and Erman, D.C. 1986. Persistence in the Size Distribution of Surficial Bed Material During an Extreme Snowmelt Flood. p. 191-197. In: *Water Resources Research, Vol. 22* (2). American Geophysical Union.

Andrews, E.D. and Nankervis, J.M. 1995. Effective Discharge and the Design of Channel Maintenance Flows for Gravel-Bed Streams, p. 151-164. In: *Natural and Anthropogenic Influences in Fluvial Geomorphology, Geophysical Monograph 89*. American Geophysical Union.

Athanasakes, J.G. and Adams, M.F. 2004. Restoration Design of Coldwater Fork Following the October 11, 2000 Slurry Spill. In: *StormCon Conference Proceedings*.

Pfankuch, D.J. 1975. Stream Reach Inventory and Channel Stability Evaluation. U.S. Department of Agriculture, Forest Service/Northern Region.

Rosgen, D.L. 1996. *Applied River Morphology*. Minneapolis: Printed Media Companies.

Rosgen, D.L. 1998. The Reference Reach – A Blueprint for Natural Channel Design. In: *Proceeding of the American Society of Civil Engineers Conference*. Denver, Co.

Rosgen, D.L. 2002. Sediment Transport Validation. Personal Communication.

Streambank Stabilization Challenges in the Glacial Lake Agassiz Sediments of the Red River Basin in North Dakota

David B. Rush, Environmental Projects Coordinator,
Red River Regional Council, Grafton, ND.
Frank W. Beaver, President, Geodynamics Inc., Grand Forks, ND
Jason Warne, Graduate Research Assistant,
Department of Geology and Geological Engineering, University of North Dakota,
Grand Forks, ND.

Introduction

For the past eight years, the Red River Basin (RRB) Riparian Project has been working to restore riparian zones and stabilize stream channels and banks in the Red River Valley. Since 1998, the project has completed 74 river miles of riparian restoration, developed eight restoration demonstration sites, and written 341 forest resource management plans for nearly 29,000 acres of riparian forest. Funded through the Environmental Protection Agency's Section 319 program, the project seeks to improve water quality throughout the watershed.

Project involvement in several slope failure stabilization efforts within the valley has ranged from providing geotechnical information to designing and implementing engineered restoration plans using technology such as soil bioengineering. Typically, the riparian vegetation has been removed or altered and the hydrology has been changed. Although the main goal of the Riparian Project is to restore a functioning riparian forest to act as a filter between urban or agricultural land use and the river, stabilization of active slope failures is frequently necessary before riparian restoration can be implemented. Riparian Project staff and cooperators recognized at the project's onset that understanding the causes of slope failure was essential to identifying riverbank stabilization solutions.

Geologic History and Present Geography

Geologic and geographic features of the Red River Valley in North Dakota, Minnesota and southern Manitoba create unique geotechnical challenges for slope and streambank stabilization efforts (Figure 1). During periods of glacial retreat, a vast lake formed over much of what is now the Red River Watershed. Material eroded from the bedrock was transported by meltwater streams and carried by icebergs to Lake Agassiz where boulders, gravel, sand, silt and clay were deposited.

The thickest deposits were along the center of the lake, the approximate location of the present day Red River of the North. These are the sediments through which the Red River flows on its way north to Lake Winnipeg and in which slope failures commonly occur. The current central valley landscape is the direct result of sedimentation in Lake Agassiz followed by recent erosion

The extremely low-gradient valley (1.5 to 0.2 feet per mile) formed less than 10,000 years ago as the glacial lake drained to the north. Large portions of this flat terrain are frequently subjected to overland and out-of-bank flooding from the Red River and it's tributaries. This relatively young river network cuts evermore sinuous channels through glaciolacustrine strata, such as the Sherack and Brenna Formations, that formed some of the richest agricultural land in the world. Although slope failures naturally occur in the Red River Valley, their frequency and severity has been exacerbated by clearing of riparian vegetation, development of riverside land, and changes in basin-wide hydrology.

Figure 1. Boundaries and shaded relief of the Red River Basin. (From UND Energy & Environmental Research Center)

Slope Stability Factors and Management Options
Several factors individually and cumulatively affect river bank instability including properties of the sediment, topographic slope, hydrology and the presence of vegetation. There exists a variety of management options to address these factors.

Earth Material Strength
Generally, dry, confined Lake Agassiz deposits have enough strength to support themselves. However, sediment strength becomes essentially zero when the material is wet and unconfined as along river banks. Where vegetation exists, sediment strength is enhanced because a live root mass provides additional strength, as well as physical buffering of the current against bank sediments. Dewatering through drainage and vegetation also adds strength.

GENERALIZED CROSS-SECTION OF FARGO-MOORHEAD

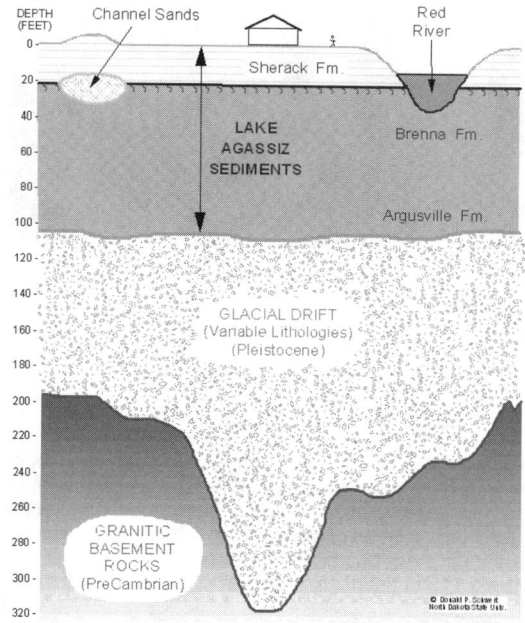

Sediment Properties

The thick smectitic clays and silty-clays derived from Cretaceous shales and glacial tills (Schwert, 2003) are particularly problematic. The Sherack Formation (Figure 2), comprising the upper 20 to 30 feet, consists of light colored, silty, laminated clays that were deposited in shallow water of the glacial lake (Harris et al., 1974). Beneath this are the dark, highly plastic, fat clays of the Brenna/Argusville Formations (Figure 2) that were deposited in deep water during higher lake levels (Harris et al., 1974). Both formations lie at or near the surface adjacent to the Red River. They are extremely plastic and exhibit little to no shear strength. Soil tests of the two formations yield high plasticity index and liquid limit results and shear strengths that are very weak (Table 1).

Figure 2. Typical cross-section of Red River Valley near Fargo, North Dakota. (From Schwert, 2003)

Table 1. Typical engineering classification results for Red River Valley clay soils.

Formation	Plasticity Index (%)	Liquid Limit (%)	Res. Friction Angle (deg)	Undrained Shear Strength (psf)

283

Sherack	38	62	27	1000
Upper Brenna	98	129	7	720
Lower Brenna/ Argusville	57	84	10	780

Data from: US ACE 2000.

Given the engineering properties of these clay soils, it is not surprising that they cause significant geotechnical problems, especially where they are unconfined. The natural process of channel migration leads to frequent slope failures along the Red River and it's tributaries as they flow though the valley soils. The two most common types of slope failure are rotational slumps and flow slumps (Figure 3), with creep and earth flows occurring to a lesser extent. As the stream banks are subjected to high water levels and drainage, it is even more probable that slope failures will occur.

a)   b)

Figure 3. a) rotational failure near Reynolds, ND, b) flow slump near Wild Rice, ND

Slump balancing
Intimately related to strength of the sediments is slump balancing. Rotational slumping results from gravity acting on a mass of material in such a manner that it rotates and moves down slope. To prevent or stop slope failure it is necessary to reduce the driving force by removing material from the top of the slump, and increase the resisting force by adding weight to the toe of the slump or by increasing the resistance along the failure plane(s). Reducing the slope by reshaping, adding fill or rock rip rap at the toe, and lowering the moisture content along the failure plane(s) accomplishes this.

Slope
The degree of slope is critical to the overall stability. A profile across a typical stream in the Red River Valley ranges from nearly flat on the uplands to vertical on cutbanks. Reducing slope grade is a viable approach to decrease the risk of slope failure. If the slope can be lowered the potential for slope failure drops. This may require as much as a 10:1 slope to ensure stability in extreme cases. A shallow slope can provide a planting site which is usually favorable for machine planting. It also provides more surface area than a vertical slope so that the shear stress generated by water flow is distributed over a greater area, subjecting the sediments to less shear stress. Deposition has occurred on surfaces which have been reconstructed from vertical (cut bank) to 3:1 under flood conditions.

Vegetation
Since the settlement of the Red River Valley in the mid to late 1800's, the riparian galley forests have been cleared to make use of the rich soils, for construction lumber, and as a source of fuel for steamboats. In recent years, expansion of the valley's urban centers have lead to the isolated removal of the remaining band of forest and native vegetation for residential and

284

commercial development. Removal of vegetation decreases shear strength of the bank materials and can increase soil moisture. Keeping root mass in place can provide slope "reinforcement" to reduce the risk of slope failure. Exposed roots along cut banks lower the flow velocity next to the bank. The shear stress on the sediments is effectively lowered. Reestablishment of riparian vegetation is a cost effective way to reduce slope failure.

Research Center)

Hydrology

Enhanced soil moisture, a critical factor of slope stability, is frequently the result development. Removal of the moisture-loving native plants and trees, installation of lawn sprinkler systems, and burying of septic drain fields may double or triple the amount of moisture these soils typically receive annually. Local hydrology is altered as homeowners seek to move water away from their houses and logically toward the river. However, water from gutters, sumps, and yard drains only helps to saturate the already wet slopes (Figure 4).

Figure 4. Slump flows caused by septic drain fields near Fargo, ND. (From UND Energy & Environmental

Regional hydrologic changes may be contributing to the problem as well. Changes in discharge and flooding have caused river channels to downcut and widen as they adjust to new flow regimes. The combined effect of the above factors has been a rapid increase in the number of slope failures across the Red River Valley. The current hydrology of the region tends to keep the Sherack and Upper Brenna Formation saturated, especially adjacent to waterways. Drainage of water from the unstable areas will tend to reduce the moisture content in the slope which increases stability. This can be accomplished by diverting surface flows away from the slope area in question. Rock filled trenches can serve to intercept groundwater and surface water and redirect the discharge; a method that has proven to be quite cost effective.

285

Slope Stabilization Case Study: The Grand Forks County Club

Efforts to stabilize a rotational failure along Cole Creek in Grand Forks County, North Dakota, make a good case study of the difficulties posed by soils and hydrology in the Red River Valley. Riparian Project staff and cooperators worked with the Grand Forks Country Club (GFCC) staff, two engineers, and three contractors to address a two acre slump that had damaged one of the club's golf cart bridges over the small creek. Two attempts were made during a three year period to stabilize the failure using a variety of techniques; some working better than others. The country club is located approximately two miles south of Grand Forks where Cole Creek confluences with the Red River. Cole Creek is a small stream draining nearly 300 square miles of agricultural land. The stream is intermittent in it's headwaters to perennial at it's mouth; flowing regularly in the spring and during summer rainfall events. It is impaired along much of it's length by a high sediment load, lack of riparian vegetation, low flows, and extreme summer water temperatures. When the course was built in 1963, trees, shrubs and native vegetation were removed. In addition to vegetation and land use changes, Cole Creek has been adjusting to increased discharge from a legal county drain that expanded the watershed by nearly a third. This change in hydrology caused the channel to downcut and become incised throughout most the GFCC. Initial assessment of the reach classified the channel as a F6c to a B6c (Rosgen 1996) depending on the entrenchment ratio in locations. Wet weather and backwater from Red River flooding in the 1990's saturated the unstable soils adjacent to the entrenched channel and triggered slope failures throughout the course.

As the 330 ft long rotational failure settled toward Cole Creek, it unearthed the wooden bridge pilings, narrowed the creek channel, and raised the channel bottom (Figure 5). The squeezing of the channel at this point was increasing flow velocities causing accelerated bank and channel scouring downstream (Figure 6). A survey of the slump area showed the channel banks to be as steep as 2:1 or greater and the slope grade to average 5:1 (Figure 7). A geotechnical report produced for the new golf cart bridge design suggested the rotational failure could be 100 ft wide and 45 feet deep (CPS 2000).

Figure 5. Looking upstream on Cole Creek at the GFCC Cart Bridge.

Figure 6. Looking downstream on Cole Creek at the scour erosion caused by the narrowed channel.

Typical efforts to stabilize this type of failure may call for keying significant quantities of rock riprap into the slope toe and channel bottom, balancing the forces causing the rotation. This method has been effective, but can be expensive, not as aesthetically pleasing, and may exacerbate downstream erosion problems. To address these concerns and balance the slump block, the Riparian Project engineer called for reshaping the existing slope to a 7 or 8:1 grade and removing an estimated 10,000 cubic yards of earth, thus weight, from the top of the slump

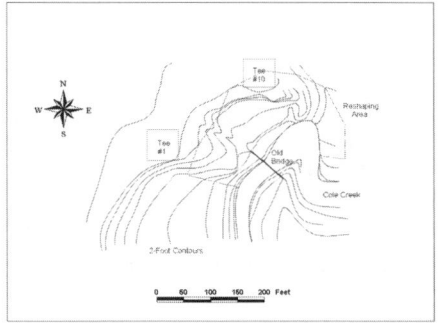
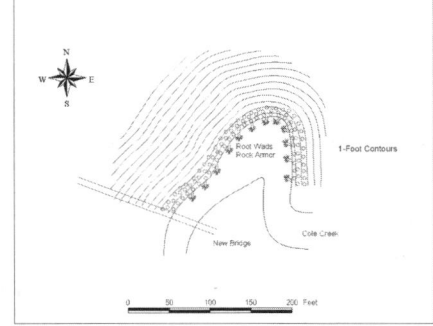

(Figure 8). Toe protection was also an
essential aspect of keeping the reshaped slope in place. Here, the engineer called for root
wads to be installed within a band of rock armor (105 cubic yards). It was expected that the root
wads would deflect energy away from the bank and the rock would add weight and protect the
toe during above-bankfull flows.

Figure 7. Topography of the slope failure   Figure 8. Design contour for stabilizing
prior to restoration.   slope failure.

Moisture management was another factor that was considered to arrest the slumping. Both
natural and human sources were contributing water to the slope. Flood waters and record
precipitation during the last decade combined with irrigation necessary for the tee box and
fairway above the slope, added weight and lubricated the slickensides. The Riparian Project
sought to solve this issue through irrigation management and the installation of deep-rooted,
moisture-wicking vegetation. The plan developed by a ND Forest Service Riparian Forester
called for 2400 dormant live sandbar willow stakes to be installed over a 7200 square foot area
(Figure 9). Below the willow stakes, 450 feet of live willow fascine would be installed for

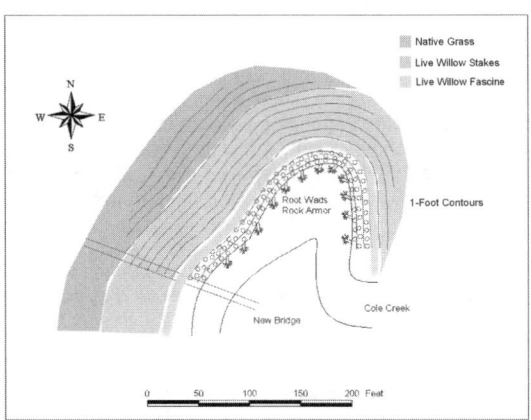

additional root mass and flow
energy deflection from the slope
toe. To maintain access to the
fairway, the top of the slope was
to be planted with a mix of deep
rooted native grasses including
switchgrass, buffalo grass, and
big and little bluestem. These
grasses would utilize more of the
existing soil moisture with their
larger and deeper root network,
and require little to no additional
watering.

Figure 9. Bioengineering and planting plan.

The design was not immediately accepted when presented to the GFCC Board of Directors. Both the earthmoving and revegetation plans indicated removing a portion of the women's tee box on hole #10. After negotiations, it was agreed that the tee box could stay, but that the final slope grading below the tee would be closer to 5:1 and therefore less stable. It was explained to the board that careful irrigation management or ceasing irrigation at the tee would be critical given the steeper final slope.

In March of 2001, the slope reshaping and rock and root wad installation was completed during the construction of the new golf cart bridge (Figure 10). The plant materials were installed in June and by August were growing vigorously (Figure 11). Total cost for the work came to

$33,400, with the excavation accounting for over half of the expense. With the exception of the native grasses being replaced with Kentucky bluegrass and irrigation continuing as before, the stabilization appeared to be holding.

Figure 10. Downstream view of Cole Creek in May of 2002 prior to planting.

Figure 11. Downstream view of Cole Creek in August of 2002.

However, by late fall, some minor slumping had occurred adjacent to the bridge where fairway drainage had not been diverted from the slope. The project engineer recommended repairs in the winter of 2002 that included stopping irrigation above the slope, diverting all drainage away, and replacing the Kentucky bluegrass with deep rooted vegetation such as alfalfa. He also recommended that rock or fill be placed at the toe of the recent slumping to balance the downward forces and additional rock be placed along the entire repaired reach. The weather experienced by the valley during the summer of 2002 prevented these repairs from being completed.

The Cole Creek watershed was struck by several storms in 2002; one dumped over 14 inches of rain in June of 2002. The torrents that flushed through the creek were followed by backwater flooding from the Red River. At four times during the summer, the repaired slump was inundated to the bridge deck. Complete sections of the rock toe had been washed away or slipped into the channel (Figure 12). Portions of the fascine and entire root wads had been pulled from the bank. Without toe protection and balancing weight the saturated slope began to move, rotating two feet and narrowing the channel by four feet. The force of the rotation had even bent the bridge piers set 80 feet into the clay.

Figure 12. Views of Cole Creek looking downstream of the cart bridge in September of 2002 (left) and March of 2003 (right). Arrows indicate the same rock in the channel.

Riparian Project staff and the project engineer met with the GFCC board in the winter of 2003 to discuss new alternatives to stabilize the failure. The engineer recommended the slope be graded to at least 8:1 and possibly 10:1, based on stable grades up and downstream of the site (Figure 13). It was also recommended that 325 cubic yards of rock riprap be placed over geotextile along the entire reach. The rock was to have a D_{50} of 10 inches and would be keyed in the toe, extending 12 to 15 feet up the slope (Figure 14). The wedge of rock would protect

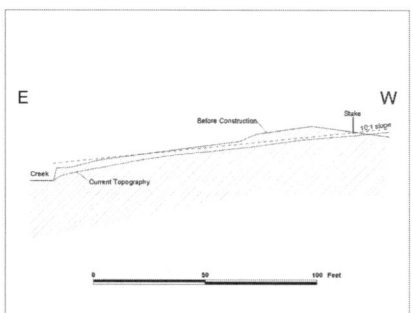

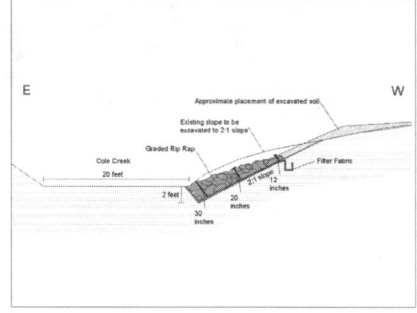

the toe to bankfull events and above and add counter-balancing weight.

Figure 13. Cross-section of the slope showing the slope grade before construction and during repairs. The planned 10:1 grade is shown for reference.

Figure 14. Typical cross-section of rock toe. Note that fill from the toe excavation will be placed directly above rock.

An aggressive revegetation and bioengineering plan was also developed for the site. To deflect energy from flood flows, increase the root mass at the toe, and improve the aesthetics, a live willow brush layer to be installed within the rock was planned for the section between the bridge and the meander. Moisture within the slump was addressed by installing three live willow pole drains among a dense planting of live willow stakes and rooted sandbar willow and false indigo conservation stock. It was expected that the willow drains would intercept surface and shallow through-flows and direct them away from the slump, while the deep rooted shrubs and trees would pull moisture from the clays. The plan again called for the upper portions of the slope to be planted with a deep rooted grass and forb mixture to include prairie cordgrass, Canada wildrye, and switchgrass. All species planned for the site were flood tolerant given the expectation of future flooding events.

The GFCC Board agreed to the plan, despite it calling for the removal of the women's tee box to

achieve a 10:1 slope. In fact, the Club carried the plan a step further by planning the installation of a six foot deep French Drain that would divert surface and subsurface water from the fairway to either side of the slope failure.

Construction began in early June and was completed by the end of the month (Figures 15 - 17). A nearly 10:1 grade was achieved by removing material from the top of the slope as well as placing fill above the rock toe. Total cost for the repairs amounted to nearly $26,000, bringing the grand total for the site to $59,400. A breakdown of the costs is shown in Table 2. Repairs to the bridge pier had been completed earlier by installing four 16 inch, concrete filled steel pipes 100 feet into the clay. Construction of the French Drain and the native grass seeding was completed during drier weather in August.

Figure 15. Aerial view of slope grading and earthwork to place rock toe.

Table 2. Costs of the slope failure repairs by practice.

Practice description	Cost
Earthwork for slope grading	$5,265
Installation of rock riprap at slope toe	$14,400
Live willow brush packing	$1250
Installation of willow stakes and rooted tree and shrub stock	$3464
Installation of live willow pole drains	$1200
Native grass seeding	$240
Total cost	$25,819

Figure 16. Rock toe being placed over geotextile. A layer of live willow branches is visible within the rock toe.

Figure 17. Looking upstream toward the cart bridge at the completed stabilization.

After approximately one year the repairs to the site appear to be functioning well. Despite a two to four inch rainfall in March 2004 and backwater flooding from the Red River, the vegetation is flourishing, channel erosion is limited, and the rotational slope failure is stable (Figure 18 a and b). Deposition was also observed on the rock toe at the apex of the meander.

a)   b)

Figure 18. Slope stabilization and bioengineering at the GFCC in May of 2004 looking a) upstream toward the new golf cart bridge and b) downstream into the meander.

Conclusion

Each slope failure must be evaluated with respect to all the critical factors that contribute to the failure. There may be multiple, alternative, cost-effective solutions to a given problem. All factors must be considered when choosing solutions including constraints placed on the project by landowners or sponsors and the possibility of extreme environmental conditions such as flooding. The GFCC case and others have taught the Riparian Project that there is a limit to how much a design can be compromised to meet the land owners objects, yet stand up to the tests of nature. The experience of the RRB Riparian Project is that a team approach is required to embrace all the factors and provide positive results. The solution that ultimately worked at the GFCC was a combination of slope adjustment, slump balancing, drainage, and revegetation. This was a "soft" soil bioengineering solution that was cost effective in a highly visible site at the GFCC. The Riparian Project will continue to apply the lessons from the GFCC and other restoration efforts to improve riparian restoration success in the Red River Basin and the region.

References

CPS, 2000. *Request for Proposal: Grand Forks Country Club Cart Bridge.* CPS, Ltd.,
    Grand Forks, North Dakota.

Harris, K. L., Moran, S. R., and Clayton, L., 1974. *Late Quaternary stratigraphic
    nomenclature, Red River Valley, North Dakota and Minnesota*: North Dakota Geological
    Survey Miscellaneous Series 52, 47 p.

Rosgen, D. L., 1996. *Applied River Morphology.* Wildland Hydrology. Pagosa,
    Springs, Colorado, 374 p.

Schwert, D. P., 2003. A geologist's perspective on the Red River of the North: history,
    geography, and planning/management issues. Proceedings: 1st International Water
    Conference, Red River Basin Institute, Moorhead, Minnesota.

US ACE, 2000. Design Documentation Report, Geology and Geotechnical Appendix,
    Alignment Refinement. United States Army Corps of Engineers, St. Paul District.

An Overview of the NRCS Stream Design Guide

Jon Fripp, Stream Mechanics Engineer, USDA-NRCS National Design, Construction, and Soil Mechanics Center, Ft. Worth, TX.
and
Kerry M. Robinson, Hydraulic Engineer, USDA-NRCS Watershed Science Institute, Raleigh, NC.
and
Jerry Bernard, National Geologist, USDA-NRCS Conservation Engineering Division, Washington, D.C.

The development of a new stream design handbook is presently underway. The Natural Resources Conservation Service (NRCS) is developing this guide as part of their National Engineering Handbook. Over 100 authors and reviewers have volunteered their time to produce the source materials for this design guide. These authors represent other Federal, State, and local agencies, universities, and private engineering firms, as well as, many NRCS employees. This pool of active volunteers represents all parts of the country, and their contributions will provide a collection of regionally sensitive design tools. The individual source documents will be synthesized into a comprehensive design handbook. A first draft of this document is scheduled for completion in 2005.

The goal of this design guide is to consolidate new and existing tools, techniques, and resources to support the design process associated with stream restoration and rehabilitation. The proposed guide is an "open-ended" document so additional methodologies can be added as they become available. Modifications will also be made to existing tools as experience in restoration brings new knowledge and insight. Formal training, preparation, and updating of supporting directives related to stream design are also anticipated.

The NRCS is actively involved in designing and implementing stream restoration projects all over the country, typically to improve habitat and stabilize the channel bed and banks. The proposed stream design handbook should be of interest to the numerous disciplines charged with putting those projects on the ground. The handbook will cover the full range of treatments from natural to structural.

NRCS STREAMBARBS, PAST AND PRESENT

Leland M. Saele

In the early 1990's, NRCS field staff in eastern Oregon began using low rock sills in stream restoration work. It was claimed these structures could redirect flow away from eroding banks and required much less rock than traditional rock riprapped banks. The structures were referred to as *"streambarbs"*. These structures offered an alternative to rock riprap (which had lost favor with state fisheries personnel), and NRCS field staff were enthusiastic because they seemed to work well with other biotechnical bank treatments.

There were no set design procedures or guidelines for installing streambarbs other than to use the largest rock available. A field evaluation in 1993 by NRCS, West Technical Center personnel resulted in the development of preliminary design guidelines for layout and installation of streambarbs. Since those first guidelines were issued there have been numerous installations of these structures in many states. Field and empirical observations have resulted in changes to the original guidelines and improvements continue to the present time. We have a much better understanding of where these structures will perform satisfactorily but there is still considerable argument as to exactly how they work. Current guidelines are much more complex than the single page initially provided. Additional design recommendations are being developed through the utilization of flow modeling results by NRCS in Oregon.

Streambarbs have been well received, and it is apparent these structures will continue to be a valuable tool for streambank restoration projects in NRCS. This paper will provide a history of NRCS experience with streambarbs over the last 10 years and describe development of our current guidelines.

Ecological Considerations for Stream Projects

K. L. Boyer. USDA Natural Resources Conservation Service, Wildlife Habitat Management Institute, Department of Fisheries and Wildlife,104 Nash Hall, Oregon State University, Corvallis, OR 97330

ABSTRACT

The NRCS Stream Design Guide provides guidance for teams of engineers, biologists, geomorphologists, hydrologists, landowners, and resource managers who are planning and designing projects intended to improve streams, and how they function. Specific stream project goals may include controlling floods or sediment sources, hastening drainage, stabilizing banks, improving fish habitat, or restoring the ecological functions and processes of a stream and its floodplain. Many approaches and techniques can be used to reach these goals, but recognition of the living and nonliving components of the stream ecosystem, its watershed, how they interact and affect each other, and the timeframes over which stream processes occur will improve the probability of desirable outcomes. The chemical and biological processes that occur within stream corridors, and between them and adjacent lands, are intricate and involve numerous linkages and feedback loops. Accordingly, this paper provides a brief overview of current knowledge regarding stream ecosystem processes and functions important to consider when designing stream improvements. Topics to be presented will include (1) ecological responses to movement of water and materials, longitudinal and lateral adjustment of channels, and floodplain alterations; (2) key ecological processes critical to aquatic community dynamics; (3) types of stream corridor habitats; and (4) the relative importance of disturbance in sustaining aquatic communities.

KEYWORDS. Aquatic, flooding, restoration, stream corridor, streams, riparian, rivers.

INTRODUCTION

Natural or minimally-altered stream corridors tend to be physically heterogeneous regardless of their size. Fluxes of energy, water and materials that occur throughout the stream corridor system create a dynamic and complex three-dimensional (length, width, depth) mosaic of different habitats and physical features, changing through time and contributing to the high level of biological diversity typical of stream corridors. The interactions occurring among the different elements of stream corridors are extensive and reciprocally obligatory for many of the plant and animal species that inhabit or use them.

Human activities in stream corridors often simplify physical structure (for example, by removing riparian vegetation) or fragment connections among their components (such as that between the stream and its floodplain), preventing or limiting natural processes important to many species. Projects designed to restore or maintain the ecological linkages of stream corridors, and their physical connections, are promising approaches for arresting the decline of aquatic and riparian species, and improving water quality.

However, the complex physical, biological, and social nature of stream corridors creates a challenge to professionals responsible for improving stream functions and conditions. Systems with more intact natural flow regimes are very complex. From a human perspective, they are also less predictable and potentially risky. Recent studies in stream ecology emphasize the importance of linkages between stream channels, riparian areas, and floodplains (Gregory et al. 1991, Stanford and Ward 1992, Brookes et al. 1996, Huggenberger et al. 1998, Molles et al. 1998). Stream corridor projects that integrate the disciplines of fluvial geomorphology, hydrology, aquatic and riparian ecology, and hydraulic and geotechnical engineering are more effective at meeting multiple objectives that address economic and ecological considerations.

1. Watersheds

A watershed captures precipitation, filters and stores water, and regulates its release through a channel network into a lake, another watershed, or an estuary and the ocean. Watersheds are characterized by different soil types, geomorphic features, vegetation and land uses. Their upland features control the quantity and timing of water and materials that make their way downstream. The environmental conditions of a stream or river corridor (such as water quantity and quality, riparian function and fish habitat) are thus linked to the entire watershed. And these linkages go both ways. Stream project designers may have little control of watershed management. Still, plans should consider past, present and future trends of watershed land use and conditions.

Landscape consideration of watersheds: spatial scales. The spatial structure of landscapes influences ecological and physical processes such as energy flow, material transport, and species dispersal within them. The spatial and temporal scales of a habitat are not fixed but rather determined by the physical and biological processes that create it, the home range of the organism, its interactions with the biotic community, and the population dynamics of the species. The habitat of a large organism or a relatively mobile one (e.g., Pacific salmon) contains within its physical boundaries the smaller scale habitats of smaller, or less mobile, organisms (e.g., aquatic insects and crayfish). How long it takes for these multiple habitats and their species to respond to stream projects depends on the project's nature and scope, and the attributes of its watershed.

A hierarchical approach to stream design identifies the main spatial scale at which each ecosystem component influences the characteristics of the stream, but it does not imply that components at lower hierarchical levels are less important than those at higher levels. In fact, the connectivity of the stream environment involves feedback mechanisms by which smaller scale components may influence larger scale patterns and processes (DeAngelis et al. 1986). Therefore, an ecologically effective stream project plan should consider factors affecting stream corridor processes at different spatial scales, from landscape to watershed to microhabitat. The plan should also consider factors that influence long-term population dynamics of aquatic species and the community of biota and physical features with which they interact.

2. Key Ecological Processes Affecting Stream Corridors

Energy flow and nutrient cycling. The flow of energy and nutrient dynamics in aquatic ecosystems occur in all dimensions, and are influenced greatly by the physical dimensions of the stream channel. In turn, these processes strongly influence the community structure of stream ecosystems and the ecological processes along their longitudinal network. In small headwater streams, channels are narrow and shallow. In forested landscapes, this means that the inputs of solar radiation to the channel are generally small and inputs of organic matter from the terrestrial ecosystem are relatively large. As streams flow downstream, channels become wider and deeper. Riparian canopy over a stream has less influence as the stream widens, and solar radiation increases, causing increased production of algae and vascular aquatic plants. Change in the longitudinal gradient of streams is also the primary factor driving hyporheic exchange flows (Harvey and Bencala 1993) which creates unique physical, chemical, and hydrologic conditions in streams and riparian zones. Considerable research over the last several decades has elucidated the importance of hyporheic zones to many alluvial stream corridor systems. Their functions include (1) moderation of stream temperature by upwelling, (2) water retention and storage which reduces peak flows and sustains baseflows during dry periods; (3) habitat creation, especially for crustaceans, and larval fishes; (4) buffering and filtering nutrients from stream flows and groundwater; and (5) aquifer recharge and nutrient enrichment.

Lateral exchange of water between a river and its floodplain is the driving force for nutrient cycling and dynamics of the floodplain biotic community. Primary productivity of floodplain habitats is closely tied to hydro-period. It is greatest in wetlands with pulsed flooding (i.e., periodic inundation and drying) and high nutrient input, and lower in drained or permanently flooded conditions and oligotrophic water. Riparian wetlands may also influence stream channel morphology and flows, buffering the stream channel against the physical effects of high flows by dissipating energy as waters spread out onto the floodplain. Alternately, as stream flows recede, riparian wetlands provide water storage, slowly releasing water back to the stream through subsurface transport, thereby influencing stream base flows.

Recruitment of large wood. Wood recruited to the stream from riparian forests provides structure and organic matter that create and enhance habitat diversity and food sources for many riparian and aquatic organisms (Boyer et al. 2003). Wood in streams also increases channel roughness and habitat complexity, triggers formation of islands, and forms dams that trap leaves, twigs and fine sediments. Fine particulate organic matter (particles smaller than 1 mm in diameter) retained by large wood pieces provides food for insects and other aquatic invertebrates. Wood is important from headwater streams to large rivers and estuaries (Maser and Sedell 1994). Small headwater streams with normal levels of wood input often contain a series of step pools formed by fallen logs that cross the channel and trap smaller pieces of trees and shrubs. At the other end of the spectrum, records exist of large rivers completely blocked by natural accumulations or "rafts" of large wood that dominated stream corridor processes (Triska 1984). Natural channel widening and bar formation associated with wood obstructions allow development of the short, braided reaches and secondary channels that are important spawning grounds for salmon and trout in the rivers of the Pacific Northwest, and smaller non-game fishes such as dace and minnows in all parts of North America where riparian gallery forests occur. In the sand-bed coastal plain rivers of the southeast United States, wood provides important habitat for invertebrates and thus provides fish with a source of food. In most if not all rivers in the US and Europe, fish abundance and diversity depend to some extent on accumulations of large wood. Wood recruitment processes are complex since they involve site-specific variables (e.g., size, species, density, and condition of riparian trees, bank geometry and erosion) and stochastic events (e.g., tree death, tree blow-down, high flows, bank failures) (Gregory et al. 2003). Removal of wood is perhaps the most widely practiced type of stream channel management, and the practice of removal ("desnagging" or "clearing and snagging") along with deforestation and removal of beaver have left many streams with only a trace of the large wood that existed previously.

3. Aquatic Habitats of Stream Corridors

Stream habitats. The dynamic nature of stream corridors and their response to floods and other disturbances create many diverse habitat types and conditions. These habitats and the processes that occur among them affect each other dramatically, adding to the complexity of the habitats and species interactions in the stream and riparian area. Thus stream corridors support a disproportionately rich biological community, relative to the rest of the landscape. Confounding this ecological richness, however, are challenges that river and stream processes (such as flooding) present to humans. Add to these the many human demands on streams as water supplies and important agricultural, recreational, and urban development sites, and managers are often compelled to take actions that compromise the ability of watersheds to sustain key ecological functions of habitats.

Stream corridors provide filtering, buffering, retention and conduit functions for water, sediment, wood, chemical compounds, seeds, and habitat for aquatic and riparian organisms. Maintaining multi-dimensional connectivity along a stream corridor is important to maintaining the species, habitats, and ecological services within them. Quality stream habitats are a mosaic of spatial diversity created by various combinations of water quality and quantity, water depth, velocity, large wood, riparian vegetation, and the organisms that inhabit stream corridors. Generally speaking, aquatic organisms need what most higher organisms need to survive: clean water, oxygen, a steady food source, a place to hide or find refuge, and a place to successfully reproduce and grow to adulthood. Some aquatic organisms such as microscopic zooplankton live almost entirely in the water column; others such as fish utilize the water column and bottom substrates. Still others rely on the interstitial spaces of hyporheic habitats below a stream's substrate.

Most species use a variety of habitats or habitat conditions during the course of their life histories, some moving upstream or downstream, others into and out of the floodplain, a few into or out of the substrate, and still others to and from the ocean, all depending on the season, their age and physiology, and the conditions they face in their habitats. The complexity of their life cycles requires comparable complexity in their habitats and connections among them to allow movement at the time they need to move. To sustain aquatic communities, stream project designers should consider the life history needs of aquatic organisms and the physical and ecological processes that provide them.

Riparian and floodplain habitats. A stream corridor is comprised of the stream channel and its riparian zone. The riparian zone forms an ecotone or transitional zone between the stream and uplands and provides value, both in productivity and biotic diversity, far greater than its relatively small area would indicate. Riparian zones may or may not include floodplains, depending on the valley form of the stream corridor. In relatively wide stream corridors, floodplains are prominent components of the riparian zone. Whereas stream channels have often been the focus of stream restoration projects over the past few decades, project designers today recognize the links between the stream channel and its riparian areas and floodplain or riparian wetlands. Thus projects that consider these linkages are becoming more common (Middleton 2002). Floodplain or riparian wetlands, which include swamps, oxbows, sloughs, ponds, backwaters, abandoned channels, and floodplain lakes, usually are remnants of historic river channels or shallower depressions created by scouring and sediment delivery associated with flooding. During overbank flows these wetland habitats are connected to the river by surface water. As water recedes, some is trapped and forms seasonal, isolated wetlands varying in size, shape, and significance for aquatic species. Low gradient streams and rivers generally have broad riparian zones and floodplains characterized by predictable hydro-periods and extensive riparian wetland systems. In these lower systems, floodplain wetlands can contribute significantly to stream ecosystem productivity and function.

The hydrological characteristics of wetlands vary from permanently flooded backwaters, to sheetwater flowing wetlands during floods, to ephemeral isolated pools. In lower gradient streams, a range of taxa including plants, invertebrates, and vertebrates have evolved life history strategies that depend on floodplain wetlands. Some macro-invertebrates complete their entire life-cycle in these habitats, persisting in seasonal wetlands in a drought resistant form such as an egg. Vertebrates (fish, amphibians, mammals, and birds) frequently make seasonal movements into floodplain wetlands (from the stream, wetlands outside the floodplain, or surrounding uplands) and time key periods of their life cycle (breeding, rearing young or migration) to flooding. Simply returning water to a floodplain is inadequate for reestablishing its function for all organisms; each may be dependent on specific timing, duration, or frequency of flooding.

Just as riparian wetlands can influence stream function, changes in stream channel morphology can influence floodplain wetland function. Riparian wetlands are often filled or isolated from the stream by levees, channel incision, or channel straightening. Physical isolation changes hydroperiod and precludes access to the floodplain by many stream obligate organisms (e.g., fish). Channelization frequently results in channel incision that will change the frequency of over-bank flows and therefore the hydroperiod of floodplain wetlands. Channel stabilization precludes avulsive processes that can form new floodplain wetlands and create complex mosaics of habitats. Stream restoration projects that facilitate stream-floodplain connectivity will improve stream functions and benefit the species that depend on them.

4. Disturbance and Response in Aquatic Ecosystems

Ecological effects of floods on stream ecosystems. In streams, floods are the most common type of natural disturbance (Resh et al. 1988). These events erode, flush, transport and deposit sediments, deliver riparian trees into the channel, transport large wood already in the channel, and transport nutrient and organic matter into streams from surrounding terrestrial ecosystems . The effects of disturbances on stream ecosystems have been reviewed extensively (Ward and Stanford 1983, Yount and Niemi 1990). Numerous studies have shown that aquatic organisms evolved to not only withstand the potential impacts of floods but also benefit from these events (Meffe 1984, Bayley 1991).

Aquatic organisms differ greatly in their life histories and their vulnerability to and recovery from disturbances (Resh et al. 1988, Yount and Niemi 1990). A review of field studies of responses to flooding reveal that, in general, algae and microbes recover in days to week, macroinvertebrates recover in less than a year, and fish recover in 1-2 years, with a few species requiring decades (Yount and Niemi 1990). The condition of the ecosystem and riparian corridors are critical factors in determining resistance to the disturbance and the subsequent rate of recovery. Refugia from disturbances are important factors in recovery and the design of stream restoration projects (Sedell et al. 1990). Floodplain rivers are larger and more complex than small streams, but the enormous power and frequency of flooding create natural processes for restoring large rivers and their floodplains (Sparks et al. 1990, Bayley 1991).

Disturbance as a restoration tool. Disturbances, such as floods, fire, and droughts, are natural processes of restoration. Design of restoration project or stream management should consider the frequency and location of disturbance events and make certain that beneficial effects of floods and other disturbances are not negated by the rush to harden stream banks, prevent channel change, and remove habitat features that provide complexity and heterogeneity (e.g., large wood, gravel bars, islands, and sloughs). In many systems, past projects that were designed to minimize the effects of disturbances (e.g., levees, rock revetments, tide gates) are being removed to restore streams, rivers, and estuaries (CALFED 2003). Restoration projects also should consider natural processes of riparian regeneration (Boyer et al. 2003). River channels may reoccupy old or abandoned side channels if revetments and other barriers are removed. Careful design and analysis of stream projects can attain a balance between taking advantage of the restorative processes of natural disturbances and the need to protect property and communities from these natural processes along the banks and floodplains of streams and rivers.

CONCLUSIONS

Fluvial systems are dynamic, changing over time and in space in response to their hydrology and geomorphology, and the interactions of these physical processes with biotic communities (bacteria, plants, animals). To protect species, habitats, and water resources managers recognize the need to incorporate environmental features into stream project designs (Shields et al. 2003). Historically, engineered solutions to stream channel "problems" featured constrained physical design systems. Today, resource managers and stream design engineers are seeking ways to modify tried-and-true designs in order to allow less constrained processes to occur. The following principles of stream engineering design incorporate ecological considerations to facilitate such modifications:

    1. *Base designs on ecological principles, as well as physical ones.* To the extent possible, restore or maintain the inherent complexities of stream corridors, ecological linkages, and their physical connections. For example: (1) incorporate native vegetation into design of flood control structures, revetments, levees, and other hard structures; (2) consider riparian restoration actions that incorporate silvicultural treatments to maximize generation of trees specifically for large wood recruitment; (3) incorporate livestock or recreational management regimes into stream design projects to protect restoration or conservation investments in riparian zones and sustain their functions; (4) remove hard structures no longer deemed necessary or functional in the watershed, due to changes in the physical and ecological conditions; (5) restore natural hydrologic regimes to a stream to the extent possible.

2. *During the design process, integrate the disciplines of fluvial geomorphology, hydrology, aquatic and riparian ecology, and hydraulic and geotechnical en*gineering. If possible, collect baseline data and post-implementation data to expedite learning and validate successful designs of innovative and new approaches to stream corridor restoration that can be used by other designers in the future.

3. *Design for site-specific response in a watershed-scale context.* Consider factors affecting stream corridor processes at different spatial scales, from landscape to watershed to microhabitat, and those that influence the long-term population status and dynamics of aquatic species and the community of species they interact with. Seek technical information about aquatic species from local university or agency biologists.

4. *Consider ecological costs and values as well as project costs and long-term costs for maintenance of engineered solutions to channel "problems".* Projects that are compatible with inherent tendencies of stream corridor systems tend to be more stable, require less maintenance, and are more ecologically sound than traditional engineered approaches; these advantages should be highlighted when evaluating design options.

Acknowledgements

I thank the colleagues who contributed to the NRCS Stream Design Guide: Bruce Dugger, Guillermo Giannico, and Stan Gregory (all of Oregon State University), and co-author Douglas Shields (USDA Agricultural Research Service).

REFERENCES

1. Bayley, P. B. 1991. The flood pulse advantage and the restoration of river-floodplain systems. *Regulated Rivers: Research and Management* 6: 75-86.

2. Boyer, K. L., D. R. Berg, and S. V. Gregory. 2003. Riparian management for wood in rivers. Pages 407-420 *in* S. V. Gregory, K. L. Boyer and A. Gurnell, editors. *The Ecology and Management of Wood in World Rivers.* American Fisheries Society, Bethesda, MD.

3. Brookes, A., J. Baker, and C. Redmond. 1996. Floodplain restoration and riparian zone management. Pages 201-228 *in* A. Brookes and J. F. Douglas Shields, eds. *River Channel Restoration: Guiding Principles for Sustainable Projects.* John Wiley and Sons, Ltd., Chichester, UK

4. CALFED.2003. Reviving Central Valley Rivers. Available at: http://science.calwater.ca.gov/pdf/Sci-in-Act-07-2003ff.pdf

5. DeAngelis, D.L., W. M. Post, and C. C. Travis. 1986. *Positive feedback in natural systems.* Springer-Verlag, New York, N.Y. 290 p.

6. Gregory, S.V., F.J. Swanson, A. McKee, and K.W. Cummins. 1991. Ecosystem perspectives of riparian zones. *BioScience* 41:540-551.

7. Gregory, S. V., K. L. Boyer, and A. Gurnell. 2003. *The Ecology and Management of Wood in World Rivers.* American Fisheries Society, Bethesda, MD.

8. Harvey, J. W., and K. E. Bencala. 1993. The effect of streambed topography on surface-subsurface water exchange in mountain catchments. *Water Resources Research* 29:89-98.

9. Huggenberger, P., E Hoehn, R. Beschta, and W. Woessner. 1998. Abiotic aspects of channels and floodplains in riparian ecology. *Freshwater Biology* 40: 407-425.

10. Maser, C. and J. R. Sedell. 1994. *From the Forest to the Sea: The Ecology of Wood in Streams, Rivers, Estuaries, and Oceans.* St. Lucie Press, Delray Beach, FL.

11. Meffe, G. K. 1984. Effects of abiotic disturbance on coexistence of predator-prey fish species. *Ecology* 65:1525-1534.

12. Middleton, B. A. 2002. *Flood Pulsing in Wetlands: Restoring the Natural Hydrological Balance*. John Wiley and Sons, New York, NY.

13. Molles, M. C., C. S. Crawford, L. M. Ellis, H. M. Valett, and C. N. Dahm. 1998. Managed flooding for riparian ecosystem restoration. *BioScience* 48: 749-756.

14. Resh et. al. 1988. The role of disturbance in stream ecology. JNABS 7:433-455

15. Sedell, J.R., G.H. Reeves, F.R. Hauer, J.A. Stanford and C.P. Hawkins. 1990. Role of refugia in recovery from disturbances: modern fragmented and disconnected river systems. *Environmental Management* 14:711-724

16. Shields, F. D., Jr., Copeland, R. R., Klingeman, P. C., Doyle, M. W., and Simon, A. 2003. Design for stream restoration. Accepted by *Journal of Hydraulic Engineering*.

17. Sparks, R. E., Bayley, P. B., Kohler, S. L., and Osborne, L. L. 1990. Disturbance and recovery of large floodplain rivers. *Environmental Management* 14:699-709

18. Stanford, J. A. and Ward, J. V. 1992. Management of aquatic resources in large catchments: recognizing interactions between ecosystem connectivity and environmental disturbance. Pages 91-124 *in* R. J. Naiman, editor. *Watershed Management.* Springer-Verlag, New York, NY.

19. Triska, F. J. 1984. Role of wood debris in modifying channel geomorphology and riparian areas of a large lowland river under pristine conditions: a historical case study. *Verhandlungen. Internationale Vereinigung Fur Theoretische Und Angewandte Limnologie* 22(3):1876-1892.

20. Ward, J.V. and J.A. Stanford. 1983. The intermediate disturbance hypothesis: an explanation for biotic diversity patterns in lotic ecosystems. Pages 347-356 *in* T.D. Fontaine and S.M. Bartell, editors. *Dynamics of lotic ecosystems.* Ann Arbor Press, Ann Arbor, MI, USA.

21. Yount, J. D., and G. J. Niemi. 1990. Recovery of lotic communities and ecosystems from disturbance--a narrative review of case studies. *Environmental Management* 14:547-569.

STREAM STABILITY INVENTORY AND EVALUATION

Wayne S. Kinney[1] and Ruth Book[2]

ABSTRACT

Inventory and evaluation of streambank erosion requires an understanding of the cause of the perceived problems. Sometimes, causes of instability are visible on-site, but many times it is necessary to consider activities in other reaches of the stream or in the watershed. Also, the problem may not be instability at all, but rather a naturally occurring process that is incompatible with the existing land use. This paper reviews the concepts of stream stability and equilibrium, along with a Channel Evolution Model (CEM), as background material. It then presents a detailed procedure for data collection and analysis to facilitate the understanding of the dynamics of a subject stream and to prepare for the design of measures to correct the problems. Published data and field-collected measurements are analyzed and compared; when all valid data match closely, the level of confidence in the analysis is high, and assessment of the situation can proceed. The suggested procedure relies heavily on a spreadsheet tool developed by Illinois NRCS to collect and compare all available relevant data, but the same analysis can be successfully accomplished without the software. Finally, parameters developed from the collected data are used in a "departure analysis" to categorize existing conditions and identify the current CEM stage, particularly applied to typical stream conditions in Illinois.

KEYWORDS. Streambank, stream stability, morphology, channel evolution model, bankfull.

INTRODUCTION

Causes of channel and bank instability can be broadly grouped into four areas of common causes: downstream, upstream, watershed-wide factors and local factors. **Downstream factors** involve lowering of the downstream base water level, which can significantly impact upstream reaches. **Upstream factors** alter the incoming discharge of water and/or sediment by installation of features such as dams and diversion channels. **Watershed-wide factors** are the result of major land-use changes; in northeastern Illinois, typical land use change would be urbanization. **Local factors** include causes such as geotechnical failures, sparse riparian vegetation and unstable planform. These local causes may be exacerbated by upstream, downstream or watershed-wide factors or they may be the primary cause.

One common misconception often encountered is the assumption that a "stable" stream should not erode its banks. The truth is that "stable" streams are not "static"; they simply migrate slowly. The difference between stable and unstable is not always distinct, as streams in dynamic equilibrium will continually migrate slowly across their floodplains. The distinction is in the rate of lateral migration being so slow in stable streams that the riparian zone remains essentially intact through the entire process. Stable streams should, however, remain essentially "static" in relation to their overall profile, i.e. they will not exhibit any degradation or aggradation.

Hundreds of years of human activity on the landscape have made significant changes in the major elements controlling stream balance. We have cleared the timber, plowed the prairie,

---

[1] Wayne S. Kinney, Streambank Specialist, Illinois Department of Agriculture. 14 Rock Hill Ct., Edwardsville, Illinois 62025, email streamdoc@charter.net.

[2] Ruth Book, PE, Ph.D., Agricultural Engineer, United States Department of Agriculture, Natural Resources Conservation Service. 2110 W. Park Ct., Champaign, Illinois 61821, email Ruth.Book@il.usda.gov.

drained the wetlands, straightened the streams, levied the floodplains, and built cities with large areas of concrete, asphalt and rooftops. Results of such activity on stream dynamics have generally had the effect of increasing runoff and stream slope and reducing flood plain width. In many watersheds, the land use changes are a significant factor leading to increased runoff. In rural areas, this may be due to more intense agriculture replacing woodland and grassland with cultivated land. In urban areas, the increase of impermeable surface within the watershed results in an increased volume of water. Additionally, the urban development of a watershed typically results in permanent land cover, either in impermeable surfaces or lawns, which produces little sediment to be delivered to the system.

Lane's Balance (FISRWG, 1998) is a tool for understanding the relationship between factors affecting channel configuration. Stability is represented when the scale is "balanced": an equilibrium condition. Instability, an "unbalanced" condition, can be represented either by degradation or by aggradation in the stream. For example, both increased runoff from impervious areas and reduced sediment loads will tend to tip Lane's Balance to channel degradation in the stream system. Increased runoff represents higher energy in the stream flow, and reduced sediment load means less work for that energy to do. The resulting excess stream flow energy is dissipated by eroding the streambanks or scouring out the bed of the channel (degradation), providing more sediment and bringing the system to a new equilibrium.

Channel modifications nearly always contribute to channel instability at some point. Some of the more obvious modifications are channelization, dam construction and levees. Some less obvious, but still significant changes include clearing and snagging, gravel mining and channel lining or paving. The changes induced can be dramatic, but more typically they appear rather insignificant to the casual observer, especially in the short-term. Time then becomes a significant element to consider in the problem identification phase, as the "lag" time between channel or watershed changes and the full effects of those changes can be decades. Because the negative impacts of channel modifications are cumulative over time, it is often difficult to identify a single modification that is responsible for an adverse condition.

The designer's most important task is therefore to be aware of the overall condition of the stream and identify trends toward or away from an identified equilibrium or "balanced" condition. Only then can alternatives be considered. The objective of this paper is to outline a method developed by Illinois NRCS for inventorying and evaluating stream conditions. A spreadsheet program, designed to assist in gathering and analyzing the data required for inventory and evaluation (I&E) of an Illinois stream segment, will be presented as a part of the suggested investigative procedure. Some of the data and analysis is very specific to Illinois: gage data and regression curves. If the spreadsheet is used outside of Illinois, the reference stream gage section and the USGS flood-peak discharge prediction section will not apply, although the user is encouraged to seek these data from local sources. The spreadsheet program can be found in its most current form on the Illinois NRCS website:
http://www.il.nrcs.usda.gov/technical/engineer/engsprdshts.html.

BACKGROUND

The inventory and analysis procedure presented in this discussion relies on an understanding of several terms and concepts that are widely used in the field. Especially pertinent are those dealing with the geomorphology of a site: what processes caused the existing conditions, and specific geometric relationships between measurable parameters.

Bankfull Discharge

A key concept in stream morphology is "bankfull" discharge (Rosgen, 1996). The bankfull parameter is a standard often used to define the stream and compare it to others. In this way, the results of existing studies on other streams can be applied to help solve problems on the subject channel. Bankfull flows are those that transport most of the sediment in the stream, thus forming the size and shape of the channel, and typically correspond to a storm return interval of 1 to 2 years (FISRWG, 1998). The 1½ year return interval is often assumed to be the approximate bankfull discharge, although in urbanized areas, it is typically much nearer to the 1 year return interval. These "channel forming" or "bankfull" dimensions represent the "stable" channel dimensions when equilibrium is achieved. Much of the inventory and evaluation procedure for streambank work is devoted to the identification of bankfull discharge and its associated channel dimensions.

While "bankfull" is important, it should be noted that it is not universally accepted as the best measure of stability. Referring back to Lane's Balance, the concept of stream stability is a balance of just the right energy level to move both water and sediment generated from the watershed with no gain or loss of energy over time. This stability can also be measured by analyzing existing channel processes, bank characteristics, sediment characteristics, channel hydraulics, and the like. In addition, recently degraded or rapidly degrading streams will not exhibit reliable bankfull indicators. However, the bankfull concept will be used in this Inventory and Evaluation procedure since it can normally be identified and measured in the field with minimal equipment and time.

Channel Evolution Model

Another aid to identifying the processes in a stream is a Channel Evolution Model (CEM). The Illinois spreadsheet tool uses a widely accepted model developed by Simon (1989). The Simon model identifies six cross-sectional stages through which a stream progresses when subjected to destabilizing influences such as the urbanization described above. Class I is the natural channel before modification; Class II represents the stream channel morphology directly after human activity such as channel straightening. Class III is then the first sign of an instability problem, with evidence of downcutting or degradation in the channel bottom. As the bottom of the channel changes elevation, support for the banks is removed and the streambanks slump, creating a widening channel shape (Class IV). At some point, a new equilibrium is being approached. The sediments from the slumped banks begin to form new, vegetated floodplains at a lower elevation (Class V) and a smaller, "natural" channel within the new banks. The new stream equilibrium (Class VI) has abandoned the former floodplain and created a new one at the lower elevation (FISRWG, 1998).

Typical streams will exhibit several of the classes defined in the CEM, depending on the location in the stream relative to the disturbance. The Simon model describes a "nickpoint": the head of an active erosion event in the stream channel, working its way upstream. Class I describes the state of the stream well above the nickpoint where the effects of the disturbance are not yet in evidence. Progressing downstream, one can see the primary nickpoint (Class III), and varying stages of bank instability in the wake of the nickpoint (Classes IV and V). If enough time has passed since the disturbance, conditions farther downstream will approach Class VI.

Geomorphic Values

A number of standard geomorphic parameters have been developed to describe the hydraulic geometry relationships in a stream. These include:

- Sinuosity, a ratio of stream length along the low flow channel to the straight-line valley length between the same two points.

- Width/depth (W/D) ratio at bankfull conditions, using a mean stream depth.

- Radius of curvature (Rc), the radius of an arc that describes the centerline of a typical stream meander bend, as identified by IEPA, 1998.

- Entrenchment ratio, a comparison of the width of stream flow during flood stage (defined at twice the mean bankfull depth) to the channel-forming, bankfull width.

There is a natural variability to these hydraulic geometry relationships, and it is important to recognize that this represents a valid range of stable channel dimensions due to such variables as geology, vegetation, land use, sediment load and grain size, and hydrology. The parameter values

305

suggested in the following procedure are based on measured observations from streams in Illinois as well as published ranges from research done elsewhere. Values for these relationships can be used as a preliminary guide to stability in stream reaches, but other techniques and local data should be considered.

DATA COLLECTION PRIOR TO FIELD VISIT

The first step in the investigation phase is to gather existing data for the stream. The information gathered will make the initial field visit much more productive. The overall objective is to compare data from as many sources as possible to develop confidence in the analysis. If the subject stream is not in Illinois, use other data as available, since several of the steps listed in this procedure are specific to Illinois streams.

- Compare recent aerial photography with older photos to determine changes in channel alignment, lateral migration rates, changes in the channel width over time, and changes in the bed features such as central bars and size of point bars. If the channel top width has gotten larger, it could be a sign of past downcutting, or excessive bedload causing aggradation.

- Identify scour patterns in the floodplain and any existing levees on the aerial photograph.

- Calculate channel sinuosity. The sinuosity of the local stream site is best determined from a recent aerial photo, with reference to a topographic map. Identify the points where contour lines immediately upstream and downstream of the project site cross the stream channel. The spreadsheet prompts the user to enter the meandering stream length and the straight line valley length between two topographic lines, to determine the sinuosity. This information is used along with the contour interval to estimate a channel slope, as well.

- From USGS topographic maps (or other suitable maps), determine the watershed boundaries of the stream reach. Calculate drainage area (if available, published data from a nearby gage can be used to help determine the drainage area).

- Regional curve "bankfull dimensions" are supplied by the spreadsheet according to the drainage area, based on work by Dunne and Leopold in 1978 (Rosgen, 1996). The data are based on relationships found in the Eastern U.S. and do not come from Illinois or from urban streams. Curve "B" bankfull widths and depths correlate reasonably well with observations of several hundred rural streams in Illinois, but should be used very cautiously (if at all) in an urban setting. Development of regional curve "bankfull dimensions" for streams in the subject hydro-physiographic area should be pursued for best results.

- Look for reference stream flow gaging data. USGS and some state and local governments may own or operate gaging equipment on the stream you are investigating (the USGS data is available online). If not, then look for the nearest gage data available in a watershed with similar soils, climate and land use to the one you are investigating. The spreadsheet has a "pull down" menu of USGS gage data in and near the selected Illinois county. The 2 yr. return interval maximum discharge, Q_2, calculated from the actual gage data is displayed for the selected gage along with the station number and its drainage area. Results of the U.S. Geological Survey regression analysis (USGS, 1987) are also displayed if available.

- To determine the USGS Flood-Peak Discharge predictions for the subject Illinois stream, the spreadsheet needs a value for "valley slope" (USGS, 1987). If desired, the user may enter topographic contour elevations and corresponding distances along the flow line of the channel, and the spreadsheet will determine the valley slope as defined by USGS. "Rainfall" and "Regional Factor" are automatically supplied from a look-up table based on the county selection, and the predicted Q_2 discharge from the regression equation will

306

be displayed, along with a typical range for "bankfull" (40% to 80% of the Q_2 discharge, corresponding to the approximate 1 to 1½-year return interval storm events commonly representing bankfull flow in Illinois).

FIELD DATA COLLECTION

With the background data gathered and an understanding of the perceived problems and risks, the designer is ready to make a field visit to the site. Actual field measurements from the subject site are used to customize the analysis. The local stream morphology section of the spreadsheet is a way to record and interpret field observations of the bankfull condition.

- Observe the roughness of the channel, affected by vegetation, obstructions, irregularities in cross section, or meandering. Select a value for Manning's "n" from the pull down menu on the spreadsheet based on channel description.

- During the field visit, walk at least two meander lengths of the stream channel, identifying "bankfull indicators". Mark the elevations of indicators with flags and use a hand level or other survey instrument to determine the height above existing flowline. The best indicators are the first "flat depositional surface", "top of washed root zone" and a "break in slope angle" on the stream bank (Steffen et al., 2000). Look for converging evidence to support your selection of indicators: when they are zeroed in to within a few tenths of a foot, take an average and use the result as your field identified "bankfull" stage. If the channel is undergoing active downcutting, as in CEM Class III or IV, there will not be any reliable bankfull indicators.

- Most streams exhibit a "pool-and-riffle" morphology. Pools (deeper areas typically occurring at the outer meander bends), are separated by riffles (steeper reaches made up of the larger bedload material) at the crossover points between one bend and the next (Biedenharn et al., 1997). Survey a cross section at the nearest **riffle** to the problem area, extending out on each side at least to the floodplain elevation. Measure the distance across the channel at the bankfull elevation. To determine a representative channel slope, survey at least several hundred feet along the stream flow line, at riffle locations. Since channel slopes are often quite flat, it is critical to take accurate measurements at a minimum of three or more riffles to determine channel slope.

- Measure the radius of curvature, Rc, of the channel bend(s) in the project area. Alternatively, this can be done using a recent aerial photograph, if desired.

- Measure the characteristics of the bedload: what size particles are moved by the streamflow. Sieve a bedload sample and do a pebble count, or estimate the D_{90} bedload size (the size mesh through which 90% of the bedload would pass). Do the same for the D_{50} bedload size.

DATA ANALYSIS AND ASSESSMENT

Analysis of the field data involves first determining the value of several standard parameters used to describe stream morphology: width/depth ratio, entrenchment ratio, sinuosity, and the ratio of radius of curvature to bankfull width. These parameters will be used to assess the condition of the stream and the potential for stabilization. Bankfull discharge and flow velocity are determined in several ways from the field data. The ultimate goal is to develop confidence in the analysis by matching discharge and velocity measurements from as many sources as possible.

- Use Manning's equation to compute velocity and flow rate through the channel cross-section taken at the riffle, at the field-observed bankfull depth. The spreadsheet accepts cross-section survey data and calculates cross-sectional area, velocity, discharge and hydraulic radius based on entries for channel roughness and slope. If the actual channel slope data is absent, the cross-section subroutine will use a slope estimate based on entries from the sinuosity determination.

- Determine width/depth ratio from the bankfull width and the mean bankfull depth. The plotted cross section in the spreadsheet helps identify bankfull width, if it was not measured directly in the field. Mean bankfull depth can be determined by dividing the cross sectional area at the field-determined maximum bankfull elevation by the stream width at the maximum bankfull elevation.

- Determine entrenchment ratio using the cross-section information to find maximum bankfull depth and the width of the channel or floodplain at twice that depth; entrenchment ratio will be automatically determined by the spreadsheet.

- Enter the measured radius of curvature; its ratio to bankfull width is automatically calculated by the spreadsheet.

- The spreadsheet allows a user-selected discharge rate "Selected Q" to be entered, representing the designer's best estimate of bankfull discharge based on all of the foregoing data (including the regression analysis and other background investigation). Typically, the discharge rate determined using Manning's equation on the surveyed data is selected, since it represents actual field conditions.

- Enter the field-determined bedload sizes on the spreadsheet. Larger cobbles indicate higher velocity flow; the spreadsheet uses this information to calculate an expected channel-forming velocity.

Finally, compare streamflow velocities calculated in a variety of ways to develop confidence. The spreadsheet will display the following:

- Velocity required to move the D_{90} bedload.

- Velocity using Manning's equation on the surveyed cross-section and slope data.

- Velocity calculated from basic field data (using a modified Manning's equation with mean depth in place of hydraulic radius).

- Velocity from the user-selected discharge rate entry, using continuity ($V=Q/A$) and a cross-sectional area determined from the basic field data section.

Velocities from all four calculations should be comparable, and should be sufficient to move the D_{90} bedload. If more than 1 ft/sec. difference is observed between these four values, it is likely that an error in data entry or bankfull identification has been made. After all the velocities compare well, then compare the bankfull dimensions with those predicted by the regional curves, and compare the "Selected Q" with the discharge predicted by the gage data and/or the regression equation. Modify entries as needed to develop confidence that the stream condition is understood. **Remember that the field indicators should be the main guide, not the regional curve data or the regression equation predictions, as the field indicators are specific to the stream being investigated.** Remember also that if the stream segment is in Channel Evolution Class III or IV, there will be no reliable bankfull indicators and the designer will be forced to rely on flow relationships developed from other similar watersheds and experience gained from previous comparisons.

DEPARTURE ANALYSIS

The final step in the inventory and evaluation procedure is to compare the existing condition of the subject stream to a predicted "stable" condition, identify the departure from equilibrium and assign a Channel Evolution class. The resulting information can then be used by the designer to determine alternatives for treatment of the identified streambank erosion problems.

Floodplain Considerations

If the floodplain elevation is at or near the elevation of "maximum bankfull depth", the channel is connected to the floodplain. Discharges larger than bankfull begin to spread out over the floodplain, slowing velocities and dissipating energy. The channel has not experienced significant downcutting. Channel Evolution Model **Class I or VI** would apply: a stable configuration. The entrenchment ratio (width at twice maximum bankfull depth/ bankfull width) will be greater than 2.5.

Conversely, when the channel is not connected to the floodplain, flood flows will remain inside the channel with little or no opportunity to spread out onto the floodplain. This is evidence of current or past downcutting. The channel evolution process is active, and its morphology is adjusting to regain equilibrium with flow characteristics. Incised channels such as this are likely to continue to erode laterally to build a floodplain. **CEM class could be II, III, IV or V**. The entrenchment ratio will be less than 2.5 (Note: Entrenchment ratio will be smallest in Class II or III channels and then increase to about 2.5 or more as channel nears a new equilibrium in Class VI. The exception to this condition will be low gradient, channelized streams with insufficient energy to erode the channel boundary, even when entrenched.)

Bedload Conditions

If the channel bed in riffle locations is comprised of bedload material, the channel is probably not actively downcutting. Bedload material is being deposited at the riffles. If the entrenchment ratio is low (less than 2.5), the channel is most likely in the widening phase of the channel evolution model, **Class IV or V.**

The channel is actively downcutting **(CEM Class III)** if the subgrade, likely a hard silt or clay, is exposed at the riffle locations. Bedload material is being swept out of this reach of channel. If the streambed is not stabilized, this reach of stream will go through all six CEM stages and the degradation will advance upstream until encountering resistance in the form of bedrock, bridge floor, culvert, or the like. (Note: Channels can be downcutting even when the entrenchment ratio is over 2.5. Streams are not considered "entrenched" until they degrade to twice the maximum bankfull depth, but degradation begins as soon as the bottom begins to be eroded.)

Stream Morphology Considerations

Width/ depth ratios can be very small (less than 10: a deep, narrow channel) in low gradient, fine grained, or sinuous channels. However, these channel types are always connected to the floodplain in stable situations. Therefore, W/D ratios less than 10, combined with entrenchment ratios less than 1.4, are good indicators that downcutting has occurred in the past, or is actively occurring at present **(CEM Class II, III, IV or V)**. If, in addition, the sinuosity is low (less than 1.2), it is likely that the stream has been channelized to create the entrenched condition.

Alternatively, if W/D is greater than 20, suspect an overwidened stream segment and sediment transport problems **(CEM Class V)**. Be aware that this condition could indicate an aggrading stream segment.

Very tight bends in the stream channel at the site indicate that the situation is outside of the normal range of planform stability. If the radius of curvature to bankfull width (Rc/W) ratio less than 1.8, it may be necessary to realign the channel or abandon the project. Natural, "stable" channel Rc/W ratios vary widely, but most commonly range from 2.3 to 2.7 or higher. With a Rc/W ratio less than 1.8, the possibility of a channel "cutoff" at this point increases dramatically.

Flow Velocity Considerations

If the velocity calculated from the Manning's equation analysis on a surveyed stream cross-section is much faster than that required to move the D_{90} bedload material, the bedload material is too small to resist existing channel-forming velocities. Therefore, downcutting is probably occurring **(CEM Class III)**. It is also likely that the riffle locations will contain residual material rather than bedload material, as described above. It should also be noted that streams with only very fine grained bedload material will exhibit excessive velocities compared to D_{90} material size. Vertical stability of these streams cannot be assessed using bedload material size estimates.

Conditions where the channel-forming velocity is much slower than that required to move the bedload could indicate an "aggrading" system where the heavy bedload generated upstream cannot be transported through the system. These conditions often occur in delta areas above impoundments or at confluences with larger streams. They also occur when channel velocities change due to slope changes (such as at the downstream end of a channelized reach), when width/depth ratios increase dramatically or when there is an exceptionally large contribution of bedload just upstream.

CONCLUSION

The underlying assumption to the designer's investigation and analysis is that every stream has a "stable" dimension, slope and planform to safely carry the water and sediment generated from its watershed under the current climate and land use. The investigative procedure is a process of determining what the "stable" conditions of each unique stream segment should be and what the current conditions are, comparing the differences between the two conditions and then attempting to understand the reasons for the differences. Only then can the designer analyze the condition of the stream and recommend action to improve an unsatisfactory condition and move the stream toward a "stable" state, or at a very minimum prevent action that would further "destabilize" the stream.

In Illinois, a procedure for inventory and evaluation of a stream has been developed, along with a spreadsheet software tool. Basic data for the stream reach is gathered from current and historical maps, as well as published gaging data and studies for the subject stream or for reference streams nearby. Field measurements are used to develop a geomorphic description of the channel, concentrating on determining "bankfull", or channel-forming dimensions.

The ultimate goal of the analysis is to develop confidence by matching discharge and velocity measurements from as many sources as possible. The spreadsheet tool determines velocities in four different ways, using field measurements, to facilitate comparisons and allow adjustments. It also summarizes the background data collected from such sources as regional curves and reference stream reaches to aid the designer in the analysis.

A Channel Evolution Model class is determined based on the results of the analysis, by observing the departure between the existing stream conditions and those of a predicted "stable" condition. Key factors include: whether the stream is connected to its floodplain, riffle bedload condition, width/depth ratio combined with entrenchment, sinuosity, and flow velocity.

REFERENCES

1. Biedenharn, D. S., C.M. Elliott, and C.C. Watson. 1997. The WES stream investigation and streambank stabilization handbook. Vicksburg, Mississippi: United States Army Engineer Waterways Experiment Station.

2. FISRWG. 1998. Stream corridor restoration: principles, processes and practices. GPO Item No. 0120-A; SuDocs No. A 57.6/2:EN 3/PT.653. Federal Interagency Stream Restoration Working Group.

3. IEPA. 1998. Field Manual of Urban Stream Restoration. Springfield, Illinois: Illinois Environmental Protection Agency.

4. Rosgen, D.L. 1996. *Applied River Morphology*. Pagosa Springs, Colorado: Wildland Hydrology.

5. Simon, A. 1989. A model of channel response in distributed alluvial channels. *Earth Surface Processes and Landforms* 14(1): 11-26.

6. Steffen, L., D. Roseboom, and W. Kinney. 2000. Fluvial geomorphology. Workshop handouts, April 17-20. Lincoln, Illinois.

7. USGS. 1987. Technique for estimating flood-peak discharges and frequencies on rural streams in Illinois. Water-Resources Investigations Report 87-4207. Urbana, Illinois: United States Geological Survey.

Designing Two-Stage Agricultural Drainage Ditches

Andy D Ward, Dan Mecklenburg, Anand Jayakaran, Larry Brown

The Ohio Department of Natural Resources and the Ohio State University have developed a two-stage ditch design procedure in collaboration with county, state and federal agencies, faculty at other institutions in Ohio, Illinois and Minnesota, and watershed groups. The goal of this work is to develop a practical procedure that can be used throughout the Midwest to correctly size the main channel, to provide a minimum bench width to ensure stability, and to size the cross-sectional capacity of the second stage to carry a design discharge to prevent over-bank flow based on a recurrence interval that satisfies local, county, watershed, or state requirements. Several demonstration designs have been installed and will be monitored to evaluate how they enhance stability, water quality, and ecological function. Highly modified channels drain extensive portions of productive agricultural land in the U.S.A. In many of these areas, most natural channels have been deepened and straightened to facilitate the flow of water from agricultural subsurface drainage outlets and to maximize conveyance. Work done periodically to maintain the drainage function typically includes removal of woody vegetation and deposited sediment, and stabilizing bank slope failures and toe scour. Drainage ditch form (pattern, profile, and dimension) was measured on ditches in Northwest Ohio. Additional measurements have been made in the Wabash River and Great Miami River Watersheds. Apparent benefits exist for incorporating fluvial process derived form into ditch construction and maintenance. To facilitate drainage, and reduce the frequency of over bank flows ditches are typically constructed such that flows as large as perhaps 5-50 year recurrence interval are contained within the ditch. The constructed ditch channel is often oversized for small flows and provides no floodplain for large flows. In response to this imbalance fluvial processes work to create a small main channel by building a floodplain or bench within the confines of the ditches. If conditions allow, these benches can reach a stable size, thickly vegetated with mostly grasses. The small main channel will often meander slightly within the ditch and is sized by nature to carry the effective discharge. Evidence and theory both suggest ditches prone to filling with accumulated sediment may require less frequent dipping out if constructed in a two-stage form. Second, channel stability may be improved by a reduction in the erosive potential of larger flows as they are shallower and spread out across the bench. Stability of the ditch bank may also be improved where the toe of the ditch bank meets the bench rather than the ditch bottom. Here the bank height is effectively reduced and the shear stresses on the toe of the bank are less. The probable dimensions of the low-flow channel can be empirically determined based on regional studies similar to those that are conducted for natural streams. A two-stage ditch has the potential to create and maintain better habitat. Two-stage ditches might also be useful in improving water quality particularly for nutrient assimilation. The primary costs of two-stage ditches are increased width and more initial earthwork.

STREAM MODULES: SPREADSHEET TOOLS FOR RIVER EVALUATION, ASSESSMENT AND MONITORING

D. E. Mecklenburg[1], A. Ward[2]

ABSTRACT

Stream physical condition is increasingly a priority for resource managers. Assessment, monitoring and restoration techniques continue to be developed and standardized. Toward these ends a suite of spreadsheet tools, the STREAM Modules, has been developed by the Ohio Department of Natural Resources and Ohio State University. This ongoing project began in 1998 and currently freely provides the following modules: 1) *Reference Reach Spreadsheet* for reducing channel survey data and calculating basic bankfull hydraulic characteristics, 2) *Regime Equations* for determining the dimensions of typical channel form, 3) *Meander Pattern* that dimensions a simple arc and line best fit of the sine-generated curve, 4) *Cross-section and Profile* that can be used to illustrate the difference between existing and proposed channel form, 5) *Sediment Equations* which includes expanded and condensed forms of critical dimensionless shear, boundary roughness and common bed load equations, and finally 6) *Contrasting Channels* that computes hydraulic and bed load characteristics in a side-by-side comparison of two channels of different user defined forms.

KEYWORDS. fluvial geomorphology, stream, channel, morphology, survey, monitoring, assessment

INTRODUCTION

The objective of this paper is to provide an overview of a suite of spreadsheet tools that the author's have developed to aid in the analysis of stream form and processes.

Stream geomorphology, or the forming of land by streams, occurs because of a series of complex processes that are not easily described by scientific theories. The most basic concept is that of force and resistance. Running water exerts force on the landscape and, in turn, the landscape offers resistance to this force. If the exerted force is less than the resistance, there is no change. If, the force is greater than the resistance, there is change to the slope or stream channel. We call this change geomorphic work. A poor understanding of these processes and inadequate consideration of the influence of changes that occur on the landscape and within the floodplain can cause a variety of adverse outcomes. Particular attention needs to be paid to the potential impact on a stream of: (a) land use changes that reduce vegetation and increase the amount of impervious area; (b) activities that modify the floodplain; (c) the construction of culverts and bridges; and (d) activities that are designed to modify the characteristics of the main stream channel. Any one of these activities might disrupt the equilibrium resulting in rapid and often undesirable adjustments. Successful stream stewardship requires combining this knowledge with sound engineering and scientific principles, together with an understanding and appreciation of the ecology of the stream and its interaction with the landscape (Ward and Trimble, 2003).

---

[1] Dan Mecklenburg, Ohio Department of Natural Resources, dan.mecklenburg@dnr.state.oh.us

[2] Andy Ward, Ohio State University, ward.2@osu.edu

Stream channels and the landscape are shaped by erosional and depositional processes that occur across a wide range of spatial and temporal scales. Therefore, if we want to develop strategies to protect, enhance or sustain these complex systems it is important to understand the origin and evolution of a stream system. It is also important to understand the scales at which changes have or will occur. This change might occur rapidly over a few weeks, months, or years. In other cases perceived instability and channel change might simply be part of the natural cycle of channel adjustment and movement that has occurred over centuries. There is also a need to understand that different scales might apply to different aspects of the ecosystem.

Perhaps because of the complexity of stream and watershed processes, and the variability of scales that can influence these processes, there is not universal agreement on theory, evolution processes, and whether stream behavior can be predicted by morphology. Rosgen (1994, 1996) proposed a hierarchy of river morphology. He divided his classification approach into the following four levels:

Level I: Geomorphic characterization that integrates basin topography, land form and valley morphology. At a coarse scale the dimension, pattern, and profile are used to delineate stream types.

Level II: Morphological description that is based on field-determined reference reach information.

Level III: Stream "state" or condition as it relates to its stability, response potential, and function.

Level IV: Validation level at which measurements are made to verify process relationships.

Many other classification methods have been developed (Ward and Trimble, 2003) but they have seen limited application in engineering design. Insight on the uses and limitations of using geomorphological stream classification in aquatic habitat restoration is presented by Kondolf (1995). Montgomery and MacDonald (2002) propose a *"Diagnostic Approach to Stream Channel Assessment and Monitoring"* and note *"Our argument is based on the observation that a particular indicator or measurement of stream channel conditions can mean different things depending upon the local geomorphic context and history of the channel in question."* The method requires the collection of a comprehensive set of information on a reach and a high level of expertise to be able to then use the data to make a diagnosis. The same statement applies to Level II, III, and IV Rosgen studies. We believe that regardless of the approach that might be used all stream assessments, designs, and management strategies should be based on extensive knowledge of stream processes. This will require the expertise and resources to make extensive measurements within the stream and watershed system and the ability to analyze the data to aid in developing an appropriate self-sustaining strategy. With this view in mind we have developed a suite of spreadsheet tools to aid in the analysis of much of the data that we believe should be obtained. Typically, it will be necessary to use several of the tools and it is important to recognize that the tools do not replace the need for an interdisciplinary team of experts to make the measurements and to analyze the outputs from the tools. Also, it is probable that it will be necessary to also use other tools in developing a self-sustaining solution.

STREAM MODULES

The STREAM modules, are a suite of spreadsheets that as the acronym implies are **S**preadsheet **T**ools for **R**iver **E**valuation, **A**ssessment and **M**onitoring. In developing these tools the author's had the following objectives: (1) to help facilitate the activities listed in the acronym by being consistent with standard or commonly used techniques; (2) to "crunch" numbers and draw plots that at times can be laborious; (3) to present some rather challenging techniques in a more understandable way; and (4) to create educational tools. Embedded in the tools are details on the equations and theory that are used to generate the reported outputs. Therefore, in this paper we will only present an overview of the purpose of each tool. Example outputs are presented for a location on Blacklick Creek in Ohio. The Blacklick Creek flows through the eastern side of

Franklin County and the western sides of Licking and Fairfield Counties Ohio. It is a tributary to the Big Walnut Creek that ultimately drains to the Scioto River. The Blacklick Creek drainage area is twenty-two square miles at the location illustrated in this paper.

The modules are grouped in two sets, the first one dealing with channel form and the second with fluvial process. Channel form (dimension, pattern, profile, and bed material) can be relatively easily seen and measured. A module is included for a reach survey and for organizing and then representing channel form from the extensive amount of data that are normally obtained during a geomorphological survey of a specific channel reach. Another approach to studying stream form is to use measured data in conjunction with relationships developed from big databases of channel form measured by others, often called the *regime approach*. Several sets of these empirical equations are organized together in a spreadsheet module. One of the more challenging tasks regarding channel form is to define and communicate not only what may exist but what are desired target or design conditions. Two modules have been developed specifically to help with this task.

The second set of modules, dealing with fluvial process, includes standard hydraulic and sediment transport equations but here too an effort has been made to illustrate and aid in communicating implications of channel process. Typically, in the United States measurements are made using English units but for applications in other countries and for research purposes the STREAM modules provide an option to use SI units.

Channel Form

Measure what is there: **The Reference Reach Spreadsheet (RRSS)**

Surveying an existing channel is perhaps the most illuminating single step toward understanding the physical aspects of a stream. *Reference Reach Spreadsheet)* is for reducing channel survey data and calculating basic bankfull hydraulic characteristics. It is generally consistent with the closest we have to a standard protocol as documented by Harrelson et al. (1994). Data may be either in the form collected with a level or a total station. In the survey, the profile serves as a framework or base line. The profile, or longitudinal slope profile, is measured from upstream down along the centerline of the bankfull channel. At each station, bed elevation in the thalwag, water depth, and bankfull elevation are recorded. If a point is measured at the beginning of each bed feature the spreadsheet will provide the average length and slope of the features. Also, by identifying the interception point of the profile with each of the cross-sections allows the instrument height and bankfull elevation information to link to the cross-section sheet. A profile of a reach of Blacklick Creek is presented in Figure 1.

Bed material is obtained a number of different ways for a number of reasons. The materials sheet of the RRSS is set up to accommodate pebble counts and bulk sieve samples. Pebble counts can be 1) individual such as the mobile riffle surface material or a zig-zag count of a reach, or 2) weighted such as between riffles and pools each representing a portion of the reach. The spreadsheet is set up to accommodate most any combination of material data collection. The standard presentation of these data is a plot of the cumulative percent, calculated D50, etc., and percentage of the various size classes (Figure 2). Because a semi-log scale is typically used the mathematical solution requires interpolation of the logs of the size values to get the various "percent smaller than" values.

Pattern is the dimension least well defined by a site survey. Often, better information can be obtained by area photos, GIS, even topographic maps, all of which allow a greater length to be assessed. This information can be entered in the RRSS. Also, while surveying the profile with a tape and level, if an azimuth is obtained and entered with each corresponding distance, then that information will be reduced and presented in plan form. In addition, the water depth information is represented on the plan view allowing an interesting perspective of pool and riffle location through the meander pattern.

Cross-sections are plotted (Figure 3) and various bankfull channel dimensions are calculated including area, width, mean and maximum depth, etc. Determining bankfull location is of course necessary but is also one of the most challenging tasks in geomorphology. The RRSS facilitates the determination of bankfull a number of ways. First, using cross-sections in conjunction with the channel profile is an established standard method for this task. The spreadsheet links the profile to the cross-section. The stage at which a bankfull-trend line from the profile intersects each cross section provides a first iteration of the bankfull stage at each cross section. Refinement of the value can be based on local trends in the profile and details of the cross-section.

Another approach to determining bankfull that is utilized in the spreadsheet is based on the idea that in gravel bed channels the particle at the threshold of motion at bankfull flow is often near the measured D50. Using Shield's parameter the spreadsheet computes the size at the threshold of motion and presents it with the D50 and D84 values for comparison.

Each cross-section has calculated values for several standard dimensions and hydraulics from width-depth ratios, discharge, shear stress and unit stream power. All the equations used in the spreadsheet are explained in comments attached to each cell. To manage the information obtained from a channel survey values are typically reduced to dimensionless ratios. This also facilitates comparisons between channels, particularly of different size. The spreadsheet provides a summary of all the sheets including an average and range of all values.

Typical values measured by others - **Regime Equations**

Another way to understand channel form is to consider typical values measured from many streams. Like dimensionless ratios, this information is usually presented in the form of relationships between variables. Several sources exist; perhaps the most extensive are the Williams equations (Williams, 1986). Richard Hey (Thorne et al., 1997) also developed a set of regime equations and Luna Leopold (Leopold and Langbein, 1966) defined some classic meander pattern relationships. These and others are in the Regime module. A refinement of this approach is a breakdown of typical values by channel types. This is not presently available in the STREAM Modules but is in Applied Stream Morphology (Rosgen, 1996). Table 1, illustrates some results for the Blacklick Creek reach. This sort of information can provide a reassuring check of measured values or be an indicator of a problem if measured values fall outside the range of typical values. At best, regime equations must be considered coarse approximations.

Defining and communicating channel form – ***Pattern and Cross-section & Profile***

The desired channel form must be illustrated and conveyed for construction documents, planning discussion, etc. Two modules have been developed to help with defining and communicating proposed condition.

The first, the **Meander Pattern** module, simply dimensions channel meander pattern. It is based on the sound but awkward sine-generated curve, which is a function of the continuously changing direction of flow. By first integrating the sine-generated curve then calculating a best fit of arcs and straight lines, standard dimensions are presented including amplitude, radius of curvature and meander length (Figure 4).

The second module, describing channel form, contrasts existing and desired cross-sections. In the module, **Cross-section & Profile**, the desired channel dimensions are represented by a compound cross-section, the proportions of which are adjustable values based on a regional curve A profile may be used for tracking elevation of proposed cross-sections but is not necessary. Survey data of existing cross-sections is entered and plotted with the desired cross-section (Figure 5). The plots are especially useful for illustrating desired floodplain form. They may be adequate as construction drawings for simple restoration involving lowering high terraces down to active floodplain. They have been proven useful in two-stage ditch design projects and in discussions of channel evolution.

Channel Process

Unlike channel form which is conceptually simple, channel process is complex, particularly the process most influential in channel form, bed load sediment movement. Two modules have been developed to make some approaches to understanding channel process more accessible. The first module, **Sediment Equations**, has three different sheets with various equations pertaining to channel process one dealing with critical dimensionless shear (Figure 6), another on relative roughness and boundary resistance and the last on bed load equations (Figure 7). The equations are presented in both an expanded format with explanations and constants shown and then in a condensed format with only input cells and answers.

The second channel process module, **Contrasting Channels**, is an application of the first. It allows a side-by-side comparison of processes given different channel forms or runoff conditions. The main strength of this module is its evaluation of an entire flow regime rather than a single surrogate stage or recurrence interval. Hydrology models developed by USGS (Koltun and Roberts, 1990; Sherwood, 1994) are built into the spreadsheet and may be used or peak discharge values from another source may be entered. Hydrographs for a series of recurrence interval storms (0.2 to 100 years) are developed. Bed load sediment is calculated by dividing the hydrographs into steps of given conditions existing for specific durations. Each runoff event is then multiplied by the number of occurrences of over a 100-year period. The calculations are performed for both channels. The results are presented by contrasting the peak flow stage of the range of runoff events as well as the calculated bed load sediment movement (Figure 8).

DISCUSSION & CONCLUSIONS

The STREAM Modules serve several needs in the area of assessing the physical quality of rivers. They provide an additional step toward standardization of methods and techniques. Also, they have application in education and in increasing the utility of methods and techniques. They can be applied to further understanding of channel form, fluvial processes and the role these play in the integrity of water resources. The existing modules are by no means a complete set of tools. The author's have developed a variety of miscellaneous tools such as a tool for determining the effective discharge (Ward and Trimble, 2003) from measured sediment data and a set of tools for evaluating different storm water management strategies. Others have developed more powerful sediment transport and bank stability models that have more extensive stream and watershed measurement requirements.

REFERENCES

Einstein, H. A., 1950. The Bed-Load Function for Sediment Transportation in Open Channel Flow. USDA SCS, Technical Bulletin 1026, Washington D.C., pp 71.

Harrelson, C. C., C. L. Rawlins, and J. P. Potyondy, 1994. Stream Channel Reference Sites: An Illustrated Guide to Field Technique. USDA Forest Service, General Technical Report RM-245, Fort Collins, Colorado, 62 pp.

Hey, R.D., 1976, Geometry of river meanders, Nature, 262, 482-484.

Koltun, G.F. and J.W.Roberts. 1990. Techniques for Estimating Flood-Peak Discharges of Rural, Unregulated Streams in Ohio. U.S. Geological Survey, Water Resources Investigations Report 89-4126, USGS Denver, CO.

Landwehr, K. and B. Rhoads. 2003. Depositional Response of a Headwater Stream to Channelization, East Central Illinois, USA. River Research and Applications, 19:77-100.

Leopold, L. B. and Langbein, W. B., 1966, River meanders; Scientific American, June, pp. 60-69.

Rosgen, D., 1994. A Classification of natural Rivers. Catena 22: 169-199.

Rosgen, D., 1996. Applied River Morphology. Wildland Hydrology. Pagosa Springs, Colorado.

Sherwood, J.M. 1994. Estimation of Volume-Duration-Frequency Relations of Ungaged Small Urban Streams in Ohio. AWRA Water Resources Bulletin 30(2):261-269.

Thorne, C. R., R. Hey, M. Newson. 1997. Applied Fluvial Geomorphology for River Engineering and Management. John Wiley & Sons.

Ward, A. and S. Trimble. 2003. Environmental Hydrology 2nd Edition. CRC Press.

Williams, G. P., 1986. River Meanders and Channel Size. Journal of Hydrology, 88:147-164.

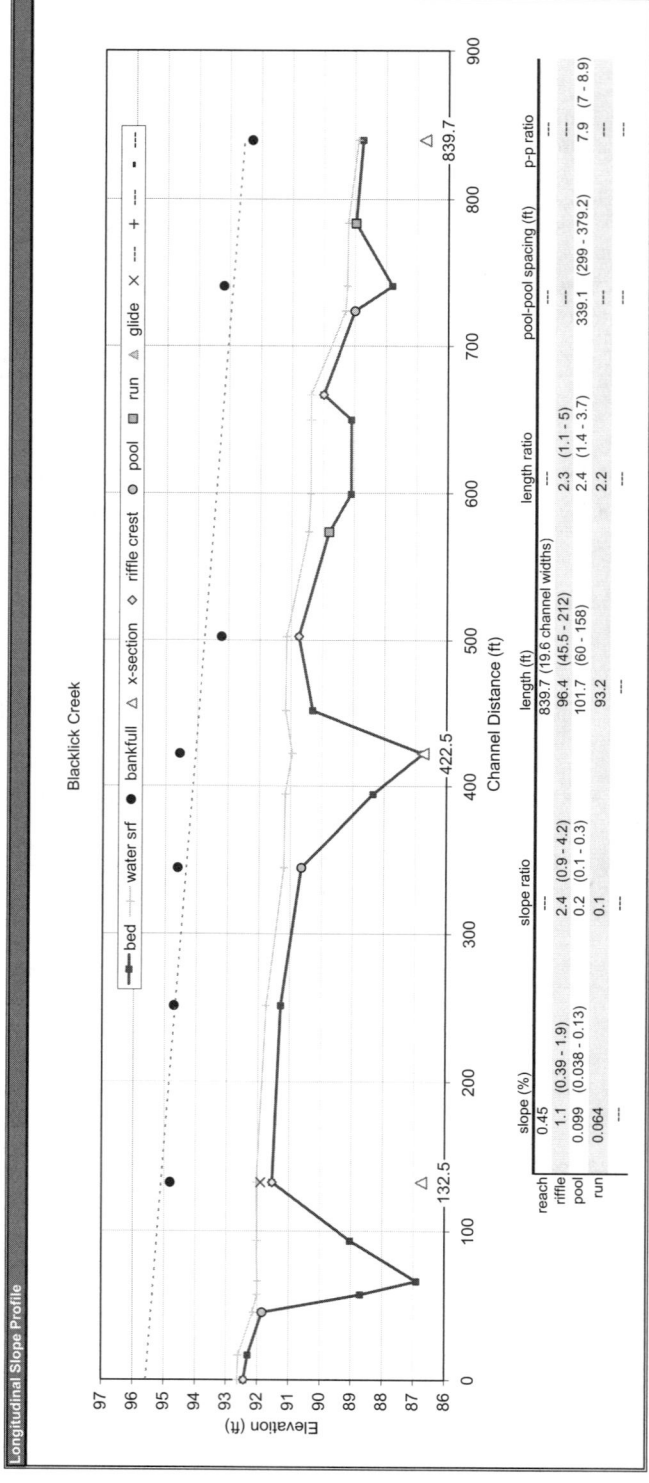

Figure1. Example profile output from the *Reference Reach Spreadsheet*.

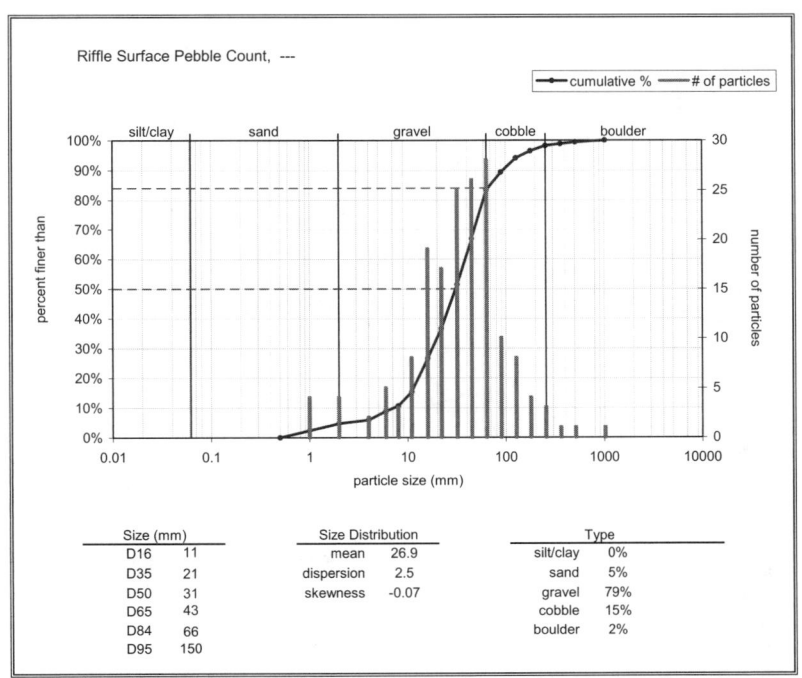

Figure 2. Example channel materials output from the *Reference Reach Spreadsheet*.

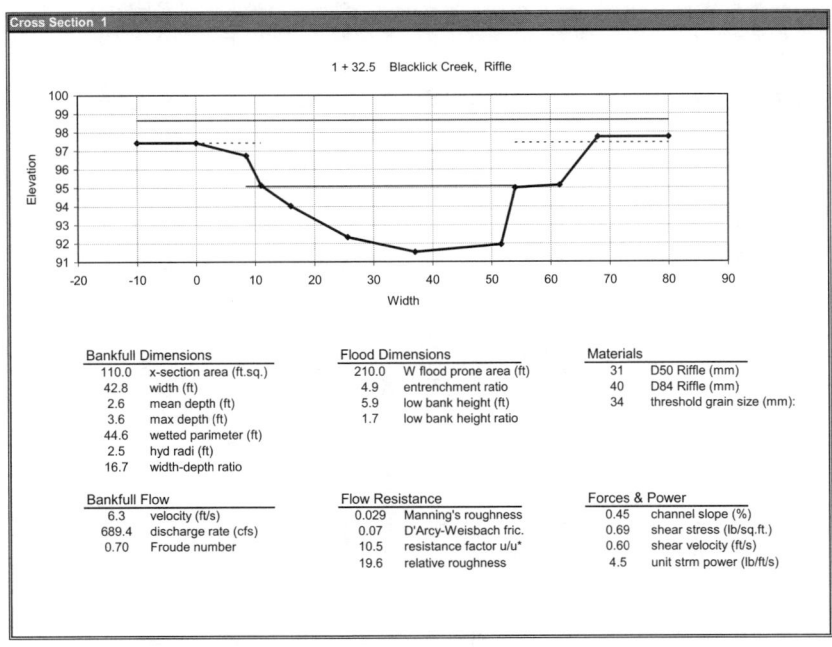

Figure 3. Example cross section output of *the Reference Reach Spreadsheet* module.

William's "River Meanders and Channel Size"

Mid values are from equations fit to 41 to 191 data points. Low and high values are from the standard deviation of residuals not the extremes.

Source: Williams, G.P., 1986. River meanders and channel size. Journal of Hydrology, 88 pp.147-164.

Meander Pattern from Bankfull Dimension	Input	meander length	belt width	radius of curvature	length bend
cross-section area	**150**	344 **545** 867	209 **327** 510	61 **106** 187	228 **400** 708
width	**50**	317 **520** 858	172 **296** 515	68 **104** 161	215 **352** 580
mean depth	**3**	281 **685** 1658	200 **425** 914	46 **122** 324	201 **457** 1041

Depth from Width or from Width and Sinuosity

	Input	meander depth
width	**50**	1.3 **2.5** 4.9
sinuosity (and entered width)	**1.3**	1.3 **2.2** 3.8

Width from Depth or from Depth and Sinuosity

	Input	width
mean depth	**3**	23 **62** 160
sinuosity (and entered mean depth)	**1.3**	68 **152** 336

Bankfull Dimension from Meander Pattern

	Input	area	width	mean depth
meander length	**500**	62.1 **126.6** 257.1	30.7 **48.0** 74.8	1.3 **2.4** 4.3
beltwidth	**300**	66.0 **129.5** 255.1	30.3 **49.7** 80.9	1.4 **2.4** 3.9
radius of curvature	**100**	56.4 **134.3** 319.7	33.2 **48.8** 72.2	1.4 **2.6** 5.0
length bend	**400**	59.9 **142.7** 342.4	34.4 **53.8** 83.9	1.6 **2.8** 4.8

Meander Pattern Relations

	Input	meander length	belt width	radius of curvature	length bend
meander length	**500**		232 **305** 400	91 **110** 133	304 **400** 528
beltwidth	**300**	372 **489** 640		74 **105** 149	294 **387** 507
radius of curvature	**100**	376 **453** 548	20 **288** 409		279 **377** 509
length bend	**400**	380 **500** 660	237 **312** 408	77 **104** 140	

Table 1. Example output from the *Regime Equations* module for Blacklick Creek Ohio.

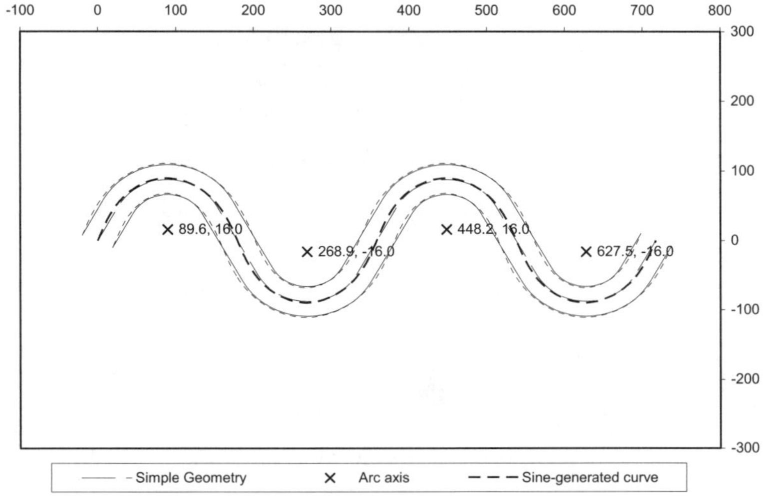

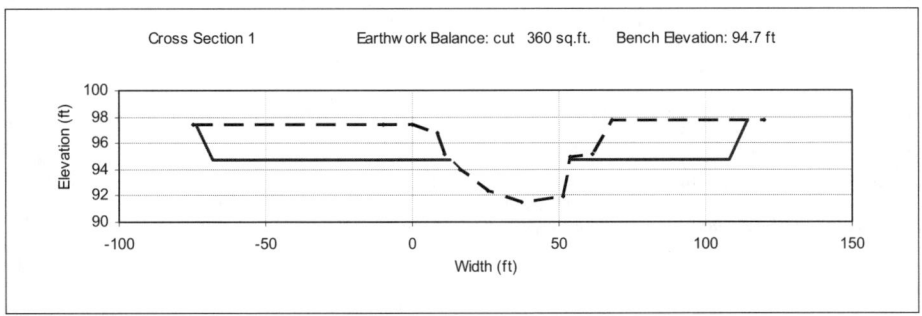

Figure 4. Example output from the *Meander Pattern* module.

DIMENSIONS			
42.8	channel width	71.7	radius of curvature
1.5	sinuosity	124	arc angle (degrees)
358.6	meander length	155.1	arc length
537.8	stream length per meander	112.3	cross over length
175.3	amplitude	62	angle of deflection (degrees)

Figure 5. Example output from the *Cross Section Module* showing existing and proposed condition.

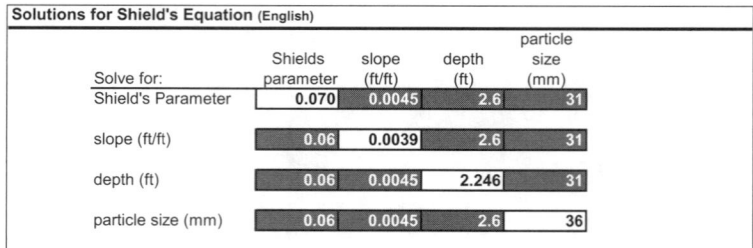

Solve for:	Shields parameter	slope (ft/ft)	depth (ft)	particle size (mm)
Shield's Parameter	0.070	0.0045	2.6	31
slope (ft/ft)	0.06	0.0039	2.6	31
depth (ft)	0.06	0.0045	2.246	31
particle size (mm)	0.06	0.0045	2.6	36

Solutions for Shield's Equation (English)

Figure 6. Example of Shield's Eq from the *Sediment Equations* Module.

Ackers and White

depth	2.6	ft
slope	0.0045	ft/ft
D_{35}	21	mm
n	0.034	
q_b	0.00067	ft³/s per foot width

Meyer-Peter

depth	2.6	ft
slope	0.45	%
D_{50}	31	mm
q_s	0.0064	ft³/s per foot width

Einstein$_{50}$

depth	2.6	ft
slope	0.0045	ft/ft
D_{50}	31	mm
Φ	0.019	
q_s	0.00439	ft³/s per foot width

Figure 7. Example of bedload equations from the *Sediment Equations* Module.

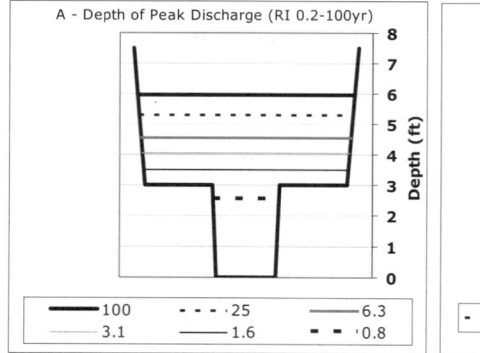

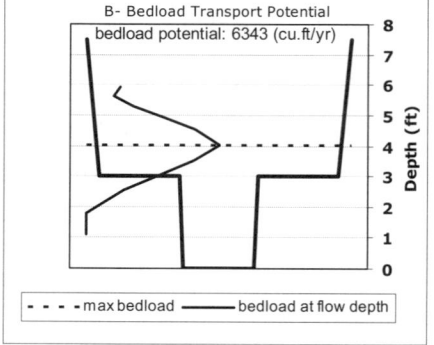

Figure 8. A. Depth of flow for different recurrence interval peak discharges suggest that the main channel can convey a discharge associated with a recurrence interval of 0.8 to 1.6 years. B. Potential bedload transport suggests the greatest bedload transport occurs at a flow greater than bankfull.

Watershed Scale Impacts of Buffers and Upland Conservation Practices on Agrochemical Delivery to Streams

T.G. Franti[1], D.E. Eisenhauer[1], M.C. McCullough[1], L.M. Stahr[1], M.G. Dosskey[2] D.D. Snow, R.F. Spalding, and A. L. Boldt[1]

ABSTRACT

Conservation buffers are designed to reduce sediment and agrichemical runoff to surface water. Much is known about plot and field scale effectiveness of buffers; but little is known about their watershed scale impact. Our objective was to estimate the watershed scale impact of grass buffers by comparing sediment and agrichemical losses from two adjacent 141-165 hectare watersheds, one with conservation buffers and one without. Rainfall derived runoff events from 2002-2003 were monitored for water runoff, TSS, phosphorous and atrazine loss. A conservation-watershed included 0.8 km of grass buffers and 0.8 km of riparian forest buffer, ridge-tilled corn, corn-beans-alfalfa rotation, terraces and grassed waterways. A control-watershed had no buffers, disk-tilled, continuous corn and grassed waterways. The same application rate and method for atrazine to corn was used in each watershed. Total rainfall during the April-June monitoring period was similar in 2002 and 2003; however, the conservation-watershed produced only 27 mm of runoff, compared to 47 mm from the control. Over two years, TSS and phosphorous losses per hectare were reduced by 97% and 95%, respectively, in the conservation-watershed. Atrazine loss per hectare was 57% less in the conservation watershed. A separation technique showed that for 2002 other conservation practices reduced TSS by 84% and buffers reduced TSS by an additional 13% compared to the control. Similarly, other conservation practices reduced atrazine losses by 29% and buffers accounted for an additional 31%. On a watershed scale buffers can add benefit to a conservation system.

Keywords. Conservation buffers, runoff, atrazine, sediment, phosphorous, watershed

INTRODUCTION

Soil erosion and subsequent sediment delivery and transport of agrichemicals, particularly atrazine, to streams continues to be a water quality problem in corn producing regions of the Midwest and Great Plains. Degradation of water quality in the Missouri River and its tributaries has been attributed to runoff contaminated with pesticides, sediment and nutrients from agricultural land in the Midwest (Clark et al., 1999; Barbash et al., 1998; Goolsby et al., 1995; Goolsby et al., 1991). In Nebraska, elevated herbicide levels in the Platte River (Snow and Spalding, 1988; USGS, 1998) and its eastern tributaries—Clear Creek, Shell Creek, Salt Creek and the Elkhorn River—are the result of a "Spring flush" in which agrichemicals are washed from treated fields shortly after application (Spalding and Snow, 1989; USGS, 1996). In particular, concentrations of 82 mg L^{-1} atrazine and 44 mg L^{-1} metolachlor have been measured in the Clear Creek tributary at its confluence with the Platte River (USGS, 1996).

---

[1] T.G. Franti, Associate Professor, tfranti@unl.edu; D.E. Eisenhauer, Professor, deisenhauer1@unl.edu; M.C. McCullough, former Research Engineer, mmccullough2@unl.edu; L.M. Stahr, Research Engineer, lstahr2@unl.edu; D.D. Snow, Assistant Professor, dsnow1@unl.edu; R.F. Spalding, Professor, rspalding2@unl.edu; A.L. Boldt, research engineer, aboldt1@unl.edu, University of Nebraska-Lincoln.

[2] M.G. Dosskey, Riparian Ecologist, mdosskey@fs.fed.us, USDA Agriforestry Center, Lincoln, NE.

Conservation buffers have long been used for erosion and surface water pollution control in agricultural watersheds. Buffers can filter out a major proportion of sediment and other contaminants eroded from row-cropped fields before runoff enters a major waterway. Research regarding the efficacy of buffers in controlling surface water contamination from agricultural runoff has been reviewed by Barling and Moore (1994), Haycock et al. (1997), Lowrance et al. (1995), Muscutt et al. (1993) and Dosskey (2000). Numerous studies have examined the efficacy of buffers to remove contaminants (Arora et al., 1996; Dillaha et al., 1989; Magette et al., 1989; Patty et al., 1997; Robinson et al., 1996 and Schmitt et al., 1999). Most research to date has examined losses during a small number of runoff events from test plots with small field area to buffer area ratios. The range of contaminant reductions varies with factors such as buffer width and field-area to buffer-area ratio which creates differences in water and sediment loading to the buffers. Very little information is currently available on actual reduction of contaminant levels in streams with the use of conservation buffers. In addition, there has been very little assessment of buffer performance at the watershed scale.

Our objective was to estimate the watershed scale impact of grass buffers by comparing sediment and agrichemical losses from two adjacent 141-165 hectare watersheds, one with conservation buffers and one without.

METHODS

Monitoring of rainfall-derived runoff was conducted in two adjacent subwatersheds which are part of the Clear Creek Watershed, a tributary to the Platte River in central Nebraska (Figure 1). The subwatersheds were situated on alluvial terrace deposits between a nearly level but dissected upland plain and the Platte River bottomlands. A 165-hectare conservation-watershed (Figure 1) was adjacent to Clear Creek and included several conservation practices (Table 1), including 0.80 km of long-term riparian forest, and 5 riparian grass buffers planted in 1999-2000, totaling an additional 0.8 km of buffer. The riparian grass buffers were designed and installed to NRCS standards, with a native grass mix, and ranged in width from 13.7 m to 18.3 m to maintain a 30:1 field area to buffer area ratio. With the addition of the grass buffers an estimated 75-80% of all cropland runoff from the conservation watershed passes through a riparian forest or grass buffer. Runoff from all corn areas passed through a grass buffer.

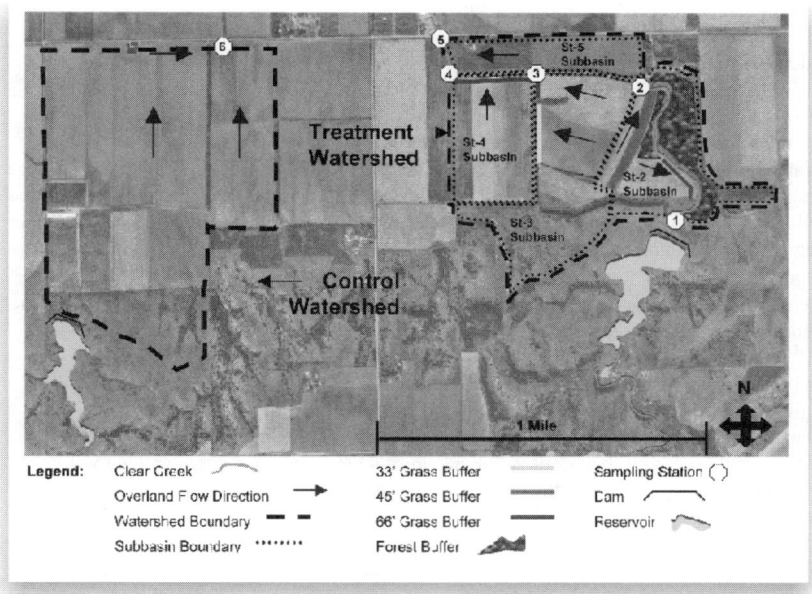

Figure 1. Clear Creek Watershed Study Site.

The control watershed had no conservation buffers. This watershed included 113 ha of continuous corn in a disk tillage system compared to 45 ha of ridge-tilled corn in a corn-bean-alfalfa rotation in the conservation watershed. Sediment and agrichemical losses were compared on a per unit area basis.

Table 1. Watershed Characteristics

Feature	Conservation Watershed	Control Watershed
Area	165 ha	141 ha
Grass Buffers[1]	0.8 km	None
Forest Buffer	0.8 km	None
Conservation Terraces	Yes	No
Grassed Waterways	Yes	Yes
Feedlot	No	11 ha
Crop Rotation	Corn-Soybeans-Alfalfa	Continuous Corn
Corn Area	45 ha	113 ha
Pasture	13 ha	8 ha
Total Cropped	155 ha	134 ha

[1] Planted in 1999 and 2000

Soils in both subwatersheds were Hord Silt loam, on 0-1% slopes. Each subwatershed was furrow irrigated, with one center pivot irrigation system in the control watershed. Each subwatershed has some area of permanent pasture. The control watershed also contains a beef cattle feedlot. (Table 1)

Atrazine was band applied to corn in both watersheds at the same rate, using the same commercial product, and was applied each year prior to when runoff monitoring began. In 2002 atrazine was applied as Bicep II Magnum at a rate of 2.2 kg ha[-1] (a.i.). In 2003 atrazine was applied as Guardsman Max at a rate of 0.81 kg ha[-1] (a.i.).

Rainfall-derived runoff was monitored at the outlet of each watershed (Station 5 and 6) during April-June of 2002 and 2003 (Fig. 1). Stream flow monitoring and water sampling was done using ISCO bubble meters and samplers programmed to sample for 24 hours after stream flow began. Samples were retrieved and samplers restarted if runoff events lasted longer than 24 hours. Water samples were tested for atrazine concentration using solid phase extraction and gas chromotrography coupled mass spectrometry (GC/MS) with ^{13}C ring-labeled internal standards for quantification of isotope dilution (Cassada et al. 1994). Method detection limits for atrazine and its degradation products in runoff samples is near 0.05 ug L[-1]. Sediment concentration was analyzed gravimetrically as total suspended solids dried at 103-105 °C (APHA, 1998). Phosphate concentration was deterimined using ion chromatography to measure orthophosphate (soluble phosphorus). The method reporting limits for phosphate are 0.10 mg L[-1].

Upstream of each subwatershed is a small flood control dam. Only one runoff event in 2002 caused flow from the dam in the conservation watershed. No flow occurred from the dam in the control watershed. Discharge monitoring and water sampling was done at the conservation watershed dam so a hydrograph and mass loss separation could be done between the total flow measured and that contributed from the conservation watershed.

Channel flow conditions at Station 6 (Figure 1) changed in the spring of 2003, which resulted in our sampler being improperly programmed and only collecting one water sample for each of two events in 2003. Discharge monitoring for other events was unaffected. Therefore, there was insufficient data to calculate the total mass loss of contaminants for these two events. To estimate the mass loss, we assumed that the ratio of total mass loss divided by total volume was proportional to the mass loss divided by volume at the time of the first sample of an event, as represented by the equation:

$$\frac{m_t}{V_t} = k\frac{m_1}{V_1}$$ [Equa. 1]

Where k = coefficient
m_t = total mass loss of contaminant (kg)
V_1 = total volume of water (L)
m_1 = mass loss at time of first water sample (kg)
V_1 = volume of water at time of first water sample (L)

We estimated the total mass loss for the events in 2003 by computing k-values using the 2002 data, and by rearranging Equation 1 as:

$$k = \frac{m_t}{m_1}\frac{V_1}{V_t}$$ [Equa. 2]

For each contaminant, the k-value was determined as the average of the k-values from four events: May 6, May 11, May 23 and May 26, 2002.

Table 2. Average k-values for 2002

	P	TSS	Atrazine
Average k-value, 2002	1.82	0.35	0.47
Standard Deviation	0.50	0.037	0.15

Total mass loss was then determined for the two events in 2003 by rearranging Equation 2 as:

$$m_t = k\, m_1 \frac{V_t}{V_1}$$ [Equa. 3]

For each event in 2003, the values for m_1, V_1 and V_t were known, and the k-value from 2002 was assumed constant for 2003.

The discharge and mass loading monitored in the conservation watershed was impacted by all the conservation practices employed. Therefore, a method was needed to separate out the impacts of other conservation practices and estimate the impact of the riparian grass buffers. The field-scale effects of the buffers were known from Helmers (2003), where he measured trapping efficiency from both rainfall runoff and irrigation runoff. Mean trapping efficiency for sediment was estimated at 80%, and the infiltration ratio for rainfall runoff events was 37%, i.e., the buffer captured 37% of the water that entered it.

Using this information on buffer performance and assuming loading is linear with runoff depth, a separation calculation was used to estimate the impact of buffers and other conservation practices on reducing TSS, phosphate and atrazine mass loss (Figure 2 and 3). The separation calculation assumed that for TSS and phosphorous the loading leaving the buffers (measured at station 5) was reduced by 80% from that entering. Because an estimated 90% of atrazine runoff is in solution the atrazine trapping efficiency was assumed to equal the infiltration ratio; therefore, loading leaving the buffers was reduced by 37% from that entering. The loading entering the buffers is the loading effected by other conservation practices. This was compared to the loading from the control watershed (measured at station 6) which was not effected by conservation practices.

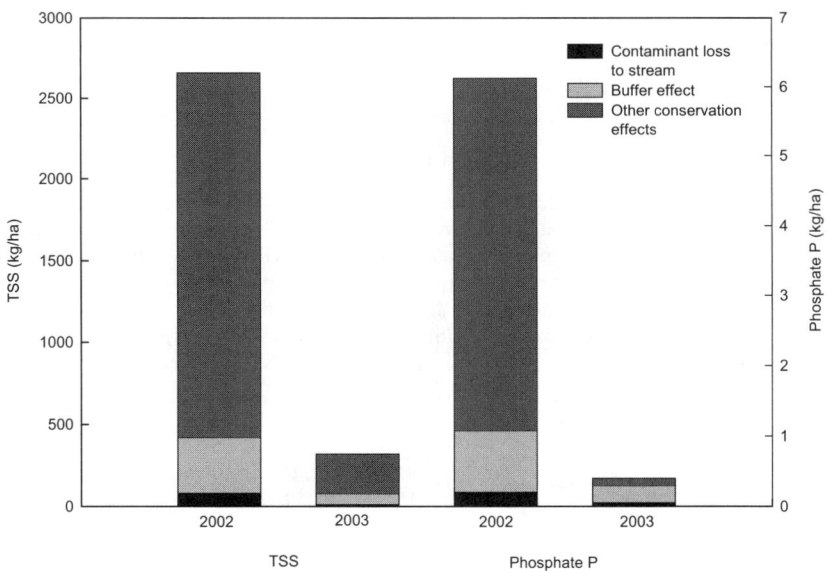

Figure 2. Effects of conservation buffers and other conservation practices on reduction of TSS and phosphorus.

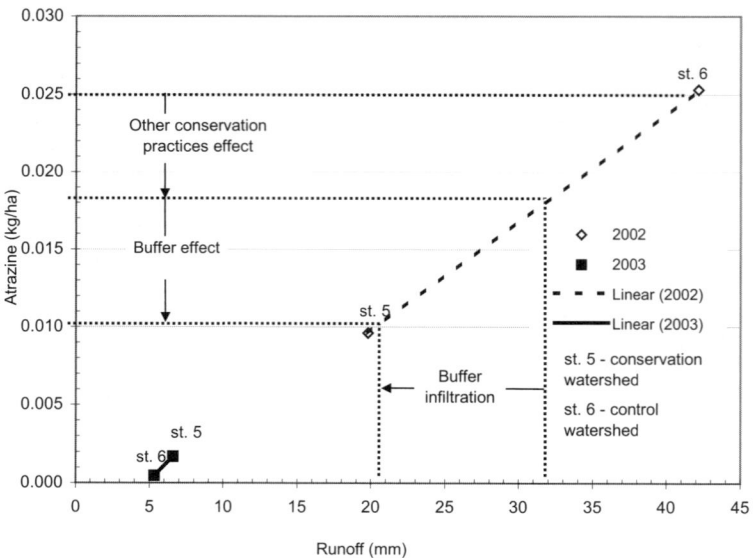

Figure 3. Effects of conservation buffers and other conservation practices on reduction of atrazine.

RESULTS

Five sampling events occurred in 2002 and four events in 2003. During the sampling period, 233-240 mm and 210-218 mm of rainfall occurred in 2002 and 2003, respectively. This resulted in a 2-year total of 27 mm of runoff from the conservation watershed, compared to 47 mm from the control watershed (Table 3). This reduction in runoff was from the combined impact of conservation tillage, conservation terraces, crop diversity and buffers.

Table 3. Contaminant Loss Results

Watershed	Year	Rainfall during sampling period, mm	Runoff mm	TSS kg ha^{-1}	P kg ha^{-1}	Atrazine kg ha^{-1}	Atrazine Percent loss %[1]
Conservation	2002	233	20	84	0.21	0.0096	0.43
(Station 5)	2003	210	6.6	15	0.059	0.0017	0.21
Control	2002	240	42	2,657	6.12	0.025	1.1
(Station 6)	2003	218	5.3	321	0.40	0.00047	0.058

[1] percent of applied

Mass loss of TSS, phosphorus and atrazine was computed on a per unit area basis. TSS and phosphorus losses were based on the watershed area and atrazine loss was based on the corn area, because atrazine was only applied to corn (Table 1). In 2002, losses of all contaminants were greater in the control watershed. In 2003 losses of TSS and phosphorus were greater in the control watershed, but loss of atrazine was less.

In 2002 and 2003, respectively, atrazine loss based on total applied mass was 0.43% and 0.21% in the conservation watershed, and 1.12% and 0.058% in the control watershed. Losses in the range of 0.21% to 1.12% are typical values; however; 0.058% is lower than expected. This suggests that atrazine loss in 2003 may be under-estimated for the control watershed. This is likely a result of the failed sampling during two events in 2003. Total mass loss was only an estimate based on equation (1) for contaminants in 2003, and for atrazine this estimate may be low.

Comparing two-year total mass loss of contaminates from the contrasting watersheds shows that TSS was reduced by 97% in the conservation watershed. Similarly, atrazine loss was reduced by 57%.

Table 4. Combined Contaminant loss for 2002-2003

Mass Loss (kg ha^{-1})	Control Watershed	Conservation Watershed	Reduction from all Conservation Impacts	Estimated Reduction by Grass Buffers[1]	Estimated Reduction by Other Conservation Practices
TSS	2,980	99	97%	14%	83%
Phosphorus	6.52	0.269	96%	17%	79%
Atrazine	0.025	0.011	57%	27%	30%

The grass buffers in the conservation watershed are part of a conservation system. The measured reduction from the impact of this system can be separated into the impact of the buffers and the impact of other conservation practices. Losses of runoff water, TSS and atrazine from the control watershed are unaffected by conservation practices. These losses represent the worst-case scenario. Using these losses for a starting point a separation calculation was used to estimate the impacts of the grass buffers alone and the impact of other conservation practices (Figure 2 and 3). Based on the separation analysis, the watershed scale impact of other conservation practices to reduce TSS was 83%, and for grass buffers, the impact was 14%. For atrazine losses, the watershed impact of other conservation practices was 30%, and the buffer impact was 27%.

For the two years of this study the impact of other conservation practices was much greater to reduce TSS (83%) and phosphorus loss (79%) compared to the impact of grass buffers (Table 4). The watershed impact of the grass buffers alone is estimated at 14% reduction for TSS and 17% for phosphorous. Buffers had a similar effect (27%) to other conservation practices (30%) in reducing atrazine runoff (Table 4). Conservation practices and buffers reduce atrazine loss by reducing total runoff amounts (45%). Within a conservation system, grass buffers can provide a significant additional impact in reducing TSS, phosphorous and atrazine

CONCLUSION

Conservation buffers are part of a conservation system that could include crop residue management, crop rotation, conservation terraces, and integrated uses of herbicides. The impact of riparian buffers to reduce TSS, phosphorus and atrazine loss was determined by comparing two years of runoff losses from a conservation watershed and a control watershed. The control watershed had few conservation practices, and continuous corn cropping, while the adjacent conservation watershed had 1.6 km of riparian grass and forest buffers, crop rotation (corn-beans-alfalfa) and conservation tillage (ridge-till). For the two years of the study, TSS and phosphorus losses per hectare were reduced by 97% and 96% in the conservation watershed compared to the control watershed. This was partially a result of a 45% reduction in the amount of water runoff from the conservation watershed. Atrazine was applied to corn at the same rate in each watershed; however, atrazine loss per hectare of corn was 57% less in the conservation watershed.

Previous plot studies (Helmers, 2003) had shown the trapping effectiveness for TSS of the grass buffers to be 80%, with an infiltration ratio (water captured in the buffer) of 37% for rainfall runoff events. Using this data, the impact of the grass buffer was separated from the impact of other conservation practices.

For the two years studied, other conservation practices (ridge-tillage, crop rotation, terraces and waterways) reduced total suspended solids by 83% compared to the control watershed, and buffers reduced TSS an additional 14%. For 2002, other conservation practices reduced atrazine mass loss by 29% and buffers accounted for an additional 31%. Thus, within a conservation system grass buffers can provide a significant benefit to reducing sediment and agrichemical losses to surface water.

REFERENCES

APHA, 1998. Standard Methods for the Examination of Water and Wastewater, 20th Edition, L.S. Clescerl, A.E. Greenberg, A.D. Eaton, (Eds.). Published by the American Public Health Association, American Water Works Assocation, Water Environment Federation, Washington, D.C.

Arora, K., S.K. Mickelson, J.L. Baker, D.P. Tierney and C.J. Peters, 1996. Herbicide Retention by Vegetative Buffer Strips From Runoff Under Natural Rainfall. Transactions of the ASAE, 39:2155-2162.

Barbash, J.E., G.P. Thelin, D.W. Kolpin and R.J. Gilliom, 1998. Distribution of Major Herbicides in Ground Water of the United States. USGS Water-Resources Investigations Report 98-4245.

Barling, R.D. and I.D. Moore, 1994. Role of Buffer Strips in Management of Waterway Pollution; A Review. Environmental Management, 18:543-558.

Cassada, D.A., R.F. Spalding, Z. Cai and M.L. Gross, 1994. Determination of Atrazine, Deethylatrazine, and Deisopropylatrazine in Water and Sediment by Isotape Dilution GC/MS. Anal. Chim. Acta, 287(1-2):7-15.

Clark, G.M., D.A. Goolsby and W.A. Battaglin, 1999. Seasonal and Annual Load of Herbicides from the Mississippi River Basin to the Gulf of Mexico. Environ. Sci. Tech. 33, 981-986.

Dillaha, T.A., R.B. Reneau, S. Mostaghimi and D. Lee, 1989. Vegetative Filter Strips for Agricultural Nonpoint Source Pollution Control. Transactions of the ASAE, 32:513-519.

Dosskey, M.G., 2000. How Much Can USDA Riparian Buffers Reduce Agricultural Nonpoint Source Pollution? Pages 427-432 in P.J. Wigington and R.L. Beschta (Eds.). Riparian Ecology and Management in Multi-Land Use Watersheds. American Water Resources Association, Middleburg, VA.

Goolsby, D.A., R.D. Coupe and D.J. Markovichick, 1991. Distribution of Selected Herbicides and Nitrate in the Mississippi River and its Major Tributaries, April through June 1991. USGS, Water-Resources Investigations Report 91-4163.

Goolsby, D.A., E.M. Thurman, D.W. Kolpin and W.A. Battaglin, 1995. Occurrence of Herbicides and Metabolics in Surface Water, Ground Water, and Rainwater in the Midwestern United States. In "AWWA," Anaheim, CA, 583-591.

Haycock, N.E., T.P. Burt, K.W.T. Goulding and G. Pinay (editors), 1997. Buffer Zones: Their Processes and Potential in Water Protection. Proceedings of the International Conference on Buffer Zones, September 1996. Quest Environmental, Hartfordshire, United Kingdom.

Helmers, M.J. 2003. Two-Dimensional Overland Flow and Sediment Transport in Vegetative Filter. Unpublished PhD diss., Lincoln, NE. University of Nebraska-Lincoln, Department of Biological Systems Engineering.

Lowrance, R., L.S. Altier, J. Denis Newbold, R.R. Schnabel, P.M.Groffman, J.M. Denver, D.L. Correll, J.W. Gilliam, J.L. Robinson, R.B. Brinsfield, K.W. Staver, W. Lucas and A.H. Todd, 1995. Water Quality Functions of Riparian Forest Buffer Systems in Chesapeake Bay Watershed. Technology Transfer Report CBP/TRS 134/95, EPA 903-R-95-004. U.S. Environmental Protection Agency, Chesapeake Bay Program, Annapolis, MD.

Magette, W.L., R.B. Brinsfield, R.E., Palmer and J.D. Wood, 1989. Nutrient and Sediment Removal by Vegetated Filter Strips. Transactions of the ASAE, 32:663-667.

Muscutt, A.D., G.L. Harris, S.W. Bailey and D.B. Davies, 1993. Buffer Zones to Improve Water Quality: A Review of Their Potential Use in UK Agriculture. Agriculture, Ecosystems and Environment, 45:57-77.

Patty, L., B. Real and J.J. Gril, 1997. The Use of Grassed Buffer Strips to Remove Pesticides, Nitrate and Soluble Phosphorus Compounds From Water. Pesticide Science, 49:243-251.

Robinson, C.A., M. Ghaffarzadeh and R.M. Cruse, 1996. Vegetative Filter Strip Effects on Sediment Concentration on Cropland Runoff. Journal of Soil and Water Conservation, 50:227-230.

Schmitt, T.J., M.G. Dosskey and K.D. Hoagland, 1999. Filter Strip Performance and Processes for Different Vegetation, Widths, and Contaminants. Journal of Environmental Quality, 28:1479-1489.

Snow, D.D. and R.F. Spalding, 1988. Soluble Pesticide Levels in the Platte River Basin of Nebraska. Proceedings of "Agricultural Impacts on Groundwater - A Conference." Des Moines, IA. National Water Well Association, 211-232.

Spalding, R.F. and D.D. Snow, 1989. Stream Levels of Agrichemicals During a Spring Discharge Event. Chemosphere, 19(8-9):1129-1140.

USGS, 1996. USGS Fact Sheets, Miscellaneous Station Analysis, process date July 18, 1996.

USGS, 1998. Water Quality in the Central Nebraska Basins, Nebraska, 1992-95. Circular 1163, 33 p.

Sediment – Water Phosphorus Equilibrium in Ozark Streams

Brian E. Haggard, Research Hydrologist, USDA – ARS PPPSRU, 203 Engineering Hall, Fayetteville, AR 72701; tele: 479.575.2879; email: haggard@uark.edu

External sources of phosphorus (P) to aquatic systems often influence P equilibrium between stream sediment and water, particularly in stream receiving municipal wastewater treatment plant effluent rich in P. Sediment – water P equilibrium is largely related to physical characteristics of sediments and may also be affected by some watershed characteristics, notably land use. Over twenty streams were sampled in Fall 2003 and Spring 2004, where three water samples and three composite sediment samples were collected at each site during each season. Water samples were analyzed for soluble reactive P (SRP), nitrate–nitrogen (NO_3–N) ammonium–N (NH_4–N), total N (TN), total organic carbon (TOC) and sestonic chlorophyll a (chl a). Sediment equilibrium P concentration (EPC_0) was estimated using a linear adsorption isotherm where EPC_0 is the x–intercept and the slope (K) of the relation is a simple measure of sediment P buffering capacity. These streams represent catchments of variable fractions of urban, pasture and forested lands. Sediment EPC_0 will be statistically compared to water chemistry, land use and other physical characteristics of the stream sediments.

Quantifying Nitrogen Loading from On-site Septic Systems to Small Watersheds: Hoods Creek Watershed

S. Pradhan, M. T. Hoover and H. A. Devine

Little is known quantitatively about the extent of nitrate-nitrogen pollution from septic systems to river basins. As a result, existing models and nutrient management plans for North Carolina's River basins have typically ignored these inputs (NCDENR, 1997). Yet the potential for septic systems to have significant impact exist. In an attempt to address this, soil and water assessment tool (AVSWAT2000) was used. The Hoods Creek watershed, located in lower coastal plain of North Carolina, was chosen as a study site due to its unique nature of all the houses being served by on-site septic system for their household wastewater treatment. In this study we considered each house as a single non-point source, and nitrogen loading from septic systems to sub-surface as daily loading. Nitrogen loss through volatilization and plant uptake during movement of effluent from septic tank to the watershed outlet was found to be insignificant. However, 50 – 75% of nitrogen loss was found to be through denitrification process. Distance of houses from the creek and the type of soil the houses were built-on had direct effect on loss of nitrogen through denitrification.

A FRAMEWORK FOR POLLUTANT TRADING DURING THE TMDL ALLOCATION PHASE

A. Z. Zaidi[1], S. M. deMonsabert[2], R. El-Farhan[3], and S. Choudhury[4]

1. ABSTRACT

The Environmental Protection Agency (EPA) encourages pollutant trading programs that help achieve Total Maximum Daily Load (TMDL) implementation goals. Such trades need to be consistent with water quality standards. For an approved TMDL, EPA recommends that the point and/or nonpoint source waste load allocations be used as the baseline for trading credits. The complexity inherent in modeling the effects of pollutants on receiving bodies makes it difficult to understand the implications associated with trading pollutant loads from different sources. The reduction of one credit from one source does not equal the reduction of one credit at another location in the watershed. Similarly, unit costs for load reductions vary considerably depending on the control strategy and the level of reduction. Though trading ratios may account for some uncertainties associated with estimates of nonpoint source loads and long-term performance of the control measures, the selection of an effective trading ratio is not very straightforward and does not fully address the environmental impacts. Although higher trading ratios may be used, this alone cannot guarantee that the water quality standard will be met and may unnecessarily increase mitigation costs. This paper proposes an alternative strategy for the TMDL trading framework. Instead of explicitly determining trading ratios, a trading scenario selection method is utilized. Water quality is simulated for the alternate trading scenarios based on various pollutant loads obtained from accepted models such as HSPF. The costs associated with the nonpoint reductions are compared for various load allocation strategies. This comparison enables an efficient screening of watersheds to identify those well-suited for a pollutant trading strategy. Watersheds with high cost variations and flexibility among allocation strategies are ideal candidates for trading. The methodology is demonstrated using the results of the TMDL allocation for the Muddy Creek WAR1 subwatershed in Rockingham County, Virginia.

KEYWORDS. TMDL Trading, Fecal Coliform TMDL, Economic Analysis, Optimization, Mathematical Modeling

INTRODUCTION

According to the United States Environmental Protection Agency (EPA) the market-based approaches such as water quality trading provide greater flexibility and have potential to achieve greater environmental benefits than would otherwise be achieved under more traditional regulatory approaches (EPA, 2003). Given the inherent scientific uncertainty in the nature of nonpoint pollutants there is a high risk involved in complying with environmental regulation while trading nonpoint pollutants. The objective of this paper is to utilize a previous model developed by Zaidi et al (2003) as a decision tool to evaluate and determine the cost-effectiveness of various trading options during the TMDL allocation phase. The approach uses a mathematical model to minimize the costs for the subwatershed load allocation to achieve the

[1] Graduate Research Assistant, George Mason University, Fairfax, VA. 22030-4444, azaidi@gmu.edu

[2] Associate Professor of Civil, Environmental and Infrastructure Engineering, George Mason University, Fairfax, VA. 22030-4444, sdemonsa@gmu.edu

[3] Senior Water Resources Engineer, The Louis Berger Group, Washington, D.C., 20037, relfarhan@louisberger.com

[4] Postdoctoral Fellow, George Mason University, George Mason University, Fairfax, VA, 22030-4444, shc758@yahoo.com

desired water quality goal while minimizing the uncertainties associated with the nonpoint pollutant trading. The costs associated with the subwatershed load allocations establish the basis for pollutant trading.

TMDL TRADING FRAMEWORK DURING THE ALLOCATION PHASE

Use of the proposed cost optimization approach can help to determine at the allocation stage whether a specific TMDL lends itself to a watershed-based pollutant trading approach. Furthermore, the approach will provide an estimate of the potential savings that may be realized by incorporating a pollutant trading approach in the TMDL load allocation phase. Water quality based trading accomplishes time and economic efficiencies. Not only are the reductions more cost-effective but also the process of achieving the reductions necessary to meet the water quality standard is accelerated.

For an approved TMDL, EPA recommends the consideration of applicable point source waste load allocations or non-point source load allocations to establish a baseline for trading credits. A baseline is defined as the level below which a reduction is made to create a pollutant reduction credit (EPA, 2003). Pollutant trading takes advantage of the control cost differentials and economies of scale between various sources of pollutants. Under EPA guidelines, watershed based trading may be considered if it results in an overall reduction of pollutant loads without violation of the water quality standard. Commonly, "trading ratios" are used to account for the expected differences in impact on water quality, for two different loadings in a watershed. These factors, may also account for the uncertainties associated with estimates of nonpoint source loads and reductions achieved through treatment options or other control measures. The relative impact of one unit of pollutant discharged from different sources from varied locations is never the same.

MUDDY CREEK WATERSHED

In this paper, nonpoint trading options are evaluated under the framework of an approved TMDL for fecal coliform bacteria in a subwatershed (WAR1) of the Muddy Creek watershed. For TMDL modeling purposes, the Muddy Creek watershed was delineated into eight interconnected subwatersheds as shown in Figure 1. The Muddy Creek watershed is located in Rockingham County, Virginia. Muddy Creek is on the Commonwealth of Virginia's 1998 303(d) list of impaired waterbodies because of fecal coliform bacteria violations. The Virginia Department of Environmental Quality (VADEQ) ambient water quality monitoring stations recorded the exceedance of the standard to indicate that the stream does not support primary contact recreation (swimming). The Muddy Creek fecal coliform target was a geometric mean of 200 counts/100ml with 0% violation. The Muddy Creek loads are comprised of direct loads (in-stream discharges due to animals, failed septic systems and uncontrolled releases) and indirect loads (surface depositions resulting from land use).

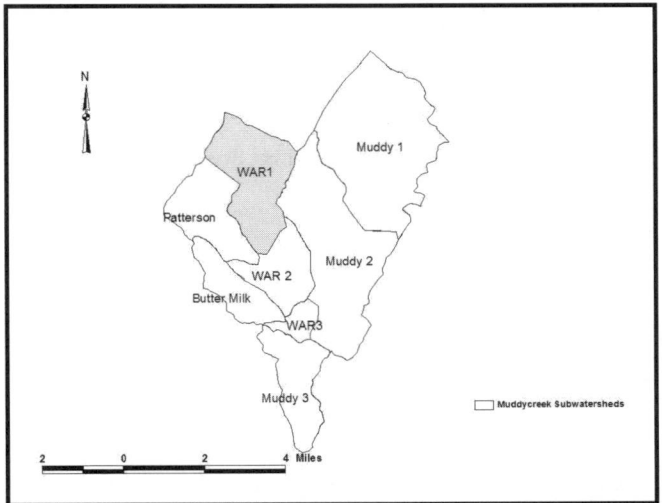

Figure 1. Muddy Creek Subwatersheds

POTENTIAL TRADING SCENARIOS

The proposed methodology identifies viable load allocation tradeoffs such that implementation costs are minimized while achieving an acceptable level of water quality in the stream. The optimization model identifies the minimum cost mitigation approach for a given allocation scenario. A comparison of the costs associated with the mitigation strategies for various allocation scenarios, provides a starting point for discussion and evaluation of trading options. For several reasons, including site characteristics (soil type, slope, etc.), fecal load deposition (direct vs. indirect), diversity in control strategies, etc. the remediation costs may differ among sources. Only scenarios that produce modeled results consistent with the TMDL standards are considered. In the case of the Muddy Creek WAR1 subwatershed, the trading scenarios were modeled using a calibrated HSPF model provided by The Virginia Department of Conservation and Recreation (DCR). The fecal coliform bacteria sources for the Muddy Creek WAR1 subwatershed are: land-based loads, wildlife loads, failed septic systems, and direct deposit as listed in Tables 1 and 2. In the final TMDL allocation report for Muddy Creek (VADEQ, 2000), the allocated TMDLs are based on variable monthly reductions for each source. For the purposes of this work, the maximum monthly reduction from each source was used in the analysis.

Table 1: Baseline Scenario

Nonpoint Source	Total Existing Load (counts/yr)	Load Reduction (counts/yr)	PV Cost[1] ($)
Land-based Loads	3.51E+11	1.68E+11	3,085
Wildlife Loads	2.67E+10	0.00E+00	30,056[2]
Failed Septic System	1.04E+11	1.04E+11	17,286
Direct Deposit	1.95E+13	1.95E+13	164,682
Total	2.00E+13	1.98E+13	215,109

1. Total Present Value (PV) costs are calculated for a 7% interest rate and 15-year planning horizon.
2. This management option reflects the cost of maintaining the existing wildlife population (assuming a 10% growth rate). As such, it does not constitute a load reduction.

Tables 1 and 2 present the baseline scenario based on the TMDL allocation report (VADEQ, 2000), and a potential trading scenario respectively. In these tables the annual load reduction from each source and the cost of achieving that reduction are given. These costs depend not only upon the level of load reduction but also upon the choices and selection of control measures. In this study the vegetative buffer strip for land-based load, wildlife management for wildlife load, system repair and installation for failed septic systems, and streamside fencing for direct cattle deposit are considered as control strategies.

Table 2: Potential Trading Scenario

Nonpoint Source	Load Reduction (counts/yr)	PV Cost ($)
Land-based Loads	8.59E+10	1,094
Wildlife Loads	1.47E+10	31,751
Failed Septic System	0.0	0
Direct Deposit	1.95E+13	164,682
Total	1.96E+13	197,527

As shown in Table 1 the total load reduction cost for the baseline scenario is $215,109, whereas for the alternate trading scenario (Table 2) the cost is $197,527. The alternate trading scenario presents a cost savings of 8% from the baseline scenario. The baseline scenario represents a slightly higher annual load reduction with higher cost as compared to the alternate trading scenario. Despite a lower annual load reduction, the alternate scenario is also a viable scenario since it complies with the state standard as modeled in Figure 2. Also the geometric means of the fecal coliform concentrations for the trading scenario are generally less than for the baseline scenario. In both cases, the direct deposit load reduction dominates the load allocation. The direct deposit is several orders of magnitude greater; for this reason, reductions in the land-based, wildlife and failed septic system loads are insignificant relative to the direct deposit reduction.

338

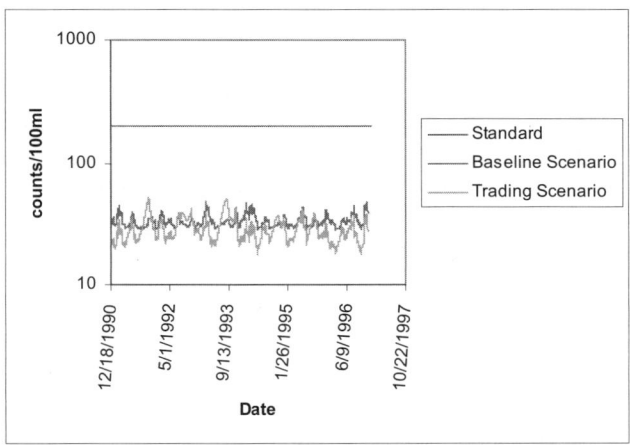

Figure 2. 30-Day Geometric Mean Fecal Coliform Loadings under Baseline and Trading Scenarios

A matrix of viable trading scenarios may be produced using the calibrated HSPF model. The calibrated HSPF model for Muddy Creek watershed obtained from DCR is used for this analysis. This matrix of options will help to identify potential tradeoffs in the reduction of pollutant loads from different categories of nonpoint sources to levels below the standard and their associated costs.

The existing load distribution in the watershed does not offer much flexibility to the Muddy Creek subwatershed WAR1 for pollutant trading because of the dominance of the direct loads. For the illustration purpose, a synthetic load distribution was used in the following section to better illustrate the strengths of the model.

TRADING SCENARIO ANALYSIS

The existing loads for the WAR 1 subwatershed were multiplied with randomly selected coefficients to provide a synthetic load distribution that would better illustrate the proposed methodology. Table 4 shows the synthetic loads used in the analysis. Table 5 presents three alternative trading scenarios. Each of these scenarios represents an option that will meet the water quality standard. Individual and total load reduction costs for each source are shown in Table 5.

Table 4: Synthetic Loads

Nonpoint Source	Loads (counts/year)
Land-based Loads	2.66E+12
Wildlife Loads	2.02E+11
Failed Septic System	2.60E+10
Direct Deposit	9.31E+11
Total Synthetic Load	3.82E+12

Table 5: Trading Scenario PV Costs

Nonpoint Source	Scenario I		Scenario II		Scenario III	
	Reduction[1] (%)	PV Cost[2] ($)	Reduction[1] (%)	PV Cost[2] ($)	Reduction[1] (%)	PV Cost[2] ($)
Land-based Loads	57 (75)	7,409	0	0	73.7 (74.2)	4,899
Wildlife Loads	67.8	249,113	87.13	253,618	49.4	244,854
Failed Septic System	100	4,322	0	0	0	0
Direct Deposit	80 (89.7)	94,706	79 (89.7)	94,706	92 (93.3)	127,942
Total	63.6 (77)	355,550	24 (26.5)	348,324	76.4 (77)	377,695

1. Values in parentheses are the actual reductions achieved after applying the control measures.
2. Total Present Value (PV) costs are calculated for a 7% interest rate and 15-year planning horizon.

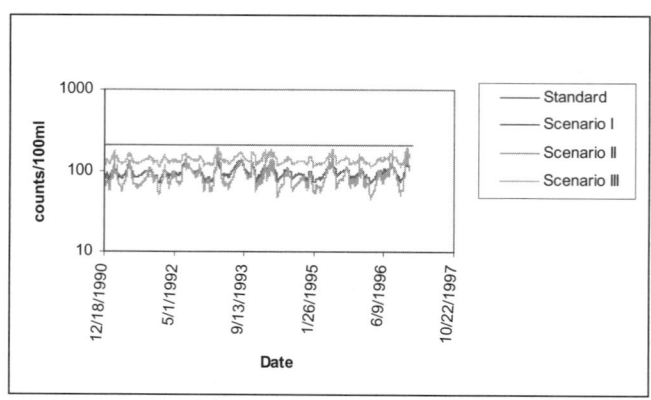

Figure 3. 30-Day Geometric Mean Fecal Coliform Loadings under Scenarios I, II and III

Discussion

Scenario II provides the best results from both an economic and environmental perspective. The modeled geometric means of the fecal coliform concentrations are generally less than those for Scenarios I and III (see Figure 3). It is interesting to note that the least cost option (Scenario II, minimum annual load reduction) provides the "best" environmental solution. This result supports the need for economic assessment during the load allocation phase of the TMDL development. In a trading environment, Scenario II would be preferred over Scenarios I and III.

Also a comparison of Scenarios I and III supports a trading solution. Scenario III yields a solution that is environmentally inferior to Scenario I, yet the total cost is greater. The total cost is 5.86% greater for Scenario III than Scenario I despite a reduced environmental load. This result suggests that trading could result in an improved solution (both environmentally and economically). Under Scenario I, the land based, wildlife and failing septic systems loads would sell the 'polluting rights' to the direct deposit loads for reduced total costs and improved water

340

quality. To better understand the concept, consider scenario III as the baseline scenario. A mutually advantageous trade may be conducted between scenarios I and III. Greater reductions in the pollutants from land-based, wildlife, and failed septic systems are needed in scenario I than in the scenario III. Similarly the costs associated with these added reductions are greater. The benefits associated with reducing the direct deposit load reductions ($33,236) far exceed the net dollar loss ($11,090) from the other three sources. Therefore, an opportunity for trading between direct deposit sources and the other sources exists. Direct deposit sources can purchase 'pollutant rights' from other sources. If the 'pollutant rights' cost more than what is required by the three sources to achieve the level of pollutant reduction in scenario I (more that $11,090), then the direct load sources will save money by purchasing these rights at a cost less than $33,236 (their cost to reduce a comparable load).

CONCLUSIONS

The trading scenario selection method provides improved assurance of meeting the water quality goal. Economic modeling during the allocation phase of TMDL development is vitally important to the environmental and economic success. Cost data, coupled with modeled geometric means of fecal coliform concentrations will enable improved allocations that accomplish both environmental and economic goals. Without economic information for the various proposed allocation scenarios, it is difficult to determine what solution would be preferred. For example, in the case study presented above, the solution with a higher load reduction produces an environmentally and economically inferior solution. This solution presents an opportunity for improved water quality at a reduced cost through the use of trading scenarios.

REFERENCES

1. USEPA. 2003. Final Water Quality Trading Policy. United States Environmental Protection Agency, Office of Water, January 13, 2003.
    http://www.epa.gov/owow/watershed/trading/finalpolicy2003.html

2. VADEQ. 2000. Revised Final Report: Fecal coliform *TMDL Development For Muddy Creek, Virginia*. The Muddy Creek TMDL Establishment Workgroup.
    http://www.deq.state.va.us/pdf/rivers/mdycrk.pdf

3. Zaidi, Arjumand Z., deMonsabert, S. M., El-Farhan, R. 2003. A Cost-Based Strategy for TMDL Allocation. Proceedings of the Second Conference on Watershed Management to Meet Emerging TMDL Environmental Regulations, ASAE, Albuquerque, New Mexico, November 8-12, 2003.

The Influence of the Water Exchange Intensity on the Contamination Processes in the Environment of Aral Sea Region

D. V. Z. [1]

[1] Institute of Hydrogeology and Engineering Geology (HYDROENGEO)
Tashkent, Uzbekistan

moaynauka@rambler.ru, hydrouz@rambler.ru
Phone: (+998 71) 1624763, Fax: (+998 71)1624763
Address: N. Khodjibaev st., 64, 700041

ABSTRACT

Nowadays many scientists and researchers emphasize solutions for the ecological problems of Aral Sea Region. In this paper attention has been paid to underground water and soil contamination processes. A study of such processes on the irrigated territory is important for decision-making and development of the efficient actions over improvement of the ecological situation in Aral Sea Region.The main sources of technogenic influences in this particular situation are agriculture, industrial enterprises, acid rains, and dusty storms. These influences are coming from the dried areas of Aral Sea bottom and from the polluted water of Amudariya River.

In this research an attention is paid to the forming factors of hydrochemical conditions and barriers that promote accumulating processes of the toxic microelements. Using the results of chemical analysis three danger classes of microelements are specified. The concentrations of them in the underground water greatly exceed the norms. The row of toxic components polluting underground water and soils was defined, and the questions of the spatial sharing of polluting metals have been considered. During the investigation process the influence, diluting the abilities of the surface water on spreading of contamination and changes in their concentration have been noticed that is also stipulated by the particularity of the water exchange. One specifies a feature of hydrogeological terms on the considered area is a shaping natural-technogenic type of water exchange. In this condition, the specific particularities of the contamination and self-clear processes of geological ambience also come into the light. The specified type of water exchange is conditioned by complex influences. There are natural and technogenic factors.

The total part of the research was an estimation of underground water quality and hazard of heavy metals spreading into adjoining areas with the provision of the natural and man-made type of water exchange.

KEYWORDS. : ecological problem, pollution processes, hydrochemical conditions, toxic components, water exchang, possibility of spreading toxic microelements.

---

[1] Put affiliations and addresses (e-mail optional) for each author in footnotes, as you would like them to be printed in the Proceedings book.

Nowadays scientists and researchers of many countries pay attention to solution of the ecological problem in Aral Sea Region. Processes of contamination of the geological ambiences (soils, priming and underground water) are considered in present paper. Studies of such processes on an irrigated territory are great importance. It is important for decision making and development of efficient action for the improvement of the ecological situation. In relation with ecological disaster in Aral Sea Region the actuality of this question is increases.

It is necessary to make a number of practical conclusions and recommendations required for the improvement of the ecological situation to solve the ecological problems. Herewith all interacting natural and anthropogenic factors should be considered. Each factor should get as qualitative, and quantitative estimation as well. The rating of the territory with view of its agricultural utilization is the necessary part of ecological studies too. The character and result of technogenic influences on geological environment is defined not only by type of this influence, but particularity of the most geological ambience as well. Such particularities of the construction of the geological ambience of the Aral Sea Region are following:
- The plain relief with very small value of the surfaces gradient (0,002).
- The prevalence of evaporation from underground water surface on waterflow doun.
- The absence of powerful humus horizon, which are the most important geochemical barrier on the migration way of the pollutants.
- Close relation of surface and underground water because the level of underground water is close towards the land surfaces.
- Low level of the first water horizonts from contamination.

One of the natural sources of microelements income to the topsoil is the maternal soilmaking rocks. The specific characteristic of microelements in the rock are practically always saved in soils. A genetic relationship between soilmaking rocks and soils is the reason of it. It is reflected in composition of underground water, which is closely connected with surface waterflow (rivers, channels, collectors of drenaige-irrigation network). The sources of technogenic influences and contaminations in Aral Sea Region are plenty enough. The basics of the last are following:
- an agriculture;
- industrial enterprises;
- acid rains;
- polluted water of Amudariya river;
- dusty boers from dry parts of the Aral Sea bottom.

During research shaping factors of the hydrogeochemical conditions have been considered. On the study territory the hydrogeochemical terms promote the process of the contamination and accumulations of the toxic microelements. High concentrations of chemical elements in the underground water of Aral Sea Region are defined by the fllowing geochemical processes:
- increasing of mineralization;
- change a correlations between concentration components limiting of the distribution elements;
- change pH and Eh parameters;
- increasing concentration of components, being catalyst for complex elements.

The microelements of the first (Cd, Hg, As, Zn), second (Li, Cr, Ni, Mo, Co) and therd (Mn, Sr) dangerous classes have been revealled in a result of the chemical analysis. These microelements are contained in soils of the aeration zone and in underground water. A presence of the geochemical barriers has been revealled. A accumulation and precipitation of determined types of elements proceeds here. In accordance to litological construction of the geological cut and geochemical particularities of the ambience the geochemical barriers are formed. The main geochemical barrier formed in loamy-sand litological difference of rocks. It is sorbtion-clayey barrier. It is stipulated with the presence of the negative charge on the surfaces of clayey minerals. The Li, Be, Zn, Cu, Cd, Pb, Hg, Co and Ni are precipitated here. The action of alkaline-carbonate barrier founded on the formation of difficult dissolved carbonate of catiogenic elements (Fe, Mn, Co, Zn, Pb). Those are the properties of soils with high level of carbonatic. The carbonatic grade in aral Sea Region soils achieves 16%. When the pH of ambiences increases an alkaline-hydrolitical barrier appears. Herewith many elements, migrating in the manner of complex (Be^{2+}, Hg^{2+}, Cr^{3+}, Fe^{3+}, Mn^{3+}, Cu^{2+}), completely or partly come in to hard phase as hydrooxids. The concentration of revealled toxic elements in underground water and soils greatly exceeds the possible standards. It once again confirms the low quality of the geological ambience and calamity of the ecological condition in Aral Sea Region.

Processes of spatial distribution of components also have been considered also. An influence of the surface waterflow, diluting abilities on spreading contamination processes has been discovered. It is conditioned of a water exchange particularities on the investigation territory. One of the features of hydrogeological conditions is a shaping naturally-man-made type of a water exchange. Such type of a water exchange actuates the processes of the selfcleaning of geological ambiences. It is formed under the influence of the complex factors. It is possible to select a natural and a man-made factors. About the natural factors have been already spoken earlier. The main man-made factor is an active exploitation of the irrigation-dreanage network by human beings. The depth of an irrigation-dreanage network channels can be achieved maximum up to 2-3 m. Water exchange in upper zone of underground water is actuated as a result of horizontal and vertical filtering of water from the surface waterflow.

In hydrogeology sciences the water exchange called the process of the unceasing change of underground water in water horizonts. It is stipulated by their supply, sewer and unload. Many researchers studied and continue to discover this process with different points of view. G.N.Kamenski as far back as 1943 was entered concept of water exchange ratio, which is relation of the annual consuption of underground water of the water horizon bassen to their total reserves. It characterizes the intensity or rates of the changing process. The process is the chenging of underground water resources by the water which comes into water horizonts from the different sources. Quantitative evaluation of the waterchange rates on separated areas can be done differently. The are many estimations made by different scientists (Korotkov, Pavlov, Sultankhodjaev, Kudelin, Bredemcamp, Vogel, Vsevolojski,Cheban, Dzilna, Zekcer and others). For example, N.A.Juravel (1979) considers a brain mineralization using waterchanche parameters. He offers the calculations determining the waterchange periods and the number of cycles. At the same time V.A.Borisov (1990) consider the same quantitative waterchanche factors as a rete, velocity and number of cycles. M.M.Krilov (1977) shows "… general and local underground sewer of the same irrigated mass often different".The velocity of general sewer of Amudariya pool is in avarage several meters annualy. Local sewer sometimes reaches considerable values. Why? Again here is the deep relation between surface and underground water.

The research area is considered as a natural difficalt water exchange. The research results of A.A.Rachinski and M.S. Merishensky (SANIIRI, Uzbekistan 1973) and "Methods of regional evaluation of drenagability of irrigated zones in arid conditions considering ground water (example oasis of Uzbekistan)", (Institute Hydrogeology and Engineering Geology, Uzbekistan, 1977) have been taken as a basis. On the basic part of the area there are loams with the capacity at 2-3 to 5-7 meters (aeration zone). Under loamy soil refoliated with a sandy soil and clay there is ia alluvial sand (fine-graned and small-grained). Subsoil water are found on the depth of 0,5-1,5 to 2,5-3,0 meters. We should note that drainage waterflow (7470 m^3/ga) has to be predominate under total evaporation of the ground water (2900m^3/ga) for more then 2 times. So, the research area is practically drainaged and the average long term yield is 25-35 thousand m^3/ga. Infilitration flow provides intensive water exchange in aeration zones. That's why in a close badding (1,5-3,0 m) subsoil water are often contain less salt (up to 3 g/l).

Besides last considered methods, during the process there were indicated quantitative analysis of water exchange factors. The analusis has been carryied out on the basis of mathematical formulas of M.M.Krilov & V.A.Borisov. The output formula for calculation was a formula, defining duration of water exchange:

$$T = (\mu * \quad / \quad)$$

$$T = \frac{\mu * F}{K_\pi * I * L}$$

where: **T** – duration of water exchange
               μ - average long term yield factor
               F – an area of the territory
               K –transmission coefficient of
                       water-bearing materials
               I – a gradient
               L – a length of the filtration way

For calculation of the water exchange duration, the research area was conditionally divided into several regions. The differenciation of the areas carried out based of the qualitative estimation of the water exchange. Herewith, on the initial stage, the natural factors of the shaping water exchange were taken into account. For each area a calculation has been made to determine the duration of water exchange. Based on the calculations data, shows that minimal duration is 115 years that prooves the character of showed down natural water exchange.

The activation of water exchange at upper layers of ground water and in aeration zone is confirmed by following facts. The velocity of ground waterflow is much bigger than the velocity at lower layers. The concentration of microelements at upper layers is well above that at the lower layers and on the surface of ground water.

Aral Sea Region is not a big urban aglomeration. The pollution scale has a local character. In this case deluted properties of surface waterflows are able to provide a cleaning of ground water. But when the values of relief gradient is low (I=0.002) and the low velocity of ground flow (6 m/year) it is categorically unacceptable increasing the scale of getting toxic substance in geological environments. It can bring the pollution in lower horizonts. Herewith the attention should be paid to location of drainage bore holes. Increasing the water rate from the bore hole and increasing the depression cone can cause a danger of appearance of polluted ground water in drinking water.

CONCLUSION

Therefore, the following can describe the water exchange influence to the pollution processes in Aral Sea Region. The water exchange has to be discribe a two basic positions. Last reflect the types of water exchange, which was determinated in Aral Sea Region:
1. natural water exchange shaping in natural condition;
2. natural-man-made water exchange shaping under influence of man activity expressed in agricultural activity (exploitation of irrigation-drenaige network)

During natural water exchange there is a danger that the accumulation of pollution components can occur. The accumulation process is described as flow less. But in relationshion with existing conditions, natural way of water exchange forming assum a natural-man-made charachter. It happens, first, because of the presence thick irrigation-drenaige network. During infiltration into the zone of aeration, water goes into ground water canals and delite the concentration of toxic microelements. Their contamination reduces.

So, in a condition of good work ability of irrigation-drainage network, zone of more active water exchange forms in upper layers at water horizont. Here we can say about selfcleaning ability of the environment. It is a collection of processes which happen in polluted water objects which are targeted on a regeneration of initial structure and properties of water. Therefore, we see double-influence of the human being not always negatively reflected on a condition of natural systems. Creation and exploitation artificial channels and collectors which are used for irrigation , has stipulated natural-man-made process of selfcleaning of polluted by humans underground water.

REFERENCES

1. **Journal Article**

2. Mamaev Yu.A. 1996. Methodic aspects of complex estimation of dangerous natural processes and phenomenons. *Geology.* 1(02): 13-21.

3. **Book**

4. Bakhireva L.V., Osipov V.I., Koff G.L., Rodina E.E. 1997. Geological and geochemical hazard as creteria of geoecological norming of territory. Moscow.

5. Akhmetjeva N.P., Lola M.V., Goretskaya A.G.. 1991. Pollution of undergraund water with fertilizers. Moscow: Nauka. 100 pg.

6. Zaikov G. E., Maslov S.A., Rubailo V. L.. 1991. Asid rains and environment. Moscow: Himija. 144 pg.

7. **Published Paper**

8. Diana V. Zakhidova. Major sources of soils and underground waters pollution in Khorezm region of the Republic of Uzbekistan.// Proceeding of the International Geotechnical Symp.-Saint Petersburg, 2003, pg. 125-127.

9. Diana V. Zakhidova. Influence of agricultural activity of human beings on environment condition.//Proceeding of Republic Scientific-Practical Conf. "Technosphere, Man and Microelements". Tashkent, Uzbekistan. 2004. pg. 242-246.

10. The 4th Ministerial Conference "Environment for Europe", States of Central Asia: Environment Evaluation. Aarhus, Denmark,1998.

11. **Dissertation or Thesis**

12. Diana V. Zakhidova. 2004. Contamination of soils and underground water by heavy metals in difficult water exchange condition (on example of the Urgench city). Unpublished PhD diss. Tashkent, Uzbekistan: Institute of Hydrogeology and Ingineering Geology.

13. **Online Source**

14. Diana V. Zakhidova, Yunir V. Gataullin. Complex approaches for solving the geoecological problems of the Aral Sea Region.// 55th International Astronautical Congress.-Vancouver, Canada, 2004. Available at: http//www.iafastro.com/congress/Vancouve%202004/Call/prabs04/Symp_B.htm. Accessed 10 May 2004.

Bankfull Channel Dimensions in Southeast Ohio

Tiao J. Chang[1], Yan Y. Fang[2], Huixian Wu[3], and Daniel E. Mecklenburg[4]

Abstract:

Relationships between bankfull channel dimensions and their associated drainage areas in the southeast region of Ohio are established. This may be used to provide the information base to assess impacts of projects for modifying channels and a possible tool to design channels for minimizing these impacts. Thirty-five sites were selected to geographically represent the unglaciated Allegheny plateaus region of Ohio and for the bankfull channel geometry determinations including width, depth and cross-sectional area relative to drainage area. A total station with an electronic sensor was used for channel cross-sectional and profiling survey. ArcGIS was used for delineating watersheds and the collected survey data were analyzed using The Reference Reach Spreadsheet Excel software. It is found that the bankfull dimensions are strongly related to the associated drainage area. No other variable showed association including further subdivision of the region.

Keywords: Bankfull Dimensions, Channel Dimensions, Regional Curve

Introduction:

Generally, the hydrologic features of a stream include channel variables of pattern, dimension, and longitudinal profile. It is believed that bankfull discharge can most effectively maintain a natural stream channel with deposition, carrying sediment with least effort, and result in average morphologic features of a natural channel. Leopold et al. (1964) showed that bankfull discharge largely controls the form of alluvial channels. Hence, bankfull discharge can also be considered as the maintaining flow because it generally transports the largest amount of sediment. Additionally, erosion and bar-building by deposition are most active at discharge near bankfull. Therefore, channel features such as patterns, channel dimensions, or longitudinal river profile associated with bankfull stage are of value for further investigation.

Bankfull is an essential term in stream morphology when describing the physical character of a natural stream. It is used to indicate the height of water that just fills the stream channel or the elevation where flooding begins (Rosgen 1996). Then, the bankfull stage is the surface elevation at which water begins to flow over the banks filling the floodplain. Furthermore, the bankfull discharge is the maximum discharge that a stream channel can carry without overflowing its natural banks. Thus, bankfull stage can also be described as the water depth at the bankfull discharge.

Channel form and stream flow are interrelated and mutually interact with each other to create the morphology of a river system and the dimensions of the bankfull channel. However, determining bankfull is often not a simple exercise particularly where the channel or watershed has been modified or the channel is unstable. Knowing approximately what size channel to expect can be a valuable aid. Bankfull channel dimensions tend to be strongly related to drainage area (Leopold et al., 1964). For a certain stream type, the relationship between bankfull channel dimensions and drainage area can be useful for estimating channel dimensions in the similar areas (Dunne and Leopold, 1978). Hence, developments of local relationships and curves based on the collected field data of channel dimensions could be valuable for stream assessment or restoration. Based on channel measurements in southeast Ohio, this study is to investigate and

---

[1] Professor, Civil Engineering Department, Ohio University, Athens, Ohio 45701
[2] Graduate Associate, Civil Engineering Department, Ohio University, Athens, Ohio 45701
[3] Visiting Scholar, Yellow River Conservancy Commission, Zhengzhow, Henan Province, China
[4] Ecological Engineer, Ohio Department of Natural Resources, Foundation Square Bldg. E-2, Columbus, OH 43224

develop regional relationships of bankfull channel dimensions and their associated watershed areas.

Site Selection:
Three areas within the un-glaciated physiographic regions of Ohio were selected for study for two reasons. The first is that they represent the three districts of Wayne National Forest which helped fund the study and second because of their distribution across the region would allow further sub-regionalization if differences were found. At all of the thirty-five sites selected, relatively stable reference reaches of approximately twenty channel widths in length, were surveyed. Eight of the sites are close to or located at stream gages maintained by USGS. Careful thoughts were taken in determining the rest of the reference sites. The guideline developed by Harrelson et al. (1994) was used to fill in the rest of the survey sites to geographically represent the Region. Sites with drainage areas that complement the USGS gaging stations to obtain a good distribution of drainage basin size were desired. The associated drainage areas for those sites close to USGS gauges were provided by USGS. For those non-USGS sites, the drainage basins were delineated using ArcGIS and their associated drainage areas were estimated. Cross-sectional and longitudinal surveys were performed at these sites to obtain channel dimension, profile and bed material data. Bankfull indicators were identified to determine bankfull stage for every site.

To determine the size of corresponding drainage areas for the selected sites, ArcGIS was applied to delineate the basin. A drainage basin is the area that supplies water to a given site. Hydrologic and geomorphic impacts of natural or man-made processes within a drainage basin or watershed are focused at its outlet (Dunne and Leopold, 1978). Therefore, the outlet location is critical to describe the drainage basin of interest.

Delineation is the process to define the boundary of a drainage basin or watershed. To delineate watershed is to map out the drainage area and determine the watershed divides, which are usually topographic highs. A watershed boundary is determined by starting from the outlet location selected. With the aid of GIS, the stream and topographic data were downloaded from the USGS geographic database. Hypsography consists of 1:24k DLG quadrangles depicting contour lines connecting points of equal elevation with reference to a vertical datum (USGS & USDA, 2003). After being converted into GIS shape files, these contour lines can be viewed in a GIS map representing a difference in elevation of twenty feet. Contour lines that are close to each other indicate a steep slope while those are farther apart indicate a gradual slope. As moving from one point to another on the hypsography map, one can decide whether it is uphill or downhill by the ascending or descending order of the elevation data. Since water flows based on the law of energy equation, the direction of water flow is from high elevation to low elevation. After the delineation, the drainage area can be estimated for each associated selected site. Figure 1 shows the locations for the sites selected their associated sub-regions.

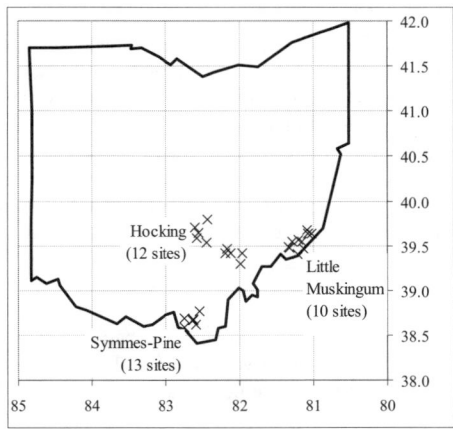

Figure 1 Site locations in areas of Ohio's un-glaciated Allegheny plateaus

Data Collection:

A reference stream reach survey is essential for natural channel design, assessment, and monitoring. The survey provides measured channel dimensions associated with the bankfull stage (Rosgen, 2000). A reference stream reach, approximately twenty times the channel width in length, was selected for each site. It is a stream segment that most likely represents the channel with evident natural features, which include effective factors in developing and maintaining the channel and floodplain. Bankfull indicators were then identified at each reach to perform cross-sectional and longitudinal measurements to collect the data of channel dimensions and profile.

A selected cross section, usually on a riffle or straight segment between bends of a stream, is the location for measuring channel geometries. The cross-sectional survey was performed using the total station with data collector. Figure 2 shows a schematic graph for establishing the cross sectional measurement (Harrelson et al., 1994). At each cross section selected, it measured and recorded all obvious breaks in slope, left and right edge of the water, and important points such as bankfull indicators and active floodplain. At least twenty vertical measurements along the horizontal line were performed at each cross section and three cross sections were surveyed for each site.

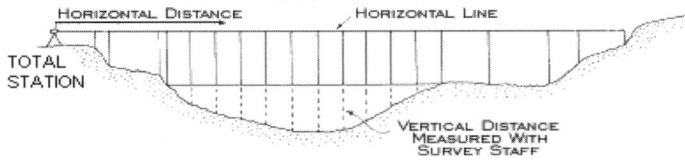

Figure 2 A schematic of cross sectional survey from Harrelson (1994)

The longitudinal profiling consists of the elevation measurements of points along the channel that describe the water surface, channel bed, and bankfull stage. It is also applied for calculating the channel slope. For each survey site, the profile extended approximately 20 times of the bankfull channel width. The elevations of thalweg (lowest point of the channel bed), water surface, center of channel, and bankfull height were measured and recorded at varied intervals of distance along the channel. For significant features, such as breaks in channel bed slope or channel meandering, more survey shots were taken at these locations.

349

It is noted that indicators of bankfull stage vary at different locations and for different stream types. An active floodplain could be useful for identifying bankfull stage, especially when the slope of the watershed is mild. A flat surface or plain adjacent to the channel, somewhat recognizable, is normally a good indication of bankfull characteristics. When the recognizable floodplain is hard to determine, other indicators may be of assistance to identify bankfull stage. According to Harrelson et al. (1994), an abrupt change of bank slope, a depositional surface top, or a change in material are possible supporting elements for bankfull indicators. The process of bankfull identification is best when using comprehensive assessment of several indicators is applied.

Analysis and Results
After the survey data were downloaded into computer, the bankfull stage was determined with the aid of The Reference Reach Spreadsheet (Mecklenburg, 1999). The data collected by the total station contain coordinates of easting and northing and elevations of the surveyed points, including centerline (center of stream), thalweg, water surface, and bankfull elevation. These data are used to develop the channel profile that can be plotted along the reach. Figure 3 shows an example of the Dogskin Run Site, which includes the points of various bankfull indicators.

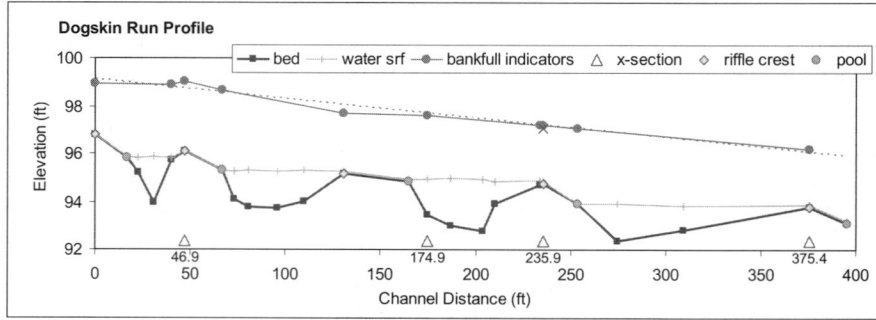

Figure 3. An example of a channle profile survey

To find bankfull channel dimensions the bankfull elevation of each cross section must first be determined. This procedure begins with the bankfull indicators surveyed in the longitudinal profile which are fitted by a linear trend line. The trend line provides a reference about which bankfull stage varies. It is checked to be generally parallel the channel slope. The channel slope is approximated by the water surface slope, S_w, between the crest of the first and last riffles of the study reach:

$$S_w = \frac{X_2 - X_1}{E_2 - E_1}$$
(1)

where X_1 and X_2 are the respective distances between the reference point and locations 1 and 2; E_1 and E_2 are the respective elevations between the reference point and locations 1 and 2.

At each cross section the corresponding elevation of the bankfull trend line is determined. For example, the profile trend line at its intersection with Cross Section 2+35.9 of the Dogskin Run Site has an elevation of 97.25. Next, by accounting for the indicators near the cross section and indicators of the cross section plot itself, the elevation from the trend line can be refined down slightly to 97.1. This is then used to define the bankfull cross section as shown in Figure 4, where the short dashed line stands for the bankfull elevation. Bankfull channel dimensions are then calculated based on this established cross section.

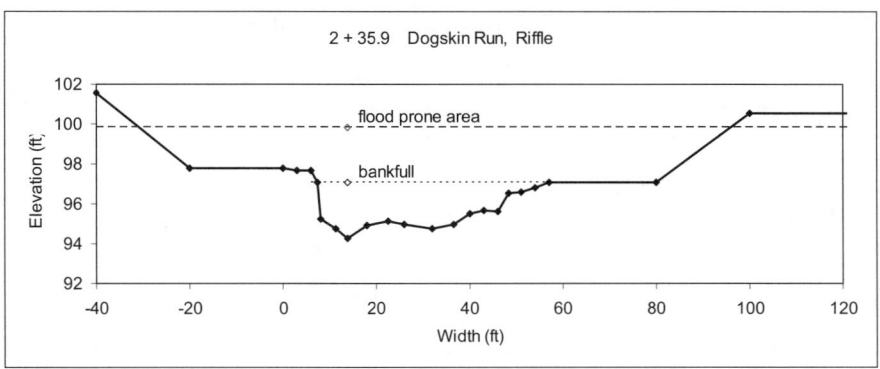

Figure 4. Example of cross section survey

The bankfull cross-sectional area, A_{bkf}, is the area of the stream channel cross-section at the bankfull stage. Bankfull width, W_{bkf}, is the width of the stream channel at the bankfull stage. Mean bankfull depth, D_{mean}, is the mean depth of the stream channel cross-section at the bankfull stage and is defined as:

$$D_{mean} = \frac{A_{bkf}}{W_{bkf}}$$ (2)

This analysis was done with each site to determine the bankfull dimensions. Table 1 lists the bankfull dimensions and their corresponding drainages and sites in the basin.

351

	Sites	Drainage Area (mi2)	Area (ft2)	Width (ft)	Mean Depth (ft)	Slope (%)	Channel Type
Hocking	Morristown	7.8	155	52.1	3.0	0.06	C
	Twphwy_129	12.8	283	62.2	4.6	0.16	C
	Rd_690	14.0	366	46.5	7.9	0.05	E
	USGS 3158000	14.8	211	42.1	5.0	0.01	E
	USGS 3158195	24.5	269	51.8	5.2	0.09	E
	Rd_329N	26.0	344	72.2	4.8	0.18	C
	USGS 3156600	30.0	381	54.3	7.0	0.18	E
	Rd_329	35.7	438	83.0	5.3	0.030	C
	USGS 3156400	48.2	714	106	8.3	0.030	C
	USGS 3157000	89.0	648	89.6	7.2	0.08	C
	USGS 3156500	90.3	582	76.5	7.6	0.11	E
	Elm_rook	116	530	92.4	5.7	0.008	C
Symmes-Pine	Easter Hollow #1	0.030	3.5	7.8	0.5	4.2	C4
	Jones Cr Trib	0.050	3.4	7.7	0.4	7.7	A4
	Easter Hollow #2	0.34	19.1	14.6	1.3	1.2	E4
	Caulley Cr Trib	0.50	14.1	12.5	1.1	1.9	E4
	Union Br	0.75	21.1	16.7	1.3	0.93	C4
	Paddle Cr Trib	1.0	37.6	32.1	1.2	1.00	C4
	Caulley Creek	1.1	39.9	27.1	1.5	0.78	C4
	Cooney Br	1.2	37.2	23.5	1.6	1.10	C4
	Little Pine Cr	2.6	57.9	26.0	2.2	0.59	E5
	Brushy Fork	3.6	57.3	22.5	2.5	0.24	E5
	Johns Cr	19.8	246	44.5	5.5	0.13	E5
	Pine Cr	25.0	269	44.9	6.0	0.015	E5
	Symmes Cr	39.0	429	66.4	6.5	0.060	E5
Little Muskingum	Ferguson Run	0.55	14.1	14.6	1.0	1.6	C4
	Jones Creek	0.63	37.7	20.2	1.9	1.4	E4
	Bakers Run # 2	1.1	55.7	31.8	1.8	1.1	C3
	Bakers Run # 1	1.1	31.9	22.8	1.4	1.2	C4
	Sycamore Fork	2.6	47.1	26.6	1.8	0.79	C4
	Dogskin Run	2.9	81.2	31.0	2.6	0.78	C4
	Dismall Creek	3.0	74.1	37.9	2.0	0.64	C4
	Sheets Run	3.2	76.3	31.0	2.5	0.94	C4
	Lt Muskingum #1	69.2	588	81.5	7.2	0.26	E4
	Lt Muskingum #2	166	1885	152	12.4	0.007	E5

To develop the relationship between the cross-sectional areas and their associated drainage areas, the graph of cross-sectional area versus drainage area is plotted and best fitted by a linear line in a double logarithmic scale as given in Figure 5. The relationship can be further expressed by a power function as follows:

$$A_{bkf} = 33.4a^{0.70} \tag{3}$$

where A_{bkf} is the bankfull cross-sectional area in square feet, a is the drainage area in square miles and the R^2 is 0.97. For example cross-sectional area for a drainage area of 50 mile^2 is found to be about 516 ft^2. Figure 5 includes as a reference the widely used regional curve provided by Dunne and Leopold (1978) for the area of Eastern United States that has a value of 240 ft^2 for the drainage area of 50 mile^2.

For the relationship between the bankfull width and drainage area, the following equation was obtained:

$$W_{bkf} = 21.6a^{0.32} \tag{4}$$

where W_{bkf} is the bankfull width in feet, a is the drainage area in square miles and R^2 is 0.91. This resulted in the bankfull width of 76 for the drainage area of 50 mile^2, compared to 65 ft obtained from the regional curve by Dunne and Leopold (1978).

Similar equation can be obtained for the mean bankfull depth relative to the associated drainage area in the following:

$$D_{mean} = 1.55a^{0.38} \qquad\qquad (5)$$

where D_{mean} is the mean bankfull depth in feet, a is the drainage area in square miles and R^2 is 0.90. The resulting mean bankfull depth corresponding to drainage area of 50 mile^2 is 6.8 ft while the value obtained from the regional curve Dunne and Leopold (1978) is about 3.7 ft.

The data were grouped by sub-region, ranges of local channel slope and by channel type. No improvement in the trends was evident either graphically or in improved R^2 values.

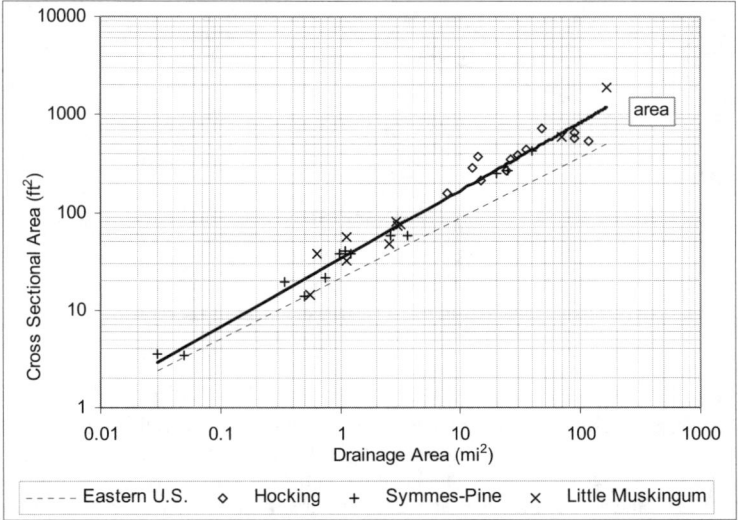

Figure 5. Cross-sectional area vs. drainage area

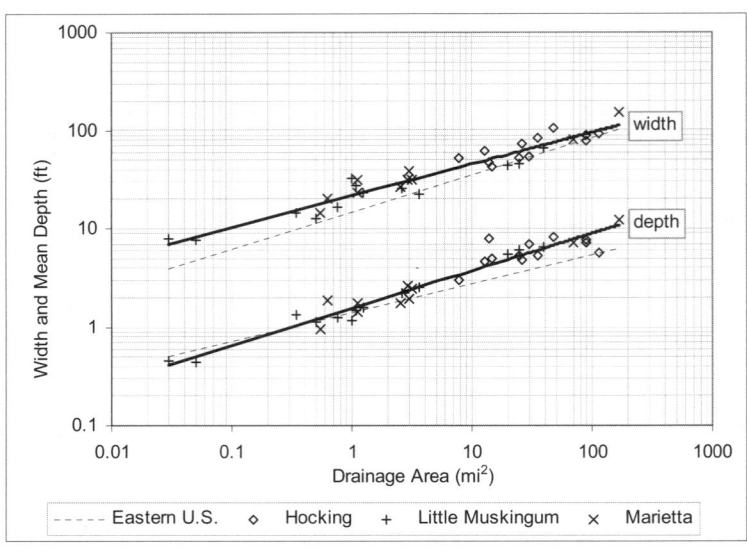

Figure 6. Bankfull width and mean depth vs. drainage area

Conclusions

Based on the analyzed results from the selected sites, it can be concluded that the bankfull dimensions of cross sectional area, width, and depth in the studied region are significantly related to their associated drainage areas. It is noted that the statistical value of R^2 for relating the bankfull cross sectional area to the associated drainage area is greater than the counterparts of relating bankfull depth and width to their corresponding drainage area. The bankfull parameters of cross-sectional area, depth, and width in the studied region are higher than the commonly referenced values published by Dunn and Leopold (1978). To select the survey sites and delineate their associated drainage areas for this study, the use of geographic information system was shown to be extremely effective.

Acknowledgments

The financial support received from the Ohio Environmental Protection Agency through a 319 Grant and from Wayne National Forest is gratefully acknowledged. The assistance for the channel survey by John Schaad and John Smith of Ohio University and Andy Ward and Dawn Farver from the University of Ohio was indispensable for the project and is deeply appreciated. We thank Lauren Lambert and Dan Imhoff of the Ohio Environmental Protection Agency for their comments and suggestions. And to Pam Stackler we are most indebted.

References

Dunne, T. and L.B. Leopold, 1978. Water in Environmental Planning, W.H. Freeman and Co., San Francisco, California.

Harrelson, C.C., C.L. Rawlins, and J.P. Potyondy. 1994. Stream Channel Reference Sites: An Illustrated Guide to Field Technique. USDA Forest Service, General Technical Report RM-245, Fort Collins, Colorado, 62p.

Leopold, L.B., M.G. Wolman, and J.P. Miller, 1964. Fluvial Processes in Geomorphology, W.H. Freeman and Co., San Francisco, California.

Mecklenburg, D. E., 2003. http://www.dnr.state.oh.us/soilandwater/streammorphology.htm.

Rosgen, D., 1996. Applied River Morphorlogy, Wildland Hydrology, Pogosa Sprong,

Colorado.

Rosgen, D., 2000. The Reference Reach – A Blueprint of Natural Channel Design. Wildland Hydrology, Colorado.

U.S. Geological Survey and U.S. Department of Agriculture, 2003. National Mapping Program Technical Instructions (Part 7, Hypsography – Standards for USGS and USDA Forest Service).

Big Darby Headwaters:
An Assessment and Discussion of Channel Morphology
November 2003

Dawn Farver[1] and Dan Mecklenburg
[1]Graduate Student, University of Arkansas, 203 Engineering Hall, Fayetteville AR 72701
dfarver@uark.edu, 479 575-5657

ABSTRACT

The main interaction between the terrestrial and aquatic systems occurs in the headwaters, the place where all great rivers begin. A stream is also tied to the land around it. Stream-floodplain interaction is potentially the most important morphological process to the long term integrity and stability of a stream. A stream system with good geomorphology and floodplain access is more likely to be tolerant and adapt to a range of watershed stressors. The Big Darby River located in Central Ohio is a high quality scenic river. This study focused on characterizing the headwaters of the Big Darby to determine what the condition of these streams was and if/how they contributed to the quality of the Big Darby River itself. Headwater streams were characterized in two different landscape areas. The northeast section of the watershed where the headwaters are located has fairly flat topography and therefore, the streams are fairly low gradient and low energy. The southwest section of the headwaters is characterized by more hilly terrain and therefore the streams are generally higher gradient, higher energy streams. The channel assessment started with field recognizance with the aid of USGS quadrangle topography maps. Once channels of interest were identified, wade-able headwater streams, cross-sections were measured to characterize levels of entrenchment, recovery and floodplain. Cross-sections were measured in the field and results were plotted in a spreadsheet.

INTRODUCTION

Welcome to the Big Darby Headwaters! This is a report on the shape of these channels, their morphology. Its purpose is to show the existing channel form, what the ideal form likely is and most importantly illustrate the difference between the two. Overall stream integrity is greatly dependent on channel morphology. Identifying the difference between the existing and ideal morphologies provides a target. Closing the gap between the two is a tangible goal to work toward.

BACKGROUND

This report describes the headwaters that are located west and northwest of Marysville Ohio. The stream system runs through Logan, Union and Champaign Counties and includes everything upstream of the Pleasant Run – Big Darby confluence. The topography is distinctly different in two parts of this area. Generally the northeast part is flat and the channels low gradient while the southwest is hillier with the channels being steeper the smaller they are. Agriculture is the predominant land use. However, agriculture is less intensive today than historically and tree cover has increased.

Figure 1: Map of the Big Darby watershed with the headwater study area highlighted

WHAT IS GOOD MORPHOLOGY?

All aspects of morphology interact and are important, however the "connection" between the channel and the floodplain is perhaps most consequential for the overall long-term integrity of the stream. It along with the amount of meandering are both drastically diminished by channelization which typically involves lowering of the stream bed causing the channel to be "entrenched" and contain larger flows instead of allowing them to spread out across the floodplain. Frequent extensive flooding is perhaps the most important process for the overall long-term integrity of the stream.

The 3, 5, 10 rule makes an excellent rule of thumb for knowing what good morphology is. It works for the vast majority streams, ones which naturally have floodplains. The 3, 5, and 10 are ratios of the width of common annual floods relative to the channel width. Floods that spread out 3 times wider than the channel are characteristic of "okay" stream quality. If they don't spread out that much, one should expect bank instability, poor habitat and negative impacts down stream. Many good streams are associated with a ratio of 5. The best streams and pristine conditions typically allow common floods to spread out 10 or more times as wide as the channel. Much of the main stem of the Darby and some of its tributaries have ratios of 10 and more.

However, many of the Big Darby headwater streams have been channelized. This has occurred over more than century and was done to increase drainage and reduce flooding. You may be wondering how the Big Darby itself is still such a high quality river worthy of protection if all of this channelization has occurred...it's due to many factors, arguably one of the most important of which is good morphology. It not that channelization has not occurred but because of Big Darby's geology it is relatively resilient and much of what has been channelized has not stayed that way but has "recovered."

The prospects for morphologic recovery depend largely on slope, or more specifically on adequate stream power (a product of flow rate and channel slope). Other factors are a supply of course sediment (gravel and cobble) and neglect. Periodic maintenance and if they are steep enough that will determine how quickly a stream can recover from channelization, if at all! Due to the low gradient of its watershed, many of the channelized stream miles of Flat Branch, a major tributary to the Big Darby, have been unable to begin the recovery process and have been left in a state of geomorphological failure. Without help, these streams may never recover to be the stable, scenic, healthy systems they once were.

Good morphology makes streams more tolerant and resistance to the sediment and other pollution inputs from the watershed. The Big Darby is able to assimilate pollutants and remove fine sediment because of its connection with its floodplain. Just like a person who takes care of themselves, eats healthy and gets regular exercise is more likely to be able to fight off a virus, a stream system with good geomorphology and floodplain access is more likely to be tolerant of the pollution it receives and other watershed stressors.

METHODS

The channel assessment started with initial field recognizance with the aid of USGS quadrangle topography maps. This allowed channels of similar form to be identified. Channel form is of course three dimensional and is typically measured with 1) cross sections perpendicular to the stream flow, 2) longitudinal slope profile, and 3) plan form showing the meander pattern. Each of these three dimensions is vital to a complete understanding of channel condition however cross sections themselves allow quick insight into the most classic issues associated with quality of the channel form. Cross-sections were taken throughout the Big Darby watershed headwaters to characterize levels of entrenchment, recovery, and to characterize the floodplain (figure 2).

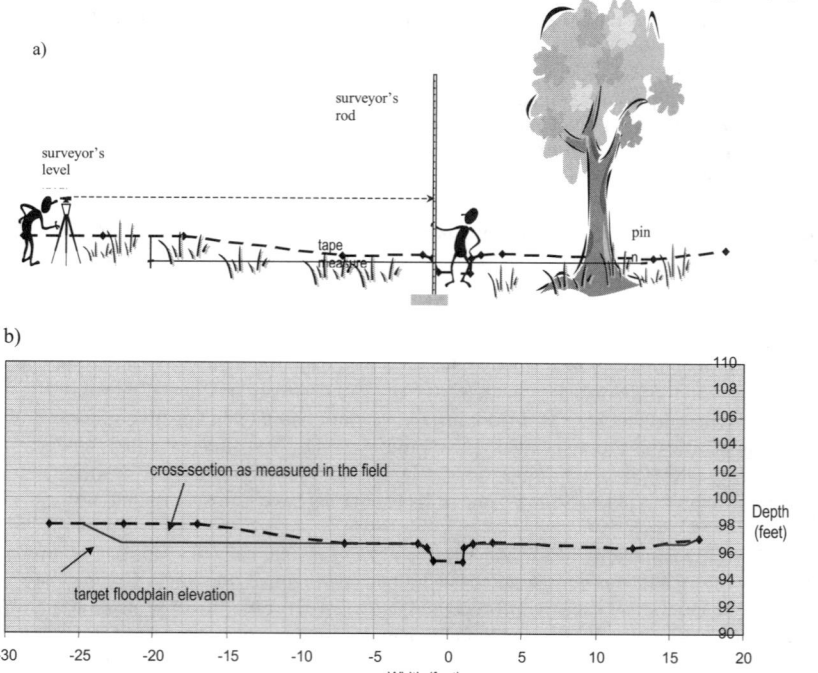

Figure 2: Surveying (a) and plotting (b) a stream channel cross-section. The elevation of the measured cross-section is shown as the blue dotted line. The solid maroon line shows what the ideal dimensions of the cross-section would be for good geomorphology to exist.

CHARACTERIZATION OF DARBY HEADWATERS

Based on the condition or quality of the stream morphology found three general characterizations are made; 1) low gradient areas found in the northeast part of the study area and much of Pleasant Run to the south, 2) the higher gradient area to the west experiencing significant recovery, and 3) high quality channels that have likely experienced little or no past channelization.

<u>Low gradient areas of NE , Flat Branch and much of Pleasant Run</u>

While some of these streams cannot recover because of the topography of the area, some streams have not recovered because of the way they are managed. These streams have been maintained as drainage ditches, deeper than ideal, and are "cleaned out" periodically when sediment deposits in the form of benches, or on the bottom of the ditch making it shallower. Often the riparian zone is mowed grass. An example is the photograph from Pleasant Run seen below:

From the photograph it is possible to see how straight the channel is, and the mowed grass around the channel. On either side of the grassway are rows of corn. From maintenance practices over time, the sediment has been cleaned out of the channel and placed on the banks, forming levees on either side of the channel. If you were to take a vertical slice of the stream and surrounding area, you would have a cross-section of the stream, fig 3. The dotted line in figure 3 represents across-section surveyed on this reach beyond the trees:

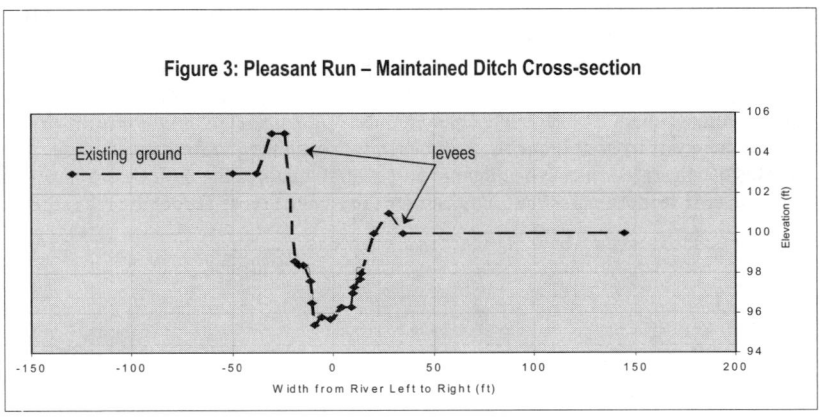

Figure 3: Pleasant Run – Maintained Ditch Cross-section

The most severely limiting factor for the quality of the stream represented by figure 3 is the area between the solid line and the dotted line. Anything that reduces the difference between the two lines must be considered. In addition to passive recovery, active restoration construction projects

359

typical take one of two approaches. The easiest to construct and which will for better and worse also cause the most flooding, is to abandon the existing channel and construct a new channel at a higher elevation. Application of approach is of course limited to areas where the lost drainage and increased flooding are acceptable. A more expensive, yet often more feasible approach to active restoration, is to lower the ground down to the active floodplain elevation. The $3 - 5 - 10$ rule of thumb can be used to determine floodplain width. The area between the existing and proposed lines plotted in figure 2 indicates the volume of earthwork required.

Channel Recovery – West and North West tributaries and the Big Darby
As you can see, regular channel work prevents natural recovery in this reach. Now you will see a channel that does not have the ability to recover due to the low gradient of the topography in the area that does not provide enough stream power to begin the recovery cycle, when no longer maintained. Below, fig 4, is a cross-section from Flat Branch where the low gradient of the landscape does not provide the stream power necessary for recovery to begin.

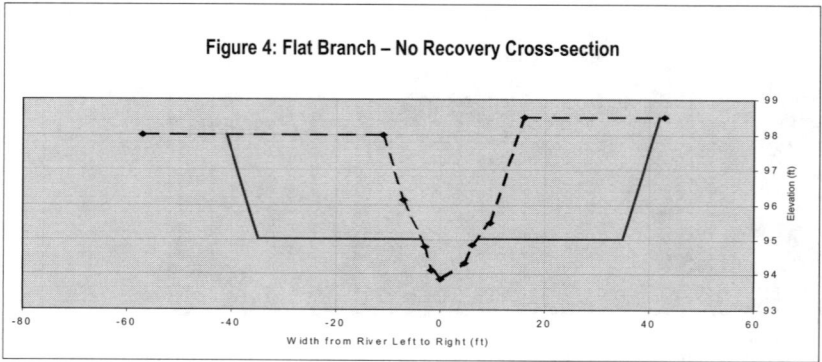

The dotted line represents the actual channel cross-section elevation/location as measured in the field, and the solid line represents a target cross-section where the stream has an active floodplain at the appropriate elevation that is five times the bankfull width. In this case, it would have to be actively provided through excavation.

An example of a stream reach in the process of recovering is on an unnamed tributary to the Big Darby. Let's call it Recovery Run. At some point, decades ago, Recovery Run was channelized. The first cross-section that was measured along this reach represents a stream in the very first stages of recovery. It is still evident that at some point the stream had been given the dimensions of a trapezoidal drainage ditch – wider and deeper than the original stream. So the stream was still straight, but there was a ray of hope! A small bench was beginning to form on one side of the channel. This bench represents the beginning of a new floodplain being built by channel processes over time. How exciting is that? You can see the channel cross-section below:

360

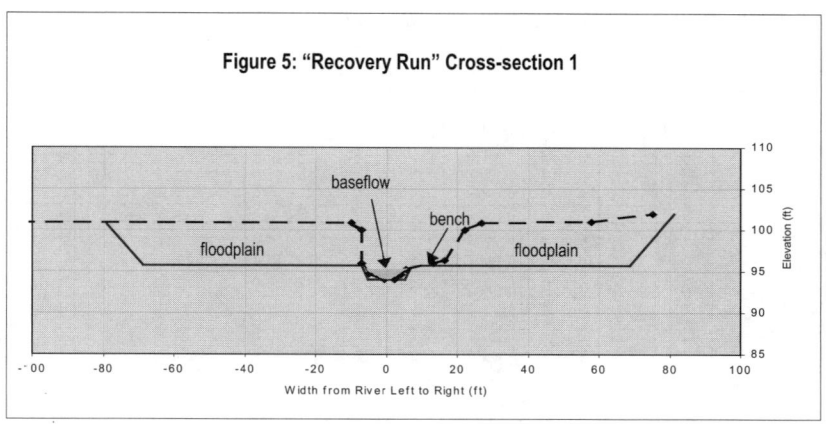

Figure 5: "Recovery Run" Cross-section 1

Keep looking, and lo and behold, there is an example, further down the stream, where there is more recovery! It just kept getting better! At this cross-section, there was a similar bench, but the stream was starting to bust out of its "drainage ditch mold" and recreate its own dimensions. As more water with more energy had been flowing down the stream, the channel started to get an itch to have some meander pattern back versus being straight ditch. Until it broke the ditch mold, there wouldn't be enough width to build the floodplain so as more and more water worked to erode the bank opposite the bench, and banks started to become unstable…as the banks failed the ditch became wider. Progress was being made! And the sediment that was now in the stream could then be used to build more floodplain. So the stream has started to move, if only a little bit, get a little sinuosity back, and continue the big job of rebuilding its floodplain. So we're moving right along with the recovery.

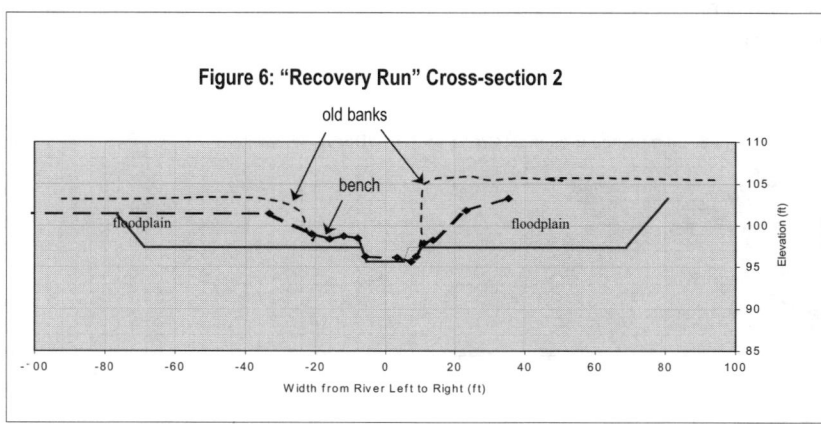

Figure 6: "Recovery Run" Cross-section 2

On this cross-section, the thin dotted line represents where the banks were on the previous cross-section where there had been less recovery. The area under between the thin dotted line and the thicker dotted line is the area where banks used to be before they eroded away.

As we wander further downstream, we observed that it continues to get even better. The next reach (the cross-section is shown below) on the same stream demonstrated even more characteristics of a recovering stream. The stream had been hard at work destabilizing the banks and showing them who was boss, continuing to build up floodplain, and becoming more sinuous. At this point, the floodplain at this site was becoming effective at helping to build up that tolerance for non-ideal conditions. ***The floodplain is the number one thing we can give back to*** *streams in the Big Darby to make them higher quality and continue to improve the overall health of the system.*

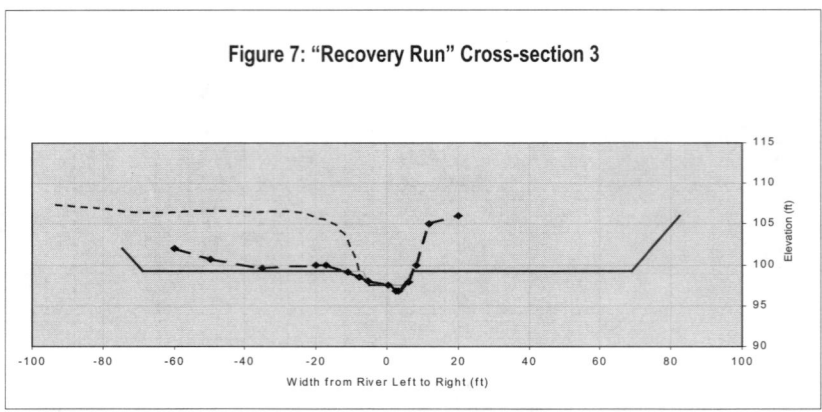

Figure 7: "Recovery Run" Cross-section 3

High Quality Channel Morphology

There are still some streams in the Darby headwaters that have had limited outside impact and are very similar to how they were hundreds of years ago. They have extensive active floodplains, are sinuous, and are stable, not aggrading or degrading. They efficiently maintain themselves and good habitat for macroinvertebrates, salamanders and fish. We have a cross-section for one of these channels, figure 8. As you can see, the channel has a large floodplain, about 10 times wider than the channel. The cross-section below is of a stream that has had little outside impact and has maintained itself as a stable, Type-E channel with a narrow, deep main channel, and a wide, frequently flooded floodplain. It is located in the very upper headwaters of the Big Darby and has a drainage area of only 0.1 mi^2.

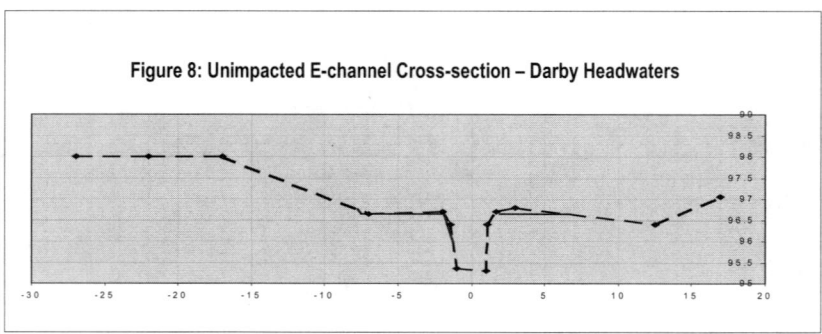

Figure 8: Unimpacted E-channel Cross-section – Darby Headwaters

Regional Curve of Active Floodplain Width
Regional Curves typically relate channel dimensions or other characteristics to drainage area size within a certain geographic region. Below is a regional curve for target active floodplain widths in the Darby watershed. Using the regional curve graph, it is possible to determine an approximate channel width and depth for a given watershed area. Figure 9 shows an example of a stream cross-section, a graph of the Big Darby regional curve, and a table of watershed areas with the associated channel dimensions.

362

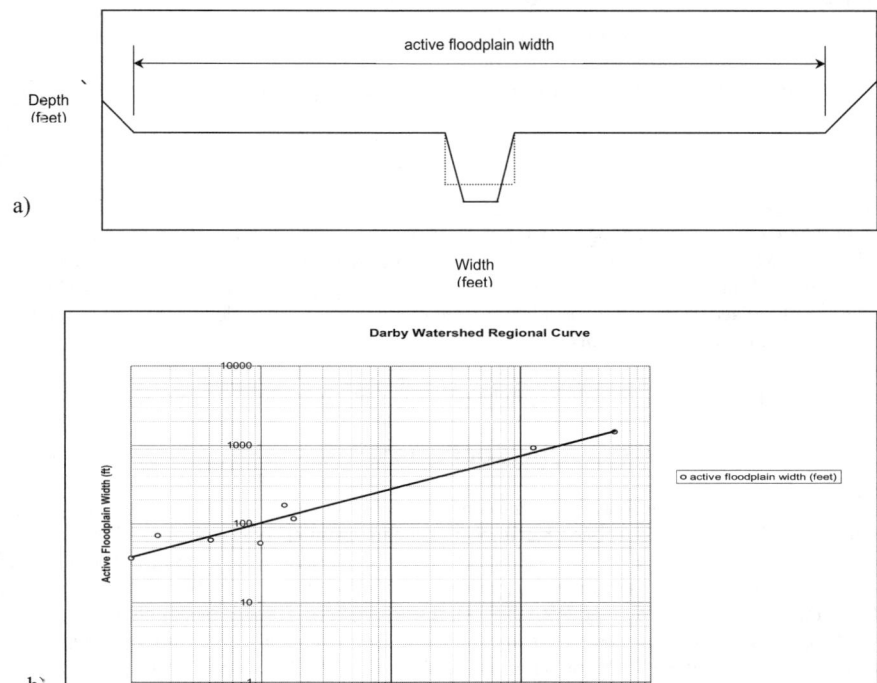

Figure 9: Target active floodplain widths for the Darby Watershed

CONCLUSIONS

So what is the first step in stream recovery in channelized streams in the Big Darby headwaters? The first step is the formation of a bench at the appropriate floodplain elevation. And what is every step of the recovery process moving toward? It's moving toward reformation of an active floodplain. If natural processes indicate that the floodplain is the most important attribute to stream recovery and health, who are we to argue? Why don't we help move the process forward instead, especially on those streams that do not have enough energy to recover on their own?

We have the capability to address the issue of the lack of floodplain in many of the headwaters of the Big Darby. From these diagrams, it is possible to see where soil needs to be excavated for floodplain at the appropriate elevation and appropriate width for the channel to begin to demonstrate improving geomorphology. A floodplain width of 3 bankfull channel widths is a minimum width necessary for a channel to begin showing signs of good geomorphology. A floodplain width around 5 bankfull widths is characteristic of a stream with good geomorphology, and a floodplain width of 10 or more bankfull widths is characteristic of a channel with exceptionally good geomorphology.

Channel Aggradation by Beaver Dams on a Small Agricultural Stream in Eastern Nebraska

M. C. McCullough[1], J. L. Harper[2], D.E. Eisenhauer[3], M. G. Dosskey[4]

ABSTRACT

We assessed the effect of beaver dams on channel gradation of an incised stream in an agricultural area of eastern Nebraska. A topographic survey was conducted of a reach of Little Muddy Creek where beaver are known to have been building dams for twelve years. Results indicate that over this time period the thalweg elevation has aggraded an average of 0.65 m by trapping 1730 t of sediment in the pools behind dams. Beaver may provide a feasible solution to channel degradation problems in this region.

KEYWORDS. Beaver dams, channel aggradation, sediment, agricultural watershed.

INTRODUCTION

In eastern Nebraska, most land, which at one time was tallgrass prairie, has been converted to agricultural land use. This conversion has impacted stream channels both directly and indirectly. One major impact of row-crop agriculture is the increase in overland runoff and peak flows in channels. Between 1904 and 1915 many stream channels in southeastern Nebraska were dredged and straightened (Wahl and Weiss 1988), resulting in shorter and steeper channels. These changes have resulted in severe channel incision and stream bank instability in eastern Nebraska and throughout the deep loess regions of the central United States (Lohnes 1997). This trend towards continued channel degradation continues to this day (Zellars and Hotchkiss 1997), creating environmental and economic concerns.

Beaver affect geomorphology of streams in ways that may counteract channel degrading processes. Beaver dams reduce stream velocities, causing the rate of sediment deposition to increase behind the dam (Naiman et al. 1986) and giving the channel gradient a stair-step profile (Naiman et al. 1988). In a forested ecosystem, Naiman et al. (1986) studied beaver dams in 4th order and smaller streams and found that small individual dams could hold $2000 - 6500 \text{ m}^3$ of sediment with an average of 10.6 beaver dams/km of stream. They calculated that approximately $10,000 \text{ m}^3$ sediment/km of channel were retained by beaver dams in two forest watersheds in Quebec, Canada.

Beaver dams may help stabilize and aggrade incised stream channels in agricultural watersheds in eastern Nebraska. However, the influence of beaver dams on channel morphology and stability in developed agricultural areas is not known. The objective of this study was to determine if beaver dams can aggrade incised streams in agricultural regions.

METHODS

The study was done on Little Muddy Creek, Otoe County, in southeast Nebraska. The study stream reach is second-order, perennial and has 430 hectares of upland watershed. Land use in the watershed is mainly agricultural, with approximately 75% of the land in cropland and 25% in

[1] M.C. McCullough, Graduate Student, University of Nebraska-Lincoln, Lincoln, Nebraska.
[2] J.L. Harper, Graduate Student, University of Nebraska-Lincoln, Lincoln, Nebraska.
[3] D.E. Eisenhauer, Professor, Biological Systems Engineering, University of Nebraska-Lincoln, Lincoln, Nebraska.
[4] M.G. Dosskey, Research Ecologist, USDA National Agroforestry Center, Lincoln, Nebraska.

pasture or Conservation Reserve Program (CRP). The predominant soils in the uplands are eroded, fine-textured loess and glacial till soils with slopes from 2 – 7 % (USDA, 1982). In the drainages, soils are clay and silt loams and silty clays.

Little Muddy Creek is a tributary in the upper watershed to the Little Nemaha River, which drains into the Missouri River. Land clearing until the 1930's increased surface runoff, soil erosion from the uplands and sedimentation in the streams, which lead to increased flooding. To reduce flooding the Little Nemaha River and its tributaries were dredged, straightened and cleared of vegetation starting in the early 1900's. This disturbance is regarded as the main cause of present channel instabilities throughout the basin (Rus 2003). In the upper reaches of the basin, Muddy Creek was straightened between 1947 and 1953. Knickpoints from channel straightening migrated upstream and most effects on streambeds had already occurred (Rus 2003). In our study area, knickpoints are still evident, working up the 1st order portions of the streams (D.E. Eisenhauer, personal communication, 2003). A recent USGS study (Rus 2003) indicated that in the Little Nemaha River Basin, streambed degradation since settlement averaged 1.9 meters for 21 bridge locations throughout the basin.

The study stream reach was 1130 meters long and was divided into upstream, affected and downstream reaches. The affected reach, from 136 to 922 m in channel distance had beaver dams since 1991 and during our survey there were 6 active dams. The upstream reach, from 0 - 136 m, had not had any beaver dams during this same period. There were not beaver dams on the downstream reach from 922 to 1121 m until the fall of 2002 when three new dams were constructed (D.E. Eisenhauer, personal communication, 2003). Since the stream survey was completed in the spring of 2003, the sedimentation effects of the new dams in the downstream reach were considered to be of negligible significance. At the boundary between the affected reach and downstream reach was a driveway culvert crossing. By dividing the stream reach into upstream, affected, and downstream segments, a comparison could be made regarding relative sediment accumulation in areas with beaver activity versus no beaver activity.

The stream reach was surveyed over a period from February to April 2003. Beaver dam locations and the stream centerline were surveyed using a mapping grade GPS unit. Stream cross-sections were measured using a surveying total station. Ten cross-sections were surveyed in the affected reach at locations above and below each active dam. A total of five cross-sections were surveyed in the free-flowing reaches upstream and downstream of the affected reach to determine channel characteristics prior to beaver recolonization. All cross-sections were measured perpendicular to stream flow direction. Cross-section survey points included stream terrace, channel profile, thalweg, and water surface.

Three point bars were selected to sample sediment deposits. Soil cores were taken at three spots at each point bar and were analyzed for bulk density and particle size distribution.

A projected streambed was developed to determine the expected cross-section had the beaver dams not been built. To determine the projected streambed, a projected thalweg elevation was calculated. A linear regression was performed on thalweg points (Fig. 1) that occurred in the free-flowing reaches (upstream and downstream). Leopold (1994) also used a linear approximation of thalweg elevation versus distance downstream. The equation of the line was used to estimate the elevation of a projected thalweg at each cross-section in the affected reach of the stream. It was assumed that the projected thalweg would have occurred at the same location across the stream as the surveyed thalweg.

The surveyed streambed was plotted on the same graph as the projected streambed and the cross-sectional area of sediment accumulation was calculated by using the trapezoidal rule. To determine volume of sediment accumulated, the distance between two successive cross sections was multiplied by the average of their cross-sectional area. The total mass of sediment collected was determined by multiplying the calculated volume by the dry bulk density.

The effect of the culvert at a downstream distance of 910 m was taken into account when calculating the sediment accumulation attributed to beaver dams. Sediment accumulation was estimated for the 736 m upstream portion of the 785 m long affected reach.

Hydraulic grade was calculated using surveyed water surface elevations at each surveyed cross section and immediately upstream and downstream of each dam.

RESULTS

The stream profile (Figure 1) shows a significant amount of sediment accumulation. The projected thalweg was calculated using a regression line between surveyed thalweg points in the upstream free-flowing reach and the downstream reach. The stream terrace slope was calculated using a regression line for surveyed terrace points through the entire study reach. In this case, the stream terrace slope is 0.0046. The projected thalweg slope was 0.0054, which is very close to the slope of the stream terrace. This is to be expected because the thalweg and stream terrace should be nearly parallel, with the thalweg having a slightly greater slope to account for the gradual deepening of the channel with distance downstream.

Channel aggradation in the affected reach was calculated as the difference between the surveyed and projected thalweg elevations. The average channel aggradation in the affected reach was 0.65 m. The shaded area on Figure 1 represents an approximation of channel aggradation.

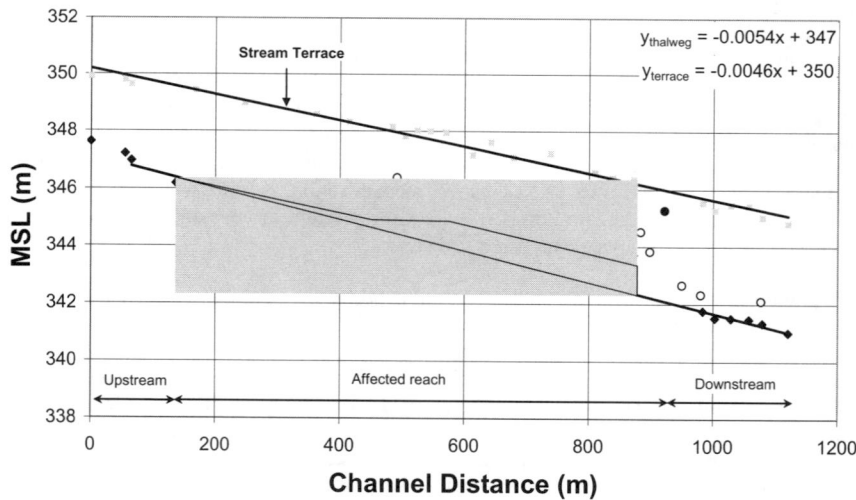

Figure 1. **Stream profile of Little Muddy Creek near Burr, Nebraska.** Squares are elevations of the stream terrace, diamonds are elevations of the thalweg, and open circles are elevations of tops of the dams. The closed circle at 910 m is a road centerline. The shaded area is an approximation of the sediment collected in the affected reach due to beaver dams.

Figure 2 shows a sample cross section. Once the projected streambed and sediment accumulation was determined for each cross section, the cross-sectional area was determined using the trapezoidal rule. The cumulative volume of sediment collected was determined by averaging two successive cross-sectional areas, then multiplying by the distance between the two cross sections. The sum of all these volumes in the affected reach was 1450 m^3 in 736 m of stream or 1970 m^3/km.

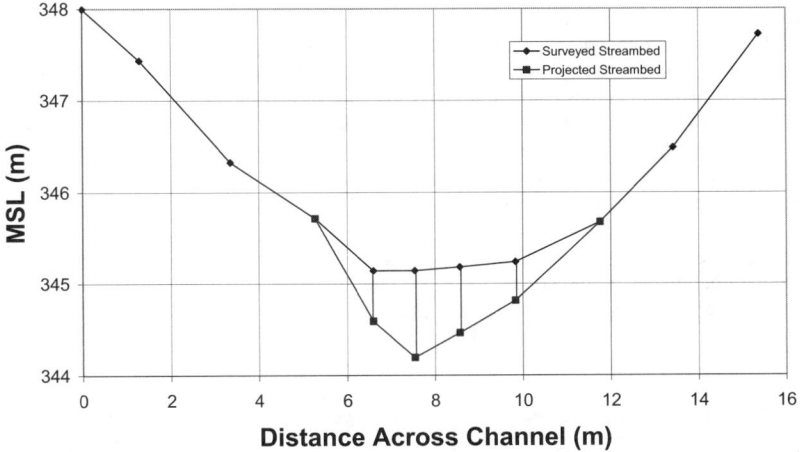

Figure 2. Sample cross section. This cross section, occurring at 550 m downstream, is a typical cross sectional profile. The top line represents the surveyed streambed, the bottom line represents the projected streambed. The trapezoids marked were used to calculate the cross-sectional area of accumulated sediment.

The dry bulk density of the accumulated sediment was determined for nine sediment samples, then averaged to determine an average bulk density. The average bulk density of all samples was 1.2 g/cm^3. The total mass collected was calculated to be 1730 t.

The hydraulic grade for the affected reach was 0.0039. This is a smaller slope than the projected thalweg (0.0056) and stream terrace (0.0046). Average water surface drop for all active dams in the affected reach was 0.45 m, creating a stair-step effect on the hydraulic grade line.

DISCUSSION

In this study, we found that beaver dams do affect sedimentation and aggrade stream bottoms in channels in agricultural streams. These results are consistent with similar studies in forest watersheds in Quebec, Canada (Naiman 1986). We found an average accumulation of sediment by beaver dams to be 1970 m^3/km, where Naiman (1986) found an average of 10,000 m^3/km. Naiman (1986) estimated an average of 10.6 dams/km stream, where we found 9 dams/km stream. Naiman's (1986) study was done in a relatively undisturbed boreal forest stream, while our study was done in an incised stream in a relatively disturbed agricultural watershed.

During the period of the topographic survey there were 6 active dams in the affected reach, however, abandoned dams were apparent. The abandoned dams also could have played a role in accumulation of sediment in the channel over the twelve years since beaver recolonized the area, no just the active dams.

The downstream reach of the stream had three new dams (less than 6-months old), the sedimentation effects of which were assumed to be of negligible impact to channel slope. Cross-sections surveyed in the downstream reach were done far enough away from new dams so as to avoid areas of recent sedimentation in order to more accurately represent the thalweg elevation of the "pre-beaver" steambed.

Beaver dams may be a viable solution to incision problems in this region and could be much less expensive than conventional engineering techniques. On this incised stream in Otoe County, beaver dams have caused channel aggradation. Where beaver have been active for the past 12 years, the stream channel bed has risen about 0.64 m and about 1730 t of sediment have been trapped in the beaver ponds (see Figure 1). By contrast, in a comparably sized watershed (405 hectares) in nearby Saline County, the cost of stabilizing 0.8 km of Sand Creek, a deeply incised stream with unstable banks, with traditional human-engineered drop structures has been estimated to cost $110,000 (Cermak et al. 2002).

Based on the stream length of 1130 m and an average of 1.0-1.2 beaver colonies per km of stream length (Chapman and Feldhamer 1982), this stream has about 1.2-1.4 beaver colonies, which is approximately 5-12 beavers. Beaver populations may have reached carrying capacity in this watershed and may go through a period of decline as stands of riparian forest are diminished. In some areas surrounding larger dams, riparian canopy was completely open.

Camp Creek, a similar stream nearby to Little Muddy Creek, was evaluated for beaver dam density and changes to hydraulic grade line. For Little Muddy Creek we found an average of 9 dams/km of stream, while Camp Creek averaged 7 dams/km. The ratio of average slope of the floodplain to hydraulic grade line was 1.2 for Little Muddy Creek and was 1.9 for Camp Creek. The problem of channel incision is regional in scale, as is the recolonization of beaver to smaller streams in agricultural watersheds.

It is expected that beaver would have inhabited the study reach to biological capacity prior to European settlement. Then, disturbance to riparian habitat and trapping pressures resulted in diminished populations and areas of local extinction by the 1930's. The native vegetation prior to settlement would have been tall grass prairie uplands with predominantly cottonwoods and willows in the riparian areas. The current riparian condition in which today's beaver are recolonizing is quite different from conditions prior to European settlement. Water flux tends less toward infiltration and more towards surface runoff, resulting in a flashy stream with more dramatic pulses of flow within shorter periods of time during the year. Observations of this study watershed indicate that beaver dams were damaged in smaller, 2-year return period storms. Dams were regularly damaged by storm flows, but beaver rebuilt them during the summer and fall to maintain their presence along this reach.

CONCLUSION

Surveying and sediment sampling were used to determine whether beaver dams cause channel aggradation. Results from this experiment showed that beaver dams did cause an increase in sediment, causing channel aggradation in an incised stream in an agricultural watershed. In a 736 m affected reach of Little Muddy Creek, approximately 1730 t of sediment were collected behind beaver dams in the twelve years since beaver recolonized the stream. Within this reach there was an average of 0.65 m of channel aggradation due to sedimentation trapping by beaver dams. Beaver dams create a stair-step effect on the hydraulic gradient line, reducing flow velocities, causing sediment deposition behind dams.

Further research is needed to determine whether beaver dams are an ecologically and economically feasible solution to channel degradation. Most research regarding beaver habitat and populations have been done in forested or mountainous areas. The carrying capacity and ecological impacts of beaver in predominantly row-crop agricultural areas could be quite different from those found in forested and mountainous regions.

Acknowledgements
The authors of this paper would like to acknowledge Alan Boldt and Paulo Luchiari , whose topographic survey work that made this paper possible. Acknowledgement also goes to Alan Boldt for his extensive compiling, organizing and analysis of the topographic survey data.

REFERENCES

1. Cermak, J., Hawley, D. and L. Stahr. 2002. Sand Creek bank stabilization project. Project report for BSEN 480: Senior design project, Biological Systems Engineering Department, University of Nebraska-Lincoln, Lincoln, Nebraska.

2. Chapman, J.A. and G.A. Feldhamer. 1982. Wild mammals of North America: Biology, Management and Economics. John Hopkins University Press. Baltimore, Maryland.

3. Leopold, L.B. 1994. A View of the River. Harvard University Press, Cambridge, Massachusetts.

4. Lohnes, R.A. 1997. Stream channel degradation and stabilization: the Iowa experience: Proceedings of the Conference on Management of Landscapes Disturbed by Channel Incision, May 1997, Oxford, Mississippi.

5. Naiman, R.J. and G. Pinay. 1994. Beaver influences on the long-term biogeochemical characteristics of boreal forest drainage networks. Ecology 75:905-921.

6. Naiman, R.J., Melillo J.M. and J.E. Hobbie. 1986. Ecosystem alteration of boreal forest streams by beaver (Castor Canadensis). Ecology 67:2365-2369.

7. National Research Council. 2002. Riparian Areas: Functions and Strategies for Management. National Academy Press, Washington, D.C..

8. Rus, D.L., Dietsch, B.J. and A. Simon. 2003. Streambed Adjustment and Channel Widening in Eastern Nebraska. Water-Resources Investigations Report 03-4003. U.S. Geological Survey. Lincoln, NE.

9. United States Army Corps of Engineers. 1996. Southeast Nebraska Streambank Erosion and Streambed Degradation Control Design Manual. Hydraulic Engineering Branch, Engineering Division, Omaha District.

10. Wahl, K.L. and L.S. Weiss. 1995. Channel degradation in southeastern Nebraska rivers. In: Ward, T.J., editor. Watershed Management: Planning for the 21st Century.American Society of Civil Engineers, New York, New York.

11. Zellars, J.A. and R.H. Hotchkiss. 1997. Channel stability adjacent to highly irrigated land: Eagle Run Creek, Omaha, Nebraska: Proceedings of the Conference on Management of Landscapes Disturbed by Channel Incision, May 1997, Oxford, Mississippi.

STREAMBANK STABILIZATION AND RIPARIAN CORRIDOR ESTABLISHMENT IN RURAL KANSAS

Phil G. Balch[1] and Brock A. Emmert[2]

ABSTRACT

The Little Blue River flows through the eastern portion of Washington County, Kansas, and has a drainage basin of approximately 3,500 square miles. In late 1999, three landowners along the river contacted the Washington County Conservation District and the District Conservationist with the Natural Resources Conservation Service (NRCS) regarding severe streambank erosion on their properties. The District Conservationist requested assistance from the Kansas State Conservation Commission (SCC). During their preliminary site visits, SCC staff determined that several stream reaches were severely over widened by excessive bank erosion, and the river had become bedload driven. Measurements of aerial photographs show a total cropland loss of 369 acres along 8.2 miles of river between 1977 and 2001. This resulted in a dry weight sediment input of 12,565,298 tons or approximately 502,611 semi-truck loads. Soil analysis showed that nutrient content of the eroded streambank soils equaled 92,270 pounds of nitrate (NO_3), 839,271 pounds of phosphorous (P), and 6,959,856 pounds of potassium (K). Bendway weirs were chosen as the primary structure for stabilization because of their ability to: help reduce width/depth ratios, reduce water velocities in the near bank region, induce sediment deposition, and maintain cost effectiveness. Additional project goals included re-establishing a riparian corridor and improving aquatic habitat. In early 2000, SCC, Kansas Department of Health and Environment (KDHE), and Natural Resources Conservation Service (NRCS) staff began conducting total station surveys of problem sites. The initial 20 project surveys, maps, and designs were developed by SCC staff and reviewed by David Derrick, U.S. Army Corps of Engineers, Waterway Experiment Station, in Vicksburg, Mississippi. The project currently involves 29 project sites on 8.2 miles of the river. Project construction began in November 2001 and was completed with the final tree planting in April 2004. This project stabilized 8.2 miles of eroding streambanks, established 110 acres of riparian habitat, resulted in the planting over 70,000 trees and shrubs, and will reduce 546,317 tons of sediment to the river annually.

KEYWORDS: stabilization, riparian restoration, habitat restoration, stream restoration, sediment reduction.

INTRODUCTION

Sediment is the most common pollutant in streams throughout the United States (U.S. EPA, 1998). Streambank erosion is a major source of stream sediment (Rosgen, 1997; Simon, 2003). As the United States Environmental Protection Agency (U.S. EPA) continues to focus on Total Maximum Daily Loads (TMDLs), stream sediment reduction via streambank stabilization and erosion control methods will become a major financial commitment for states attempting to comply with sediment standards. A cost-effective solution to streambank erosion must be developed to resolve this water quality problem and restore America's degraded stream corridors to a healthy condition.

[1]P.G. Balch – The Watershed Institute, Inc., Tetra Tech EM, Inc. – 1200 SW Executive Drive – Topeka, Kansas 66615. 785-272-2252 – FAX 785-272-7349- philip.balch@ttemi.com

[2]B.A.Emmert – The Watershed Institute, Inc., Tetra Tech EM, Inc. – 1200 SW Executive Drive – Topeka, Kansas 66615. 785-272-2252 – FAX 785-272-7349 – brock.emmert@ttemi.com

The Little Blue River Stream Stabilization and Riparian Corridor Restoration Project is the first such attempt in Kansas to remedy large-scale streambank erosion on large rivers with limited funds. Other project goals were to:

> - Reduce excess stream sediment
> - Improve stream channel dimension, pattern, and profile
> - Improve aquatic habitat
> - Establish a riparian ecosystem
> - Improve terrestrial habitat
> - Improve water quality
> - Reduce nutrients and chemical pollutants.

STUDY AREA

The Little Blue River flows through the eastern portion of rural Washington County, Kansas, and has a drainage basin that ranges from 2,752 square miles at Hollenberg, Kansas to approximately 3,500 square miles at Waterville, Kansas. More than half of the river basin is in south-central Nebraska. The bed material is predominantly sand and small gravel (.0024 – 2.5 inches in diameter) (Figure 1-1). Bankfull flow ranges from 5,646 cubic feet per second (cfs) at Hollenberg to 11,230 cfs near Barnes, Kansas with return intervals of 1.29 years and 1.44 years respectively.

The bank material composition varies from silts and clays (<. 0.0024 inches in diameter) to sand (0.0024 – .0787 inches in diameter). The Little Blue River is not impounded by large reservoirs and does not contain areas of major levee construction. The river is slightly entrenched with natural riparian vegetation that includes three species of willow (*Salix sp.*), eastern cottonwood (*Populus deltoides Marsh.*), silver maple (*Acer saccharinum L.*), box elder (*Acer negundo L.*), elm (*Ulmus sp.*), burr oak (*Quercus macrocarpa Michx.*), American linden (*Tilia americana L.*), black walnut (*Juglans nigra L.*), hackberry (*Cetlis occidentalis L.*), red mulberry (*Morus rubra L.*), and green ash (*Fraxinus pennsylvanica Marsh.*).

Riparian understory vegetation is dominated by wild ryes (*Elymus sp.*), poison ivy (*Rhus radicans L.*), reed canary grass (*Phalaris arundinacea L.*), buckbrush (*Symphoricarpos orbiculatus Moench*), and wild gooseberry (*Ribes missouriense Nutt.*).

Most fields along the stream were under cultivation within a few meters of the streambank edge each year (Figure 1-2). Among the stabilized areas, only site number 3 had any permanent riparian vegetation.

SURVEY AND DESIGN

Each site was surveyed by Kansas State Conservation Commission (SCC) and Natural Resources Conservation Service (NRCS) staff with a total station. Data points were downloaded into a computer and topographic maps were produced for each site. The maps were then used for measurements and project stabilization design (Fig. 1-3).

Initial site assessments recognized that the Little Blue River had severe bedload problems. Numerous sites contained mid bars and the stream was extremely shallow. Areas surveyed with water depths greater than 18 to 25 inches were upstream of a few isolated, large, woody, debris piles. SCC and Kansas Department of Health and Environment (KDHE) staff designed all projects, choosing bendway weirs for the primary stabilization structure because of the stream's high width/depth ratio (Figure 1-4).

Bendway weirs redirect water flowing over them which slows water velocities along the near bank region (Derrick, 2001). A weir also moves the thalweg away from the bank to the in-stream end of the weir. The design height of all bendway weirs was 1 to 1.5 feet above the water surface at low flow. David Derrick, U.S. Army Corps of Engineers, reviewed the initial 20 project designs.

On sites # 8 and # 21 the radius of curvature was very low. In order to keep from pushing the thalweg a great distance from the bank and keep from radically redirecting stream flow, rock vanes were chosen as the stabilization method for these sites. (Figure 1-5).

PROJECT FUNDING

The SCC's Riparian and Wetland Protection Program (RWPP) was originally targeted as the main source of project funding. Increasing numbers of landowners enrolling in the project rapidly grew beyond the RWPP's financial capabilities. Fortunately, KDHE was able to provide $265,000.00 of U.S. EPA 319 funds to the project. Additional financing came from the SCC's Non-point Source Pollution Control Program, the Kansas Governor's Water Quality Initiative, the Kansas Alliance for Wetlands and Streams (KAWS), and the Kansas Chapter of the National Wild Turkey Federation. Combining federal and state funds provided 100% funding for the stabilization portion of the projects. This project required participating landowners to enroll a 100-foot wide strip into the United States Department of Agriculture's (USDA) Continuous Conservation Reserve Program (CCRP). Costs associated with planting and maintenance of the CCRP strip were not included in the listed construction cost. Tree planting costs for the riparian area between the CCRP strip and the edge of water were included in the construction cost, or shared with the Kansas Forest Service's (KFS) Forest Land Enhancement Program (FLEP) and RWPP. Total construction costs for the Little Blue River Stabilization Project and repairs are estimated at $550,000. This equals $13.02 per lineal foot and does not include any cost associated with the CCRP plantings.

STRUCTURE INSTALLATION AND RE-VEGETATION

On early projects, weirs were constructed by excavating ramps into the streambanks, dumping rock on the ramp, and then pushing the rock into the stream with bulldozers (Figure 1-6). After the first few projects, rock was dumped directly over the streambank and then moved into place with an excavator (Figure 1-7).

Following construction of the bendway weirs, the vertical banks were reshaped to a 3 feet horizontal to 1 foot vertical slope (Figure 1-8). On all sites utilizing bendway weirs for stabilization, the near vertical banks were shaped by pushing them into the river channel (cut and fill method). This accomplished three things: it eliminated the need to "key" the weirs into the bank, reduced construction costs by reducing required equipment time, and reduced the amount of valuable cropland required achieve the desired slope.

After reshaping the vertical banks, winter wheat or oats was sown on the slopes and mulched with native prairie hay on all projects in phase 1. Projects constructed in phase two and three were sown to wheat or oats but not mulched (Figure 2-9).

The riparian area between the CCRP strip and edge of water was planted with live willow stakes and bare-root cottonwood seedlings. Live willow stakes were planted on 4 x 4 ft. spacing. Cottonwood seedlings were planted on 6 ft. x 6 ft. spacing. In the CCRP strip, trees were planted on 8 ft. x 8 ft. or 10 ft. x 12 ft. spacing. All shrubs were planted on 6 ft. x 6 ft. spacing. A 25-foot wide strip of native grasses and forbs between the shrubs and the cultivated crop field completed the CCRP (Figure 2-10).

All trees on the 3:1 slope were planted by hand. Trees on the field portion of the buffer were planted with farm tractors and tree planters (Figure 2-11). Agency personnel with NRCS, SCC, and the KFS measured and flagged the tree rows (Figure 2-12).

Prior to planting, willows were soaked for a minimum of 10 to 14 days (Figure 2-13). Research has shown that the survival rate for live willow stakes doubles when the stakes are soaked for this amount of time prior to planting (Schaff et. al., 2002). Student members of area Future Farmers of America chapters harvested all willow stakes used in the 2002-planting season. Live willow stakes for the 2003-planting season were purchased from the Kansas Forest Service (KFS).

In the spring of 2002, landowners, agency personnel, and conservation district personnel planted the trees on 12 sites with volunteer help from several Boy Scouts of America troops. In the 2003 planting season, landowners and agency personnel planted all trees on 12 additional sites. Landowners planted native grasses with a no-till drill provided by the Washington County conservation District. Over 70,000 trees and shrubs were planted during the springs of 2002, 2003, and 2004.

FROM DROUGHT TO FLOOD

The project area experienced a severe drought during the late spring and summer of 2002. Rainfall throughout the project area totaled less than 7 inches during the summer. Because of the drought, trees were replanted on some sites dominated with sandy soils and south facing slopes in April 2003.

Projects completed in early 2002 were inundated with two flows that approached the bankfull magnitude -- in June and September 2002. Another bankfull flow event occurred in early May 2003. Minimal erosion occurred at the stabilized sites during these flows. Slight erosion from the moderate flows required the addition of one structure on two sites. In late June 2003, severe weather and torrential rainfall in south central Nebraska resulted in substantial flooding along the Little Blue (Figure 2-14).

U.S. Geological Gage data logged flood flows of 42,600 cfs at the Hollenberg, Kansas stream gage. Downstream, at the USGS gage near Barnes, Kansas, flows peaked at 32,800 cfs. No damage occurred at 20 of the 24 completed sites (Figure 2-15) from this flood. On the four sites that incurred slight erosion, the damage was limited to a small portion of each. All problems appeared on the lower one third of the project sites and were corrected in the fall and winter of 2003 by installing Longitudinal Peaked Stone-Toe Protection (LPSTP) or adding additional rock to the weirs.

Sediment deposition occurred on several sites. This was evident between the weirs and on the banks (Figure 2-16).

RESEARCH

Various types of research projects were conducted at project sites. One of the first was installation of bank erosion pins. In April 2001, 4 ft. and 6 ft. long bank (erosion monitoring) pins were installed at six sites (Figure 2-17). At another site, obvious severe erosion warranted placing two benchmarks 28 feet and 29.1 feet away from the bank edge. Five weeks later, an inspection trip discovered all pins lost at the six sites due to streambank erosion. At the other site, only 8 feet of the original 29.1 feet remained between the bank and one remaining benchmark.

Soil samples were also taken at each site. On most sites, one sample was taken for every 3 feet of bank height (Figure 3-18). The Kansas State University soils laboratory analyzed all soil samples for nutrient content. Total nutrient input, associated with the bank erosion, was calculated using the resulting data and soil loss calculations.

Dr. Charles Barden, Kansas State University Research and Extension - Professor of Forestry, assisted with tree planting design and also conducted research on various types of tree shelters (Figure 3-19).

Fisheries biologists with Kansas Department of Wildlife and Parks (KDWP) conducted fish sampling surveys at several sites prior to project construction (Figure 3-20). These sites will be re-sampled in subsequent sessions to determine any changes in fisheries species composition and biomass.

Similar studies in Mississippi have shown a greater increase in total biomass and species diversity at sites with rock weir type structures than at sites with other stabilization methods (Shields et. al., 2000).

In October 2001, researchers with the USDA Agricultural Research Station (ARS), National Sediment Laboratory in Oxford, Mississippi, conducted research on the root strength and density of various species of willow and eastern cottonwood (Figure 3-21).

The ARS National Sediment Lab also conducted soil tension strength analysis on limited sites and is now investigating possible causes for the systemic bank instability throughout the river basin.

Three sites were chosen for comparison studies of riparian planting methods. Two sites will look at riparian area natural regeneration. The other will compare direct seeding and nut plantings to sites planted with bare-root tree seedlings.

Kansas State University Department of Agricultural Economics was enlisted to conduct a socio-economic study of the project. The results of this study showed the average landowner gained an additional $810 annually from participating in the project. Gains were realized by the value of cropland acres not lost to streambank erosion, income from the acres not lost, and income from the Continuous CRP payments. Furthermore, the assessment showed a positive net present value to the landowner for establishing a riparian buffer in CRP and a negative net present value if removing an existing riparian buffer.

CONCLUSION

The Little Blue River stabilization and riparian corridor establishment project has proven that large-scale streambank stabilization can be constructed in a cost- effective, "river friendly" manner. Bendway weirs on sand bed streams can diversify fisheries habitat and assist in restoring a stable fluvial geomorphology to streams. A cost comparison between Bendway weirs and rip-rap was conducted for site 28. The cost estimate for rip-rapping this site was $165,000. The actual construction cost to install 6 bendway weirs, reshape the 1,200 lineal feet of streambank, and plant trees was $17,789.43. This project not only reduced the amount of sediment entering the stream due to bank erosion, but also removed excess sediment from the stream during high flow events as evidenced by sediment deposition in several locations.

The Little Blue River Stabilization and Riparian Corridor Establishment Project has reduced loss of valuable cropland to bank erosion, extended downstream reservoir life, increased wildlife habitat, increased fisheries habitat diversity, and improved water quality.

REFERENCES

Derrick, D.L., 2001. Unpublished Research.

U.S. EPA. *305b Report*. 1998.

Schaff, S. D., Pezeshki, S. R., and Shields, F. D., Jr. 2002. "The Effect of Pre-planting Soaking on Growth and Survival of Black Willow (Salix nigra) cuttings." *Restoration Ecology* 10(2):267-274.

Shields, F. D., Jr., Knight, S. S., and Cooper, C. M. 2000. "Warmwater Stream Bank Protection and Fish Habitat: A Comparative Study." *Environmental Management* 26(3):317-328.

Simon, A. (2003). Personal Communication.

Rosgen D. (1997). Personal Communication.

(Figure 1-1) Little Blue River Streambed

(Figure 1-2) Site 22, Prior to Construction

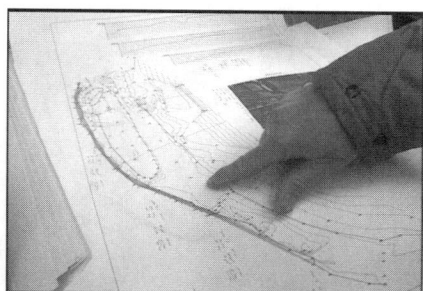

(Figure 1-3) Typical Project Map and Design

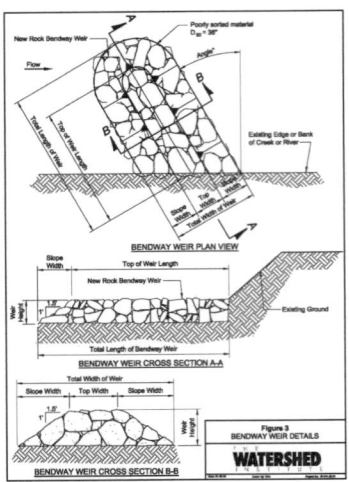

(Figure 1-4) Bendway Weir Detail

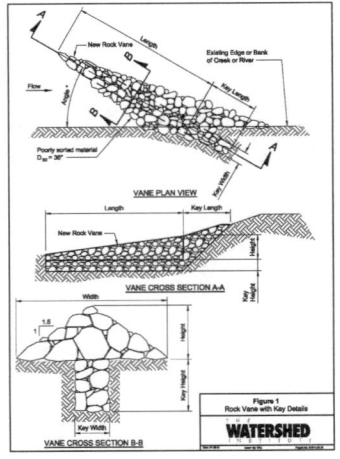

(Figure 1-5) Rock Vane Design Detail

(Figure 1-6) Building Weir on Site 28.

376

(Figure 1-7) Weir Construction on Site 22.

(Figure 1-8) Bank Shaping with Bulldozer

(Figure 2-9) Site 1, 4/02/03 Oats Beginning to Grow.

377

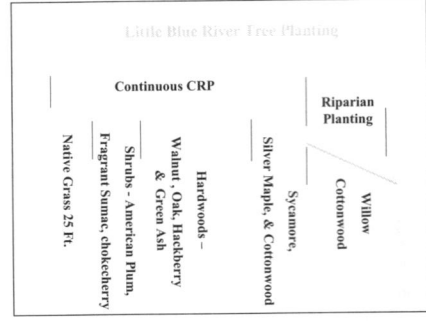

Little Blue River Tree Planting

Continuous CRP

Riparian Planting

Native Grass 25 Ft.

Fragrant Sumac, chokecherry

Shrubs - American Plum,

Walnut , Oak, Hackberry & Green Ash

Hardwoods –

Silver Maple, & Cottonwood

Sycamore,

Cottonwood

Willow

(Figure 2-10) Planting Diagram

(Figure 2-11) Tree Planting on Site 11

(Figure 2-12) Measuring and Flagging Tree Rows

(Figure 2-13) Willow Cuttings Soaking in Ponds.

(Figure 14) Site 1, 6/24/03 Prior to Peak Flood

(FIGURE 2-15) SITE 1, 7/02/03, FOLLOWING 42,600 CFS FLOW

(Figure 2-16) Sediment Deposition

(Figure 2-17) Installing Bank Erosion Pins

(Figure 3-18) Taking Soil Samples Site 6

(Figure 3-19) Tree Shelter Test on Site 11

 (Figure 3-20) Fish Sampling on Site 22

 (Figure 3-21) Root Strength Testing

Restoring Wetlands for Nitrate Removal from Subsurface Drainage: Research Foundation and Program Elements of the Iowa Conservation Reserve Enhancement Program

William G. Crumpton, Department of Ecology, Evolution, and Organismal Biology,
Iowa State University

Surface water nitrate loads in the Midwest Corn Belt are among the highest in the Mississippi River Basin, and in addition to local water quality impacts, are suspected as a major contributor to hypoxia in the Gulf of Mexico. Wetlands have the potential to significantly reduce nitrate loads, if wetlands can be sited and designed to intercept nitrate carried in subsurface drainage tile. For the past 10 years, Iowa State University and the Iowa Department of Agriculture and Land Stewardship have cooperated on a targeted research program addressing the hydrologic and water quality functions of wetlands in agricultural watersheds. This effort supported the development and application of performance forecast models for siting, design and assessment of wetland restorations in agricultural watersheds. This work elucidated the benefits and limitations of wetland restorations for nutrient reduction and provided the research foundation for the Iowa Conservation Reserve Enhancement Program (CREP). The Iowa CREP provides financial incentives to restore wetlands that are strategically located to intercept tile drainage and reduce nitrate export from agricultural watersheds of the Des Moines Lobe. A unique aspect of the Iowa CREP is that expected water quality benefits will not simply be assumed based on wetland acres enrolled, but in fact be calculated based on the measured performance of CREP wetlands. As an integral part of the Iowa CREP, the Agricultural Drainage & Wetlands Research Team at ISU is monitoring a representative subset of wetlands to document nitrate reduction. This will allow further refinement of modeling and analysis tools used in siting and design of CREP wetlands. Research at Iowa State University has shown that wetlands meeting CREP requirements will remove 40-90% of the nitrate received, which for sites currently constructed or under design equates to 36,900 tons of nitrate over the design life of the wetlands.

An Investigation of Strategies to Predict Flood-Peak Discharges on Variously-Sized Watersheds in Ohio

Jessica L. D'Ambrosio, Andy Ward, Lance Williams

This study (1) evaluated how flood-peak discharges are associated with several common approaches for determining the recurrence interval (return period) and annual frequency of storm events, and (2) analyzed and tested the United States Geologic Survey (USGS) rural and urban flood-peak discharge regression equations for Ohio to determine if they can provide effective estimations of peak discharges at lower recurrence intervals (less than 2 years) and using simple, practical data collection methods. The USGS equations were tested on 79 watersheds in Ohio with various drainage sizes. An analysis of measured discharge used annual series streamflow data from USGS gaging stations and related it to recurrence interval using the Weibull method. An analysis of predicted discharge involved solving the USGS equations and comparing the results to measured discharge values. This relationship was evaluated based on $y = (1)x$. Slope and intercept terms were evaluated for significance (alpha = 0.05) for each trend line equation. In general, on smaller watersheds (less than 35 mi^2) the USGS rural and urban equations provided reasonably good estimates of measured discharge. On watersheds greater than 100 mi^2, the USGS urban equation provided good estimates of measured discharge. Different methods of collecting data for the rural equation variables included the use of USGS 7.5-minute topographic quadrangle maps and Geographic Information Systems (GIS) technology. Comparing the two methods showed that for the watersheds used in this study, simple practical methods of data collection provided similar estimations of peak discharge to data gathered using GIS. Manipulating the coefficients in the urban equation through regression analysis provided even better estimates indicating that some modification to the current urban equation may be useful. The basin development factor (BDF) variable was not significant (alpha = 0.05) in the urban equation across all watersheds sizes and recurrence intervals suggesting that the watersheds used in the study may not be urbanized enough for it to be important in the equation. However, this variable is needed because it is the only indicator of urban development in the USGS equations.

STUDY OF RELATIVE TRANSPORT ABILITY AND OTHER STREAMFLOW FORMATION FACTORS

Volodymyr Dumenko

Chernivtsy Institute of Agroecology and Biotechnology

This research is dedicated to examination of streamflow formation factors which influence on formation and modification of streamflow processes types of Ukrainian lowland rivers. It is conducted a search of streamflow formation factors which cause the forming of several various types of streamflow processes under various conditions of their display. Such factors help to find reasons of existence of some types of streamflow processes and their mutual transformation under change of ratio of balance between component parts of streamflow formation factors. Exploitable principles of system approach demand independent reveal of streamflow formation activity of factors on various system levels. It is considered relative transport ability (ratio between transport ability and drifts entry) as one of the streamflow formation factors. It is proved the necessity of relative transport ability calculation not only on system levels "river – reservoir" and "water - soil" but also on system level "stream – river-bed" as one of the reasons of formation meandering and branchy by type of river-bed multi-branch streamflows. Relative transport ability influence is expressing as in both river longitudinal profile ("river – reservoir") and local erosion and inwash sections ("water - soil") as on system level "stream – river-bed". In the research it was established multifactor model of streamflow processes which leads to joint analysis of independent streamflow formation factors. In the research it is presented bi-factorial tables which are projections of multidimensional space of determining factors of streamflow processes types to determinative streamflow forming axis of relative transporting ability and axes of other streamflow forming factors. Actuality of study the reasons of streamflow processes types change is explained that regularities of streamflow changes within the bounds of certain types of streamflow processes are well-known enough and are used for streamflow deformation forecast. However reasons and regularities of types' changes under change of ratio between streamflow formation factors are studied insufficiently. The main purpose of this research is substantiation of selection of relative transporting ability as a factor that determinates formation and change of streamflow processes types on the system level "stream – river-bed" in following consistency: meandering – straight streamflow – multi-branch streamflow. It is examined joint influence of this streamflow formation factor and other factors on type changes and it is also offered bi-factorial table of streamflow processes types.

30 Year Historical Trends in Coralville Reservoir Water Quality

Claudia O. Espinosa-Villegas[1*], Dr. Craig L. Just[2] Dr. Jerald L. Schnoor[1]

1 * CGRER, University of Iowa, Iowa City, IA, 52242, USA

2 Dept. of Civil and Environmental Engineering, University of Iowa, Iowa City, IA, 52242, USA

Iowa's water resources are perhaps its single most valuable natural asset, yet the number of water bodies and river miles impaired by nutrients and siltation has increased in the last years. With agriculture as the leading industry in Iowa, yet a major contributor to water resources degradation, it is difficult to protect our water resources without negative economic and social impacts to farmers. In watersheds that are primarily agricultural, it has been estimated that over 80% of the total nitrogen added in to the watershed comes from commercial fertilizers use related to agriculture. Many times, the amounts of in-stream nutrients can be correlated to the land-use of the watershed.

In this study a 30 year historical water quality record for the Coralville reservoir has been used in order to determine the water quality trends of the reservoir and the river that flows into it. Preliminary analysis of the data set shows that the amount of nitrate-nitrogen in the reservoir waters has a positive trend over the 30 year time span. The expansion of suburban developments has increased the impervious land cover and has affected not only the runoff patterns within the watershed, but has also increased the amount of anthropogenic nitrogen contribution due to the changes in land use, installation of industrial enterprises and increases in population waste load. The amount of total suspended solids in the water appears to decrease over 30 years and there appears to be greater variability in the amount of sediment in the water prior to 1980. Another nutrient of concern in the Coralville watershed is phosphorus. The waters of the reservoir and the incoming river usually contain less than 1 mg/L of orthophosphate (as phosphorus) but much of the time, 4-5 times more than the recommended 100 ppb. Examination of the land use changes, the urbanization patterns and evaluating the efficiency of BMP's will help elucidate the factors contributing to the long-term data trends.

* Corresponding author : cespinos@engineering.uiowa.edu

An Assessment Of The Instream Flow Requirements:
The Case Of Mazowe River Catchment

Tichatonga Gonah
House Number 221, Mkoba 1, Gweru, Zimbabwe
tichgonz@webmail.co.za/tichgonz@yahoo.co.uk

ABSTRACT

In a bid to determine the status of instream flow requirements and if they are being met, the researcher carried out a case study comprising of the Mazowe river basin, falling under Mazowe catchment, one of the hydrological catchments of Zimbabwe. As the concept of instream flow requirements is still new in Zimbabwe, this justifies the need for carrying out such a research. The major objectives were to enlighten the status of instream flow requirements in Zimbabwe and to establish the effect of dams on flow regimes. This study was carried out by considering a sample of 15 sub hydrological zones falling under this river catchment. It is mainly runoff data that was integrated into the Hydrological Database and Analysis Tool (HYDATA) model to come up with the most important input parameters of the Hughes and Munster desktop methodology that was used to estimate the instream flow requirements. The researcher quantified the instream flow requirements and reconciled these with the river catchment water resources potential revealing that most of the river systems are having their environmental water needs being met. Dams have been found to regulate flow regimes through increased or decreased flows and the base flow index was discovered to be linearly related to the instream flow requirements. The researcher concluded that there is a potential of carrying out environmental rehabilitation particularly river restoration and expanding other water uses. The researcher has recommended the need for more research and investment into the issue of environmental flows with the overall goal of achieving sustainable allocation and utilisation of our scarce water resources.

Key words: instream flow requirements, river rehabilitation, naturalization, sustainable watershed management, flow regulation

Protecting Surface Water from Military Activity with Riparian Buffers and Low Water Stream Crossings

Hutchinson, S.L., P.L. Barnes, J.M. Shawn Hutchinson,
C.G. Oviatt, J. Steichen, and N. Zhang

Non-point source (NPS) pollution has been called the nation's largest water quality problem, and its reduction is a major challenge facing our society today. As of 1998 over 290,000 miles of river, almost 7,900,000 acres of lake and 12,500 square miles of estuaries failed to meet water quality standards. Military training maneuvers have the potential to significantly alter land surfaces in a manner that promotes NPS pollution, resulting in the inability of military installations to meet water quality standards and the decline of training lands.

Military readiness depends upon high quality training. Effective maneuver training requires large areas of land and creates intense stress on this land. Environmental protection requirements place additional restrictions on land use and availability. Because military training schedules are set well in advance to make the best use of installation training facilities and National Training Centers, there is little flexibility to modify training events and maintain readiness. In order to avoid maneuver restrictions, proactive management plans must be developed giving commanders the information they need to assess the environmental cost of training and management practices that reduce the environmental impact.

The objective of this work is to identify sources of NPS pollution resulting from military activities, to assess the impact of this pollution on surface water quality and to provide information for commanders to lessen the impact of training on water quality. Investigators are assessing the impact of two major sources of NPS pollution, 1) erosion from upland training areas and 2) channel erosion at stream crossing sites, on surface water quality at Fort Riley, KS. Project objectives will be met through a comprehensive analysis of military activities, climatic factors and environmental response.

Researchers are using watershed water quality models in conjunction with remotely sensed information and a geographic information system (GIS) to assess the impact of training on water quality, in particular on the amount of soil erosion. A matrix of training intensity and weather will be created for assessing the environmental cost of training maneuvers. In addition, researchers are collecting surface runoff at three buffer sites to determine the effect of vegetated buffers for controlling NPS pollution and using new real-time data collection systems to assess the impact of vehicle crossings on stream water quality and erosion dynamics at Low Water Stream Crossings (LWSCs).

IMPACT ASSESSMENT OF CONSERVATION MANAGEMENT PRACTICES ON WATER QUALITY IN UPPER BIG WALNUT CREEK WATERSHED

Kevin W. King, Ph.D.
USDA-ARS
590 Woody Hayes Dr.
Columbus, OH 43210

Recent source water protection initiatives and legislative mandates housed in the 2002 farm bill require watershed scale assessments of conservation practices. The objective of this study is to share an approach being proposed and utilized in the Upper Big Walnut Creek watershed in Ohio for attaining this goal. Much of the Upper Big Walnut Creek Watershed is in agricultural row crop production and is extensively tile drained. Understanding the partitioning (quantity and quality) of surface runoff, true tile flow, and combined surface/tile flow is critical. The study consists of both modeling and field research using a multi-scale, multi-land use, nested, paired watershed approach. Water quality instrumentation includes control volume structures and automated samplers. Conservation management practices such as water table management (controlled drainage) as well Core4 practices will be evaluated. The approach could be adopted or modified for similar assessment studies.

The NEMO for STREAMS Program

Timothy Lawrence, Jessica D'Ambrosio, the Ohio State University

NEMO for Streams combines land use issues and stream functionality in a program to provide educational opportunities for local decision makers in improving water quality. Traditional storm water management strategies have primarily focused on water quality and/or reducing and managing the impacts of floods. Many land development strategies modify the stream and floodplain system. These practices only indirectly consider fluvial processes and often fail to prevent stream channel stability problems. aims to: assist in the establishment of an infrastructure and interdisciplinary groups necessary to develop viable ecological solutions for stream and watershed management; improve the function and quality of streams through education on the use of natural channel design concepts; provide educational programs that focus on the adoption of preventative and self sustaining strategies of restoration and aquatic management; and encourage recognition that for stream systems it is necessary to consider a range of discharge conditions and that high frequent events might be more important than extreme rare events that are often used in design. While concern lies in having water that is drinkable and swimmable, flooding and property damage due to erosion are more common complaints in urban neighborhoods. The NEMO for STREAMS Program encourages landowners and local officials to recognize that water quality and water quantity go hand-in-hand and managing them together makes economic and environmental sense. Management strategies employ site design and land use regulations that decrease impervious surfaces, employ natural channel design, include wetlands and bioretention areas, and maintain adequate stream buffer zones.

History and Status of Physical Habitat Assessments of USA Streams

Derek Martin and Robert T. Pavlowsky, Department of Geography, Geology, and Planning, Southwest Missouri State University, 901 S. National, Springfield, MO 65804, djm242s@smsu.edu.

Field assessment techniques are needed to quantify the physical characteristics for stream restoration and ecological monitoring purposes. This paper describes the origins, history, and present status of physical habitat assessment protocols currently in use in the USA. While the driving force for the development of standard procedures for classifying and quantifying stream properties has been to support ecological analyses from the biological perspective, the roots of the methods and their scientific rationale are more multidisciplinary. For example, the concepts of scale-dependent variables, watershed linkages, dominant/bankfull discharge, and channel-sediment relationships were developed by workers in geography, geology, hydrology, and engineering. A 50 state survey shows that most stream assessment protocols are modifications of those published by Federal agencies, including the Rapid Bioassessment Protocol (EPA, 1989, revision 1999), the Environmental Monitoring and Assessment Program (EPA, 1998, revision 2001), and the National Water-Quality Assessment Program (USGS, 1993, revision 1998). The EPA's original RBP (1989) was based on Stream Classification Guidelines for Wisconsin (Wisconsin DNR, 1982) and "Methods for Evaluating Stream, Riparian, and Biotic Conditions" developed by the USDA Forest Service (1983). EMAP and NAWQA were developed in response to the need to incorporate broader scale habitat assessments within water resource programs. About one third of the states responding to the survey claimed to use protocols designed specifically for physical habitat assessment. Of those responses, only a small portion implements those assessments as part of a long-term monitoring program. The origin, overall organization, and logistics of these methods for field studies are discussed. The need for effective and scientific stream assessment techniques is underscored by recent trends toward the development of long-term monitoring programs, stream classification systems, human impact studies, and linked studies of field-based and remotely-sensed data.

Dam Removal as a Solution to Increase River Water Quality

Matthew Nechvatal
Civil and Environmental Engineering Department
The Ohio State University

Dr. Timothy Granata, PhD
Civil and Environmental Engineering Department
The Ohio State University

Abstract

Dams have historically been constructed as public works projects for water supply, hydropower, recreation, flood control, and transportation, however, dams disconnect the aquatic ecosystem, degrade water quality, and alter aquatic habitats. Dam removal has recently become a new restoration tool for restoring stream and river ecosystems. St. John's Dam was a concrete low-head dam constructed in the 1930's on the Sandusky River in north central Ohio. The dam was 2.2 m high, 4.6 m wide, and its reservoir served as a water supply for the city of Tiffin. The St. John's Dam was removed on November 17, 2003. Studies are being performed to determine how dam removal is affecting water quality above and below the dam. As part of this work, a YSI water quality sonde was attached to a canoe to sample surface water quality longitudinally from upstream of the dam to the dam. These longitudinal transects show spatial variation in oxidation reduction potential, dissolved oxygen, temperature, specific conductivity, turbidity, and pH. Transects were performed before and after dam removal. Water samples were also taken at four sites below the dam and four sites above the dam to determine how nutrient concentrations (PO_4^{3-}, NO_3^-, NH_4^+) change with removal. Results from preliminary studies showed increasing temperature and turbidity starting 17.4 km above the dam to just above the dam. After removal there was no increase in temperature or turbidity as you moved down the former impoundment. Prior to dam removal, nitrate levels increased toward the impoundment while after removal nitrate was less variable. Phosphate and ammonium levels were undetectable above and below the dam. Removal did not seem to affect the levels of nitrate below the dam. Time series data showed a significant increased (spike) in turbidity below the dam during removal, a decrease in ORP, and slow increase in temperature as the water was released downstream.

Watershed-scale Assessment of Bank Stability in an Urban Watershed, Springfield, Missouri

Presenter: Aaron Nickolotsky, Southwest Missouri State University

Co-Authors: Ron Miller, Southwest Missouri State University
Robert T. Pavlowsky, Southwest Missouri State University
Adam Coulter, USDA-NRCS, Ozark, MO

Stable stream banks reduce sedimentation problems, improve water quality, and enhance the aesthetic character of riparian lands. The purpose of this study is to assess the influence of urbanization on stream bank stability in the Ward Branch watershed (drainage area is 28.41 km2). The study watershed drains a karst area in the Ozarks of southwest Missouri. The headwaters are located in Springfield, the third largest city and one of the fastest growing urban areas in the state. Over the past decade, Ward Branch has been subjected to more frequent flooding and channel erosion. The objectives are to: (1) describe the spatial distribution of reach-scale bank conditions using the USGS National Water Quality Assessment Program's bank stability index; (2) evaluate the patterns of geomorphic disturbance indicators by GPS mapping; and (3) provide baseline information for planning of stream restoration projects. Overall, 22 reaches were studied in the watershed with 15 on the main stem and 7 along Workman Branch, a major tributary. The most unstable banks are located in the upper portions of the watershed where urban runoff is concentrated and released to natural sections and along lower, more sinuous reaches, with higher fine-grained banks. Local bank erosion problems are also associated with poorly designed bank improvements, road culverts and bridges, and woody/construction debris jams. In general, bank erosion at the watershed-scale is related to proximity to urban areas or impaired riparian corridors.

Demonstration Project for Channel Restoration in Urbanizing Streams of the Ozark Plateaus, Greene County Missouri

Timothy W. Smith, P.E., Kevin R. Barnes, P.E. and Marc R. Owen,
Greene County Resource Management Department.

Channel instability in Ozark streams due to urbanization has impacted streams in the Springfield, Missouri metropolitan area resulting in flooding, bed scour, bank instability and increased sediment yields. Presently, the geomorphic processes related to urban channel erosion are poorly understood and no tested channel restoration guidelines are yet available in the Ozarks. This poster describes recent efforts by Greene County to identify management options for restoring and stabilizing urban streams in the region. Preliminary results involve a nonpoint source pollution 319 grant aimed at developing a demonstration site along Ward Branch for channel restoration measures. The Ward Branch of the James River is located on the south side of Springfield. The study reach is located in an area where Greene County purchased and removed homes affected by flooding from recent developments. The objectives of the project are: (1) Develop, install and monitor alternative stream stabilization techniques. (2) Evaluate stream stabilization techniques based on cost, availability of materials and effectiveness. (3) Educate area residents, developers, contractors and community leaders on the importance of stream channel erosion in the transport of nutrients linked to water quality in the James River Basin. Initial channel surveys show the 325 meter reach is deeper, has longer riffle-pool spacing and lower sinuosity relative to other urban streams in the area. This project is a collaborative effort between the City of Springfield, Southwest Missouri State University, the Watershed Committee of the Ozarks, Missouri Department of Conservation and Greene County that will facilitate a watershed approach to urban stream management.

Using Geomorphology to Assess and Manage the Risks Associated with Clean Sediments in the Little Miami River Watershed (Southwestern OH)

Christopher Schultz (NRMRL), Joseph Schubauer-Berigan (NRMRL), Matthew Morrison (NRMRL), Bernie Daniel (NERL), Michael Troyer (NCEA) and Michael Griffith (NCEA)

We are evaluating the use of stream geomorphology and related measurements in the assessment and management of channel risks associated with stream impairment associated with clean sediments. The relationships between various geomorphological variables have been used by Rosgen and others to classify streams into groups that can be related to sediment transport or to bed and bank stability. These methods may be used to predict and evaluate the sensitivity of stream reaches to altered hydrologic regimes that subsequently result in bank destabilization and excessive sediment transport. Factors, such as increased erosion and stream channel destabilization, can result in impairment by excessive clean sediment. This project is intended to determine (1) the most effective, timely and cost-efficient methods for collecting channel morphology data, and (2) which variables may be used as indicators of increased risk of impairment from suspended and bedded sediments. Previous research in the Little Miami River (Southwestern Ohio) established thirty-five sites where data was collected for several years using the U.S. EPA Environmental Monitoring and Assessment Program protocols. Eight of these sites were selected for intensive surveying to determine the level of effort required to obtain representative data for stream classification. Slope and sinuosity measurements from the surveys resulted in the same Rosgen stream classification as that determined from measurements from aerial photographs and topographic maps. This indicates that these types of data may be used in place of extensive surveying at least in Southwestern Ohio. Other measurements such as pebble counts, entrenchment and stream profile still require field visits. The extent to which a qualitative rating system can be used is also being evaluated. The most important parameters for determining the stability of stream geomorphology and near-stream erosion risks appear to be slope, bed and bank material stability, incision and vegetative cover/type in the riparian zone.

Measuring the Effectiveness of an In-stream Passive Bedload Sediment Removal System in Mountain Streams

Michael B. Shaffer, EI and Greg Jennings, PhD, PE
North Carolina State University, Raleigh, NC
Streamside Systems LLC, Findlay, OH
Balsam Mountain Preserve, Waynesville, NC
mbshaffe@unity.ncsu.edu

There many existing approaches to preventing sediment impact to streams but very few means of removing sediment from a previously impacted stream. Most techniques have significant negative impacts as well as positive.

Many of the mountain streams of North Carolina have been significantly impacted by sediment from historic logging, mining and development. Aquatic habitat and stream hydrological functions have both been affected.

A new system of equipment patented by Streamside Systems, LLC offers promise for providing a tool to selectively by size, capture and remove in-stream bedload sediments without significantly disrupting aquatic habitat. By being fully passive, the equipment could be installed and should operate continuously without attention, making it much more likely to be utilized. The collected fine sediments may have commercial value as well.

North Carolina State University has set up experiments within the Balsam Mountain Preserve that will measure the effectiveness of this equipment, the impact of this equipment and develop specific installation and application guidelines and techniques. The equipment was installed in 2003 and preliminary findings and results will be presented.

Development of the Ecological Unit Model for the Sandusky River Watershed

Cynthia Smith, Jay Martin
The Ohio State University
Department of Food, Agriculture, and Biological Engineering
Ecological Engineering Group

In coastal settings, including the Great Lakes, policies are being proposed to address human impacts on water quality and habitat change. However, few tools exist to quantitatively evaluate the impacts of these plans across time and space. A spatial watershed model will be created to predict and quantify the effects of land use policies on water quality and habitat change within the Sandusky River watershed (Lake Erie, Ohio). The spatial watershed model will incorporate three cellular models: hydrodynamic, ecological, and economic across a spatial grid. This project focused on the development of an ecological unit model that incorporated the important processes involved in three different ecosystems; agricultural, forest, and wetland, found throughout the watershed.

The ecological unit model used similar relationships and mathematics with ecosystem-specific coefficients to simulate edaphic, vegetative, and nutrient processes characteristic of each ecosystem. Ecosystem-specific coefficients related to vegetative processes that were not found in the literature were parameterized to represent the dominant vegetative species of each ecosystem. Calibration of unknown coefficients for the ecological unit model produced biomass, nutrient uptake, and water movement that compared well with studies of similar ecosystems. Most values produced were within 30% or less for all three ecosystem types. The validation of the forest, wetland, and agricultural ecosystem produced good comparisons despite changes in initial storages to match the validation data. The ecological unit model was within 20% of the validation data for the water and biomass submodels. For the nutrient submodels for the three ecosystem types the ecological unit model was less comparable to the validation data. However, often the validation data for the nutrient submodels were different from the calibration data. After validation, three types of scenarios were performed on the ecological unit model; climate change, water level decrease, and reduced fertilizer. The three ecosystems responded predictably to temperature and precipitation increases. For the lower water level scenario the wetland simulated reduced growth. The reduced fertilizer scenario for the agricultural ecosystem had little effect on nutrient runoff and biomass.

The ecological unit model was designed to be a simple model that simulated water, biomass, and nutrient dynamics. Though simplified processes reduced accuracy when compared to literature data, the model performed well. Nutrient dynamics were the most impacted by the simplification processes, but these are often variable in the literature and still not thoroughly understood.

SUCCESSIONAL MANAGEMENT IN RESTORED OLD-FIELD WETLANDS

Joshua L. Smith
Environmental Science Graduate Program
The Ohio State University

Exotic plants are known to have invaded various ecosystems throughout the United States. Wetlands in particular have a number of invasive exotic plants known to detrimentally affect both floral and faunal communities and ecosystem function. Loss of species diversity and ecosystem function are concerns shared by managers of both natural wetland preserves and created treatment wetlands alike. In northeast Ohio, exotic species such as purple loosestrife (*Lythrum salicaria*) and common reed grass (*Phragmites australis*) are of particular concern. The George Jones Memorial Farm, in Oberlin, Ohio, is the site of six created wetland cells being used to test different management strategies for not only the control of invasive plant species, but also to determine what management strategies provide the best ecological services while still maintaining a high plant species diversity. Three treatments are replicated at the study site: two planted wetland cells being actively managed for invasive and exotic species; two cells being planted and allowed to vegetate without management; and two cells being left as unplanted controls. The objectives of this project are to see not only which management strategy best reduces invasive and exotic plant species in this part of Ohio, but also to determine which strategy can provide the most cost-effective way of preserving diversity as well as restoring ecological function. The project was initiated in the fall of 2003 with wetland cell creation and plant installation; the collection of field data will begin in early spring of 2004.

Hydrology, Morphology and Water Quality in Constructed Open-Ditch Channels

J.S. Strock, R.T. Venterea, J.A. Magner, W.B. Richardson,
and P.H. Gowda

Open-ditch channels are potential transporters of considerable loads of nutrients, sediment, pathogens, and pesticides from agricultural land to small streams and larger rivers. An open-ditch research facility incorporating a paired design was constructed near Lamberton, MN during 2002. Our objective was to compare hydrology, ditch channel morphology, and water quality between two experimental open-ditch channels both spatially and temporally. Channels were constructed with a 0.1% slope. A 200-m reach of existing drainage channel was converted into a system of four parallel channels. The facility was equipped with water level control devices and instrumentation for flow monitoring and water sample collection on upstream and downstream ends of the system. Hydrographs from base- and storm-flow periods are presented along with water quality data. Longitudinal profiles along the horizontal length of each open-ditch were made every 3 m. In addition, eight longitudinal cross-sections were measured in each open-ditch. Hydrographs from simulated flow during 2002 indicated flow between channels at upstream and downstream monitoring locations were similar. Cumulative outflows were 60 and 85 m^3 for channels A and B, respectively. Channel cross-sections near the upstream end of the facility generally exhibited a trapezoidal shape whereas cross-sections near the downstream end tended to exhibit a parabolic shape. The parabolic shape coincided with an area of increased gradient (0.7%) in the channel bottom. Base- and storm-flow water samples were analyzed for sediment, nitrate plus nitrite, ammonia, orthophosphate, and total phosphorus. Stable isotopes of nitrate-nitrogen, $\delta^{15}N$ and oxygen, $\delta^{18}O$, were used to assess the source of nitrate-nitrogen and the magnitude and mechanisms of nitrogen transformation processes in these open-ditch channels. Results from two years of data collection are presented, and our plans for future experimental work at this facility are described.

MODELING OF EROSION PROCESSES AND ESTIMATION OF INTERRILL ERODIBILITY SOIL PARAMETERS IN FOREST-STEPPE ZONE OF UKRAINE

Snizhana Tsopa, Environmental Research Scientist
Department of Bioinformatics, Chernivtsy Institute of Agroecology and Biotechnology, 88a, Kashtanovaya Street, Chernivtsy 58026, UKRAINE
snizhana_tsopa@yahoo.co.uk

Erosion models which are able to describe erosion processes correctly under a number of given parameters in the specified point of space and time are an important apparatus of erosion-resistant agrolandscapes design and creation. The Water Erosion Prediction Project (WEPP) model is used to study erosion processes of soils of Ukraine as water flow physical laws to incline plane are in its basis.

Since 1999 till 2003 it has been conducted the researches of the grey forest and chernozem soils erodibility in forest-steppe zone of Ukraine on the Chernivtsy, Khmelnytskyy and Vinnitsa regions territory. According to WEPP methodic field experiments were fulfilled to define parameters of rill and interrill zones erodibility. The process of water erosion was simulated by water filling method without raindrops energy bringing. Both rill and interrill erosion research experiments were held on the same objects of Chernivtsy and Khotyn experimental stations of Chernivtsy Institute of Agroecology and Biotechnology in the Prut and Dnister River basins respectively and also on the territory of three agricultural farms. To conduct these experiments it was used the sprinkling simulator worked out by scheme described in publication of Karlos. It was designed four rain regimes different by intensity. The reiteration of the experiment was double.

As a result of the experiments in contrast to earlier received data it has been determined appearance of the nonlinear dependence of interrill erosion value D_i from $I_q S_f$ (where I – rain intensity, S – surface down gradient) under high rain intensity. The received data indicate the necessity in unification to realize field experiments to define coefficient of interrill soil erodibility K_i. In view of WEPP model structure and general requirements to physical modeling (highest possible approximation of modeling conditions to natural one), according to the research experiments, it is necessary to model "typical" downpour for research region. With the above purpose it has been conducted territory zoning of south-western Ukrainian territories, including Chernivtsy region by downpour risk based on downpour probabilistic estimation. The research data were received with all four intensions use of its "typical" for region research downpour. The interrill soil erodibility K_i coefficients calculated by means of different models were analyzed. The models estimations, which are characterized by coefficients of correlation between interrill erosion D_i value and flow determinative indexes, were fulfilled. The analysis of the research data showed that interrill erodibility coefficient value strongly vary depending on formula used. According to the above, soil losses calculated by WEPP model based on the generally accepted formula of K_i determination significantly differ from the received data. As a result of the research formula for under consideration soils on Ukrainian territory has been calculated. It has been also determined the conditions when WEPP model methods incorrectly evaluate the interrill erosion value and interrill erodibility coefficient.

It has been also determined coefficient K_i value by a different way – by obtaining regression dependence K_i from soil main parameters based on the experimental data. In view that it is impossible to use dependences received on soils in the USA, according to model WEPP calculations, it have been conducted researches to define interrill erodibility coefficient by experimental methods for maximum quantity of soils various by its genesis in Ukraine. As a result of data processing by method of direct single-step regression it has been constructed regression model of interrill erodibility coefficient estimation that gives possibility to estimate coefficient K_i according to data of granulometric structure and organic carbon contents without active experiment realisation. It has also been estimated correctness of this model on the territory of Ukraine.

Designing Two-Stage Agricultural Drainage Ditches

Andy D Ward, Dan Mecklenburg, Anand Jayakaran, Larry Brown

The Ohio Department of Natural Resources and the Ohio State University have developed a two-stage ditch design procedure in collaboration with county, state and federal agencies, faculty at other institutions in Ohio, Illinois and Minnesota, and watershed groups. The goal of this work is to develop a practical procedure that can be used throughout the Midwest to correctly size the main channel, to provide a minimum bench width to ensure stability, and to size the cross-sectional capacity of the second stage to carry a design discharge to prevent over-bank flow based on a recurrence interval that satisfies local, county, watershed, or state requirements. Several demonstration designs have been installed and will be monitored to evaluate how they enhance stability, water quality, and ecological function. Highly modified channels drain extensive portions of productive agricultural land in the U.S.A. In many of these areas, most natural channels have been deepened and straightened to facilitate the flow of water from agricultural subsurface drainage outlets and to maximize conveyance. Work done periodically to maintain the drainage function typically includes removal of woody vegetation and deposited sediment, and stabilizing bank slope failures and toe scour. Drainage ditch form (pattern, profile, and dimension) was measured on ditches in Northwest Ohio. Additional measurements have been made in the Wabash River and Great Miami River Watersheds. Apparent benefits exist for incorporating fluvial process derived form into ditch construction and maintenance. To facilitate drainage, and reduce the frequency of over bank flows ditches are typically constructed such that flows as large as perhaps 5-50 year recurrence interval are contained within the ditch. The constructed ditch channel is often oversized for small flows and provides no floodplain for large flows. In response to this imbalance fluvial processes work to create a small main channel by building a floodplain or bench within the confines of the ditches. If conditions allow, these benches can reach a stable size, thickly vegetated with mostly grasses. The small main channel will often meander slightly within the ditch and is sized by nature to carry the effective discharge. Evidence and theory both suggest ditches prone to filling with accumulated sediment may require less frequent dipping out if constructed in a two-stage form. Second, channel stability may be improved by a reduction in the erosive potential of larger flows as they are shallower and spread out across the bench. Stability of the ditch bank may also be improved where the toe of the ditch bank meets the bench rather than the ditch bottom. Here the bank height is effectively reduced and the shear stresses on the toe of the bank are less. The probable dimensions of the low-flow channel can be empirically determined based on regional studies similar to those that are conducted for natural streams. A two-stage ditch has the potential to create and maintain better habitat. Two-stage ditches might also be useful in improving water quality particularly for nutrient assimilation. The primary costs of two-stage ditches are increased width and more initial earthwork.

Olentangy River Watershed, Ohio Total Maximum Daily Load (TMDL) Project

Jon Witter, Andy Ward, Kevin King, Lance Williams, Jessica D'Ambrosio, and Cynthia Smith, the Ohio State University

Participants from a wide variety of organizations including local governments, university, industry, and environmental groups have been working to develop a water resource inventory and management plan for the Upper Olentangy watershed as part of a 319 Watershed Grant. The Ohio EPA and Ohio State University are taking results and information from those efforts and developing TMDL's for subwatersheds in the Olentangy basin. In addition to computer modeling activities this project will use concepts of fluvial geomorphology to determine the role of stream processes for influencing stream health. Currently this role is relatively unknown and has generally not been considered in TMDL studies around the nation. Also, Ohio is one of a few states that include biology as part of Water Quality Standards. As part of this research study we will examine the relationships between biology, habitat, geomorphology and water quality using multivariate statistical methods.